CHEMISTRY RESEARCH AND APPLICATIONS

NEWS IN CHEMISTRY, BIOCHEMISTRY AND BIOTECHNOLOGY

STATE OF THE ART AND PROSPECTS OF DEVELOPMENT

CHEMISTRY RESEARCH AND APPLICATIONS

Additional books in this series can be found on Nova's website
under the Series tab.

Additional e-books in this series can be found on Nova's website
under the e-book tab.

CHEMISTRY RESEARCH AND APPLICATIONS

NEWS IN CHEMISTRY, BIOCHEMISTRY AND BIOTECHNOLOGY

STATE OF THE ART AND PROSPECTS OF DEVELOPMENT

GENNADY E. ZAIKOV
GRZEGORZ NYSZKO
LARISA P. KRYLOVA
AND
SERGEI D. VARFOLOMEEV
EDITORS

nova publishers
New York

For permission to use material from this book please contact us:
Telephone 631-231-7269; Fax 631-231-8175
Web Site: http://www.novapublishers.com

Library of Congress Cataloging-in-Publication Data

ISBN: 978-1-63117-273-1

Published by Nova Science Publishers, Inc. † New York

CONTENTS

PREFACE

"The most practical thing in the World is a good theory"
Albert Einstein, USA

"The main lifeline is a fight against boredom"
Henry Reznik, lawyer, Russia

"If the government doesn't wish to invest money in education,
it will then be compelled to invest even bigger
money in the construction of prisons"
An opinion of the Russian scientists

"Experimental data is everything that was obtained
by you and your colleagues,
and then was published by your boss
without any reference of you"
Russian proverb

"The one who knows nothing is blissful.
He doesn't risk being unclear"
Confucius, Ancient China

In this volume, we included information about the preparation, characterization and applications of magnesium stearate, cobalt stearate and copper stearate, and the water sorption of polyvinyl chloride–luffa cylindrica composites. The control of the particle size and purity of nano zinc oxide, hyaluronan – an information rich messenger reporting on the physiological and pathophysiological status of synovial joints are also discussed. Further information is included as well, such as: the surface properties of polyimide copolymers; polyvinylchloride antibacterial pre-treated by barrier plasma; new types of ethylene copolymers on the base nanocomposite; the interaction of hybrid antioxidants – ichphans with erythrocyte membrane; and changes in the structural parameters and molecular dynamics of polyhydroxybutyrate–chitosan mixed compositions under the external influences and antifungal activity of animated chitosan against three different fungi species.

We collected the reviews and original papers about the collagen modified hardener for melamine-formaldehyde adhesive for increasing the water-resistance of plywood, including: the possibile applications of collagen colloid from secondary raw materials as a modifier of polycondensation adhesives; preparation and properties of animal protein hydrolysates for optimal adhesive compositions; a novel supramolecular hyaluronan/polyborate system for tumour treatment with boron neutron capture therapies; the analysis of the common factors of inactivation and stabilization of glutathione peroxidase I with the use of polyacrylic acid as a way of receiving preparations for curing diseases of the central nervous system and a comparison of two bioremediation technologies for oil polluted soils.

Many interesting results in the field of phase behaviour and structure formation in aqueous solutions of bovine serum albumin are also discussed, such as: phase transitions in a water-in-water BSA/dextran emulsion in the presence of strong polyelectrolyte, the crucial role of milk xanthine oxidoreductase in the conversion of toxic nitrate and nitrite to physiologically important nitric oxide and the Prostor and Ferm KM complex probiotic additives – innovations in biotechnological preparations for enhancing the quality of domestic fish mixed feed.

We also included information about common licorice glycyrrhiza glabra as an example of the use of plant extracts and biological components obtained from the plants of an arid zone and the study of morphogenetic peculiarities of winter rape (*brassica napus* L.) primary explants in vitro culture. Additionally, cytological changes in the spermia of the russian sturgeon (acipenser guldenshtadti b.) after cryopreservation based on the composition of the cryoprotective medium, the development of nontoxic methods of rodent population control as an alternative approach for big cities and antioxidantive activity of forest and meadow medicinal herbs are also discussed.

The editors and contributors will be happy to receive some comments from the readers which can be taken into account in their future research.

Editors

Gennady E. Zaikov
Head of Polymer Division,
N.M. Emanuel Institute of Biochemical Physics
Russian Academy of Sciences
4 Kosygin str., 119334 Moscow, Russia
chembio@sky.chph.ras.ru

Grzegorz Nyszko
Deputy of Director,
Military Institute of Chemistry and Radiometry
105, Al.gen.A. Chrusciela
00-910 Warsaw, Poland
Grzegorz.nyszko@wichir.waw.pl

Larisa P. Krylova
Member of N.M. Emanuel Institute of Biochemical Physics,
Russian Academy of Sciences,
4 Kosygin str., 119334 Moscow, Russia
LPKrylova@sky.chph.ras.ru

Sergei D. Varfolomeev
Director of institute,
N.M. Emanuel Institute of Biochemical Physics
Russian Academy of Sciences
4 Kosygin str., 119334 Moscow, Russia
SDVarf@sky.chph.ras.ru

In: News in Chemistry, Biochemistry and Biotechnology ISBN: 978-1-63117-273-1
Editors: G. E. Zaikov, G. Nyszko, L. P. Krylova et al.

Chapter 1

HYALURONAN: AN INFORMATION RICH MESSENGER REPORTING ON THE PHYSIOLOGICAL AND PATHOPHYSIOLOGICAL STATUS OF SYNOVIAL JOINTS

***Ladislav Šoltés*[1,*] *and Grigorij Kogan*[2]**

[1]Institute of Experimental Pharmacology and Toxicology,
Slovak Academy of Sciences, Bratislava, Slovakia
[2]Directorate Health, Directorate General for Research and Innovation,
European Commission, Brussels, Belgium

ABSTRACT

Hyaluronan-degrading enzymes in synovial fluid, if any, is extremely low. Thus, the high rate of this glycosaminoglycan turnover in synovial fluid, around 12 hours, should result from a cause different from enzymatic catabolism. An alternative and plausible mechanism is that of oxidative-reductive degradation of the biopolymer chains by a combined action of oxygen, transition metal cations, and ascorbate.

Reactive oxygen species, which are generated during the oxygen metabolism, may participate in physiological catabolism of native high-molar-mass hyaluronan within the joint synovial fluid. However under pathological circumstances, such as the inflamed joint, the free-radical oxidative hyaluronan decay should prevail.

Keywords: Glycosaminoglycans, hyaluronan catabolism, reactive oxygen species, synovial fluid, transition metals

* Fax (+421-2)-5477-5928; Email: ladislav.soltes@savba.sk.

Introductory Remarks

On average, a healthy person living in the developed countries currently reaches lifespan of ca. 80–85 years. Women often live longer than men. This fact could be associated with their enhanced redox load during the reproductive phase of their life. Physiological bleeding (with a periodicity of ca. 4 weeks) is accompanied by changes in the concentration of iron ions. Pre-menopausal women are believed to have a lower risk of common diseases because amounts of iron in their body are unlikely to be excessive at this time [1].

Fe ions are regarded as one of the most important catalytical agents that contribute to the augmented generation of the reactive oxygen species (e.g., $^{\bullet}OH$ radicals). However, such "radical training" of female organism lasting on average 40 years (i.e., in a period between ca. 15 to 55 years) can have a positive effect on females in the sense that their organism is better adjusted to the oxidative stress. In the "free radical theory of ageing" oxidative stress is considered to be a risk factor that is usually associated with such negative consequences as serious diseases or even premature death [2,3].

Life can be in a simplified way divided into three periods: childhood, maturity, and senescence. Maturity is the longest lasting part of human life. It lasts from the end of development and growth of a skeleton (around ca. 20 years) till the old age, which start can be marked as at ca. 70–75 years. Thus, maturity lasts about half a century. During this period, human skeleton can be considered invariable regarding the number of bones (206), their size, and mass.

The human skeleton consists of both fused and individual bones supported and supplemented by ligaments, tendons, and skeletal muscles. Articular ligaments and tendons are the main parts holding together the joint(s). In respect to the movement, there are freely moveable, partially moveable, and immovable joints. Synovial joints, the freely moveable ones, allow for a large range of motion and encompass wrists, knees, ankles, shoulders, and hips.

The Structure of a Synovial Joint

Figure 1 illustrates a normal healthy synovial joint indicating its major parts.

Cartilage

In a healthy synovial joint, heads of the bones are encased in a smooth (hyaline) cartilage layer. These tough slippery layers – e.g., those covering the bone ends in the knee joint – belong to mechanically highly stressed tissues in the human body. At walking, running, or sprinting the strokes frequency attain approximately 0.5, 2.5 or up to 10 Hz.

Cartilage functions also as a shock absorber. This property is derived from its high water-entrapping capacity, as well as from the structure and intermolecular interactions among polymeric components that constitute the cartilage tissue [5]. Figure 2 sketches a section of the cartilage – a chondrocyte cell that permanently restructures/rebuilds its extracellular matrix.

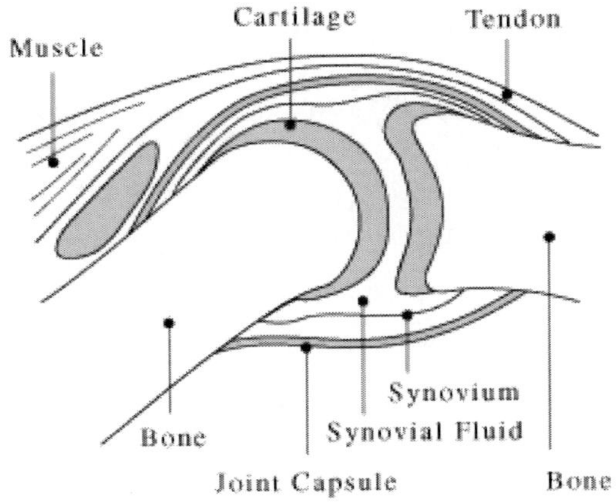

Figure 1. Normal, healthy synovial joint [4].

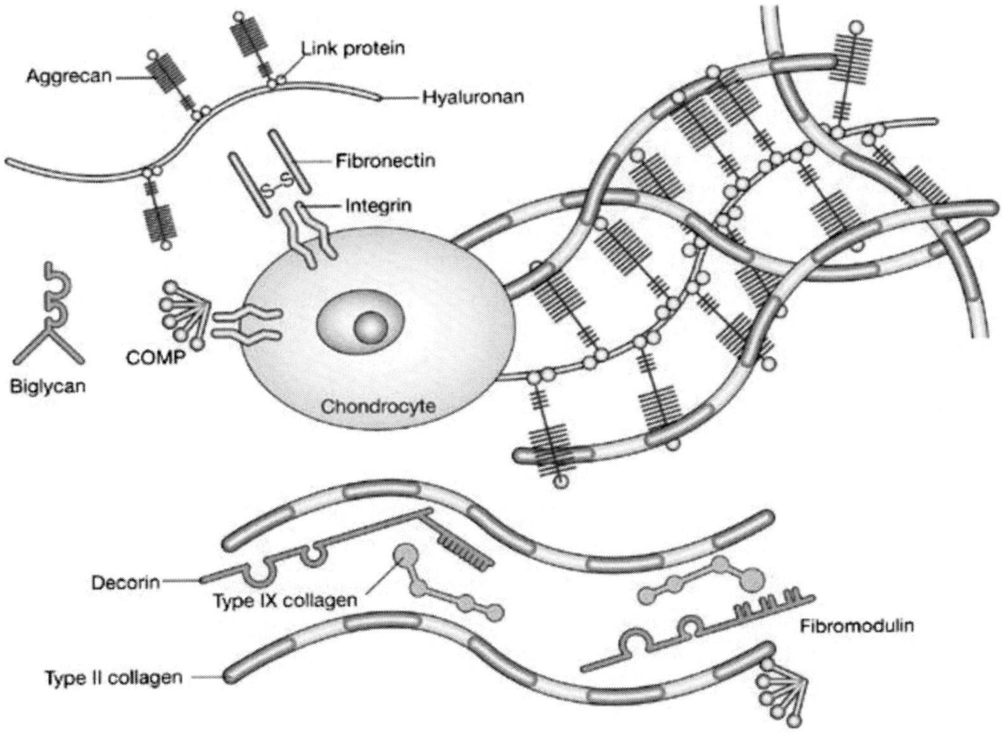

Figure 2. Articular cartilage main components and structure [6].

Three classes of proteins exist in articular cartilage: collagens (mostly type II collagen); proteoglycans (primarily aggrecan); and other noncollagenous proteins (including link protein, fibronectin, COMP – cartilage oligomeric matrix protein) and the smaller proteoglycans (biglycan, decorin, and fibromodulin). The interaction between highly negatively charged cartilage proteoglycans and type II collagen fibrils is responsible for the compressive and tensile strength of the tissue, which resists applied load *in vivo*.

Synovium/Synovial Membrane

Each synovial joint is surrounded by a fibrous, highly vascular capsule/envelope called synovium, which internal surface layer is lined with a synovial membrane. Inside this membrane, type B synoviocytes (fibroblast-like cell lines) are localized/embedded. Their primary function is to continuously extrude high-molar-mass hyaluronans (HAs) into synovial fluid (SF).

Synovial Fluid

The synovial fluid, which consists of an ultrafiltrate of blood plasma and glycoproteins, in normal/healthy joint contains HA macromolecules of molar mass ranging between 6–10 megaDaltons [7]. SF serves also as a lubricating and shock absorbing boundary layer between moving parts of synovial joints. SF reduces friction and wear and tear of the synovial joint playing thus a vital role in the lubrication and protection of the joint tissues from damage during the motion [8].

The nutrients, including oxygen supply, upon crossing the synovial barrier, permeate through the viscous colloidal SF to the avascular articular cartilage, where they are utilized by the embedded chondrocytes. On the other hand, the chondrocyte catabolites (should) cross the viscous SF prior to being eliminated from the synovial joint [9]. It can thus be concluded that within SF, the process of "mixing" at the joint motion, significantly affects the equilibrium of influx and efflux of all low- and high-molar-mass solutes. It appears that the traffic of solutes is determined by molecular size, with small polar molecules being cleared by venular reabsorption, while high-molecular-sized solutes are removed by lymphatic drainage [10].

Hyaluronan

Figure 3 represents the structural formula of hyaluronan (also called hyaluronic acid, hyaluronate) – regularly alternating disaccharide units composed from *N*-acetyl-D-glucosamine and D-glucuronic acid. HA is a polyelectrolyte component of SF; the concentration of HA in healthy human knee SF is 2.5 mg/ml on average [11]. While in the articular cartilage matrix HA is firmly associated via a link protein with proteoglycans (cf. Figure 2), in SF the HA macromolecules are, if at all, only loosely interacting/bound to proteins.

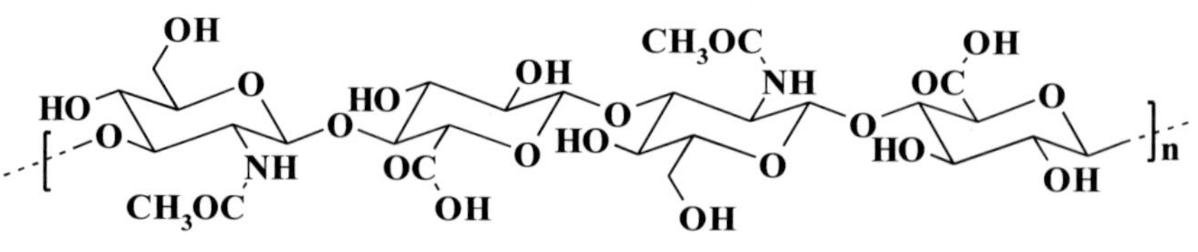

Figure 3. Hyaluronan – the acid form.

HA is a linear non-branched non-sulfated glycosaminoglycan (bio)polymer. In aqueous solutions, HA is represented by negatively charged hyaluronate macromolecules (p*K*a = 3.21 [12]) with extended conformations, which impart high viscosity/viscoelasticity, accompanied also by low compressibility – the characteristic property of SF [13].

REACTIVE OXYGEN SPECIES IN ARTICULAR CARTILAGE

Articular cartilage is an avascular, acidic (pH 6.6–6.9) and hyperosmotic tissue dependent on diffusion of nutrients supplied mainly from SF (and perhaps partly from subchondral bone [14]) to provide for the metabolic requirements of chondrocytes. The oxygen levels in this tissue are low, ranging between 1 and 6% (cf. Figure 4). While reduction in O_2 tension to 6% in all other tissues is already hypoxic, for chondrocytes such oxygen level is normoxic.

In the mitochondria of the eukaryotic cells, not all O_2 is fully reduced to water. A small fraction of oxygen is reduced incompletely yielding reactive oxygen species (ROS), which are assigned to the defense of the organism against viral/bacterial invaders [15]. It has been established that while ROS content within the articular cartilage tissue remains normal at 6% O_2, it decreased at 1% O_2 [14].

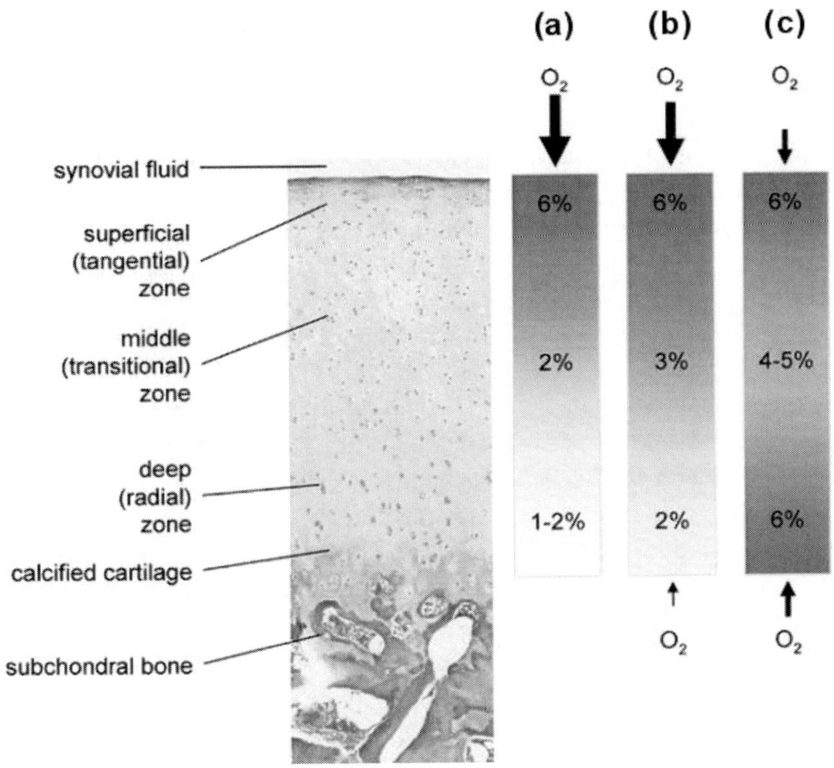

Estimated levels of O_2 within the cartilage tissue are shown for three scenarios:
(a) penetration of O_2 exclusively from SF;
(b) O_2 supply mostly from SF with a small contribution from subchondral bone;
(c) supply of O_2 in equivalent amounts from SF and subchondral bone.

Figure 4. The structure of articular cartilage and its oxygen supply (adapted from [14]).

Since hydrogen peroxide generated within the mitochondria of chondrocytes can freely permeate through the chondrocyte cell wall, one should admit the presence of H_2O_2 in all (deep, middle, and superficial) zones of the articular cartilage (cf. Figure 4). The higher the O_2 tension, the greater is the content of H_2O_2 and vice-versa.

The ROS within the cartilage tissue could serve both as intra- and inter-cellular signaling devices and a reactant participating in the so-called Fenton reaction

$$H_2O_2 + Me^{n+} \rightarrow {}^{\bullet}OH + Me^{(n+1)+} + OH^{-} \qquad (1)$$

where Me^{n+} and $Me^{(n+1)+}$ represent a (biogenic) transition metal ion in reduced and oxidized state. Among these metals, primarily iron and copper are usually ranked, however, several further trace/biogenic metals can be taken into account as well [1,16].

ROS in SF and Their Function Thereof

The capillaries within synovium continuously provide a plasma filtrate supplying in this way nutrients to the joint tissues (the arterial blood O_2 tension is 13% [17]). This is particularly important for homeostasis of the avascular articular cartilage [10]. As recently stated [16], taking into consideration that articular cartilage does not contain any teleneurons, chondrocytes should perform their autonomic (metabolic) regulation most plausibly using a chemical process, in which both O_2 and ROS play significant roles [17]. To understand this tenet, one should take into consideration that in the joint relaxed state – for example, at night – chondrocytes experience a decreased oxygen supply (a status termed "hypoxia"). However, when the status changes to an enhanced mobility in the morning, joint SF receives elevated supply of O_2 (a situation termed "re-oxygenation"). Such increased content of oxygen can be, however, deleterious for the homeostasis of the chondrocytes – the cells that in adults lack mitotic activity.

Let us assume that Me^{n+} ions in a given concentration are "entrapped" by (highly) negatively charged cartilage glycosaminoglycans (GAGs) within the superficial (tangential) zone of the articular cartilage (cf. Figure 4). During the utilization of O_2 – respiration – by chondrocytes, a limited amount of H_2O_2 liberated from their mitochondria can react with the entrapped transition metal ions generating hydroxyl (${}^{\bullet}OH$) radicals. Due to extremely short half-life of these species (picoseconds), they react *in situ nascendi* with GAGs – chondroitin sulfate (CS) and/or keratan sulfate (KS). The C-type radicals of CS or KS can, however, instantly undergo a reaction of hydrogen radical transfer onto the neighboring HA macromolecules within the SF. In such a way, free C-(macro)radicals of hyaluronan appear nearby the superficial zone of the articular cartilage. And it is this very C-(macro)radical (denoted later as $A^{\bullet}$), which reacts and in this way reduces the (free "hyperoxic") O_2 tension within and nearby the superficial zone of the articular cartilage – according to the reaction presented in the following scheme:

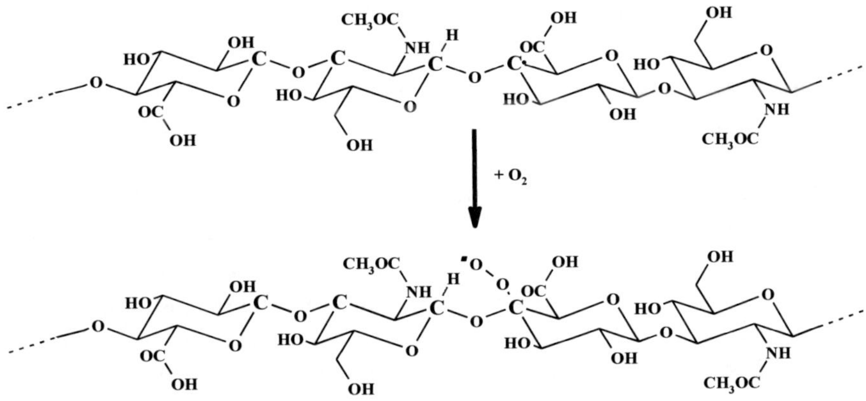

Scheme 1. Entrapment of oxygen by the hyaluronan C-(macro)radical ($A^{\bullet}$) yielding a peroxyl (macro)radical (A–O-$O^{\bullet}$).

or briefly

$$A\text{–}H + {}^{\bullet}OH \rightarrow A^{\bullet} + H_2O \quad (2)$$

$$A^{\bullet} + O_2 \rightarrow A\text{–}O\text{-}O^{\bullet} \quad (3)$$

where A–H represents the intact hyaluronan macromolecule (cf. Figure 3 and Scheme 1).

Subsequently, this A–O-$O^{\bullet}$ peroxyl (macro)radical can transform simply by an intramolecular 1,5-hydrogen shift to another C-(macro)radical – $A^{\bullet}$ (cf. Scheme 2). By participation of another O_2 molecule, this $A^{\bullet}$ radical can yield two fragments of the HA biopolymer: (i) the fragment, which possesses an aldehyde terminus, and (ii) the fragment bearing a hydroperoxide functional group. It is naturally evident that both fragments differ in their chemical structure from the initial HA macromolecule, not only due to the included novel substituents (–C=O; –O-OH) but above all by a reduced molar mass of both polymer fragments compared to that of the parent biopolymer.

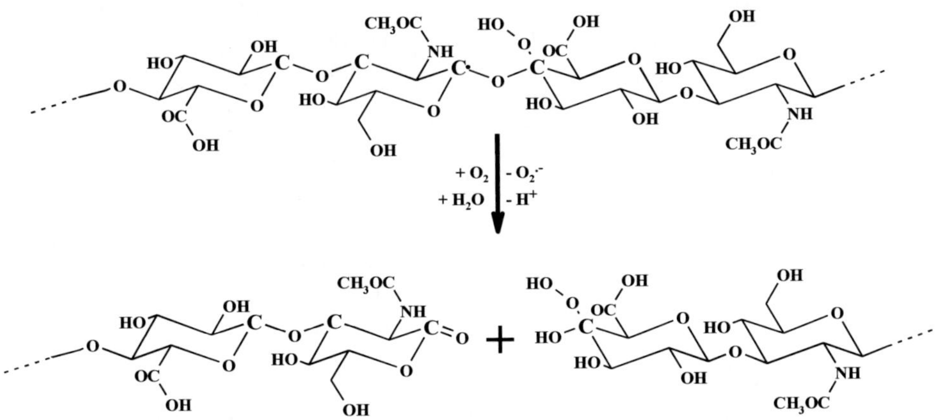

Scheme 2. Strand scission of the C-(macro)radical ($A^{\bullet}$) yielding two fragments.

Since the intermolecular reaction between the CS and KS radicals and the native HA macromolecule could yield various $A^{\bullet}$ radicals – formed for example at C(4) of the D-glucuronate/D-glucuronic acid (GlcA) unit (cf. Scheme 1) or at C(1) of GlcA unit, as well as at C(1) or C(3) of *N*-acetyl-D-glucosamine (GlcNAc) [18] – various biopolymer fragments are produced.

Very recently Kennett and Davies [19] reported the data obtained with both the C(1)- and the C(2)- ^{13}C-labeled *N*-acetyl-D-glucosamine, and the apparent highly selective generation of radicals at the C(2) position of the isopropyl group of the β-isopropyl glycoside, which allow the authors to rationalize the specific banding pattern observed on oxidation of hyaluronan: The lack of reactivity at C(1)/C(2) of the *N*-acetyl-D-glucosamine monomers and the specific formation of radicals on the isopropyl group, which models the C(4) glycosidic linkage site of the glucuronic acid, implicate attack at C(4) of the glucuronic acid subunits and subsequent β-scission of this radical as a major route to cleavage of the hyaluronan backbone (Scheme 3). A contribution from reaction at C(1) of the glucuronic acid and subsequent cleavage of the alternative glycosidic linkage cannot be discounted; however, it is clear that an alternative route involving C(3) on the *N*-acetyl-D-glucosamine monomer is less favored, as only low levels of initial hydrogen atom abstraction seem to occur at this position as judged by the low yield of radicals that did not have additional ^{13}C couplings observed with the two labeled *N*-acetyl-D-glucosamine species. It should be pointed, however, that the products of the hyaluronan strand cleavage depicted in Scheme 3 do not take into account that the ubiquitous oxygen participate within the strand scission reaction and thus, analogously to Scheme 2, the involved O_2 molecule with the $A^{\bullet}$ radical yields two fragments of the HA biopolymer: (i) the fragment bearing a hydroperoxide functional group, and (ii) the fragment, which possesses an aldehyde terminus. As stated above, both fragments naturally differ in their chemical structure due to the included –C=O or –O-OH substituent and, above all, by the reduced molar mass of both polymer fragments compared to that of the parent HA biopolymer.

Along with the fragmentation reactions shown in Schemes 2 and 3, the radical attack on the GlcA and GlcNAc moieties can also lead to the ring opening without breaking the polymer chain [11, 18, 20, 21].

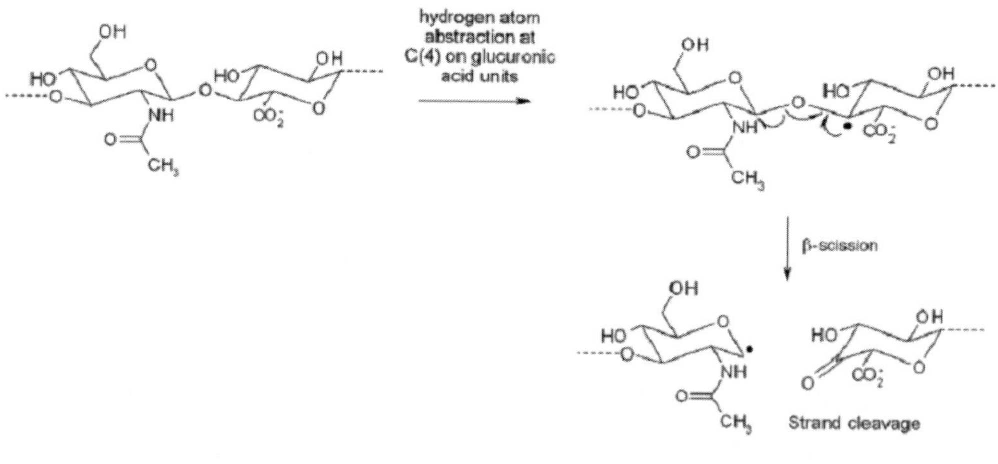

Scheme 3. Potential mechanism of hyaluronan strand cleavage as a result of hydrogen abstraction and radical formation on C(4) of the glucuronic acid unit (adapted from [19]).

There exists, however, a remarkable phenomenon of *in vivo* free-radical oxidative degradation of hyaluronan: Under physiological conditions, the SF viscosity does not undergo any changes since the content of "native" hyaluronan remains constant due to permanent *de novo* production of megaDalton HA macromolecules by (stimulated) type B synoviocytes. Thus, the self-perpetuating oxidative (non-enzymatic) HA "catabolism" in SF represents a rather delicate and properly balanced mechanism that presumably plays significant role in regulating the physiological – normoxygen – homeostasis for chondrocytes. At the same time, the produced polymer fragments, which are probably cleared from the joint by drainage pathways, serve most likely as chemical messengers/feedback molecules. These play role in the adjustment of the optimum mode of functioning of the synovial membrane and of the HA-producing cells, B synoviocytes, localized within. In other words, during physiologic joint functioning, the hyaluronan in SF plays the role of a "scavenger antioxidant", whereas the produced polymer fragments can subsequently serve as messengers mediating information on the changes occurring in the homeostasis of the joint [16].

High "protective/scavenging efficiency" of hyaluronan against the *in vitro* action of $^{\bullet}OH$ radicals has been earlier pointed out by some authors [22, 23]. Presti and Scott [23] described that high-molar-mass hyaluronan (megaDalton HA) was much more effective than the lower-molar-mass HAs (hundreds of kiloDaltons HAs) in scavenging $^{\bullet}OH$ radicals generated by a Fenton-type system comprising glucose and glucose oxidase *plus* Fe^{2+}-EDTA chelate.

Hypoxia and Re-oxygenation of the Joint

As SF of healthy human exhibits no activity of the hyaluronidase enzyme, it has been inferred that oxygen-derived free radicals are involved in a self-perpetuating process of HA catabolism within the joint [24]. This radical-mediated process is considered to account for ca. twelve-hour half-life of native HA macromolecules in SF.

To understand how to maintain a radical reaction active/self-perpetuating, its propagation stage should first be analyzed. If a peroxyl-type (macro)radical ($A–O-O^{\bullet}$) exists within SF, due to the relatively high reactivity of the unpaired electron on oxygen, the following intermolecular reaction can be assumed

$$A–O-O^{\bullet} + A–H \rightarrow A–O-OH + A^{\bullet} \quad (4)$$

In the case when $A^{\bullet}$ is a C-type (macro)radical, it is this very reactant that traps the dioxygen molecule, dissolved in SF, according to the reaction

$$A^{\bullet} + O_2 \rightarrow A–O-O^{\bullet} \quad (5)$$

Hence, by combining the reactions 4 and 5, the net reaction

$$A–H + O_2 \rightarrow A–O-OH \quad \text{(net reaction)}$$

corroborates the statement that one particular function of (a high-molar-mass) HA is to trap the oxygen excess during the phase of joint re-oxygenation [16].

Physiologic Oxidative Catabolism of Hyaluronan: Participation of Biogenic Transition Metal Ions

As stated in Scheme 2 and reaction 4, A–O-OH hydroperoxides are generated during the self-perpetuating – propagation – stage of the hyaluronan oxidative catabolism. The fate of A–O-OH type hydroperoxides, however, is significantly dependent on the presence or absence of the transition metal ions within SF. In the former case, the following reactions could be suggested for decomposition of the generated A–O-OH hydroperoxides

$$A\text{–}O\text{-}OH + Me^{n+} \rightarrow A\text{–}O^{\bullet} + HO^{-} + Me^{(n+1)+} \qquad (6)$$

$$A\text{–}O\text{-}OH + Me^{(n+1)+} \rightarrow A\text{–}O\text{-}O^{\bullet} + Me^{n+} + H^{+} \qquad (7)$$

As can be seen, while the "propagator" that participates in reaction 4 is (re)generated by reaction 7, reaction 6 produces an alkoxyl type (macro)radical A–O$^{\bullet}$. The ratio of the A–O-O$^{\bullet}$ to A–O$^{\bullet}$ radicals is, however, governed by the present transition metal ions, or, more precisely, by the ratio of $Me^{(n+1)+}$ to Me^{n+}. To answer the question, which transition metals may be present in SF and cells or tissues of healthy human beings, one should take into account the data presented in Tables 1 and 2.

Table 1. Contents of transition metals in blood serum of healthy human volunteers and in *post mortem* collected SF from subjects without evidence of connective tissue disease

Element	Mean concentration in blood serum [μg/100 mL][a]	Mean concentration in synovial fluid [μg/100 g][a]
Iron	131.7 (23.6)[b]	29.0 (5.19)[b]
Copper	97.0 (15.3)	27.5 (4.33)
Zinc	115.4 (17.7)	17.6 (2.69)
Manganese	2.4 (0.44)	2.4 (0.44)
Nickel	4.1 (0.70)	1.2 (0.20)
Molybdenum	3.4 (0.35)	1.0 (0.10)

[a]Reported by Niedermeier and Griggs [25].

[b]Data in parentheses are the values in μM calculated in assumption that 100 g of SF has a volume of 100 mL.

Table 2. Average relative abundance of some biogenic transition metals in the mammalian blood plasma and cells/tissues

Element	Blood plasma [μM][a]	Cell/Tissue [μM][a]
Iron	22	≈ 68
Copper	8-24	0.001-10
Zinc	17	180
Manganese	0.1	180
Nickel	0.04	2
Molybdenum	-	0.005

[a]Adapted from [26].

Based on the data listed in Table 1, iron and copper are the two prevailing redox active transition metals in SF. It should be, however, pointed out that the respective concentrations of ca. 5.2 μM of iron ions and 4.3 μM of copper ones do not represent those, which are (freely) disposable to catalyze the oxidative catabolism of hyaluronan within SF. As has been reported, the availability of iron to stimulate *in vivo* generation of $^{\bullet}OH$ radicals is very limited, since concentrations of "free" iron, are seldom larger than 3 μM in human samples [27].

Let us now deal with the oxidation states of iron within SF of a healthy human. By accepting that the concentration of ascorbate in SF of healthy subjects reaches the values close to those established in blood serum, i.e., 40–140 μM [28], it must be admitted that the transition metal ions in SF of a healthy human being are in the reduced oxidation state, i.e., Me^{n+}. Thus, in the case of the ascorbate level, which many times exceeds the concentration of transition metal ions, the actual concentration of ferrous ions should exceed that of ferric ones, and thus A–O$^{\bullet}$ radicals should prevail. These radicals could, similarly to the A–O-O$^{\bullet}$ ones, propagate the radical chain reaction as follows

$$A\text{–}O^{\bullet} + A\text{–}H \rightarrow A\text{–}OH + A^{\bullet} \qquad (8)$$

Yet, due to the redox potential of the pair $RO^{\bullet},H^{+}/ROH$ = +1.6 V, which surpasses significantly that of $ROO^{\bullet},H^{+}/ROOH$ = +1.0 V, the actual content of A–O$^{\bullet}$ in SF is practically nil; the half-life of the A–O$^{\bullet}$ radicals is much shorter than that of A–O-O$^{\bullet}$ ones – microseconds *vs*. seconds.

OXIDATIVE/NITROSATIVE STRESS

Oxidative and/or nitrosative stress are terms used to describe situations, in which the organism's production of oxidants exceeds the capacity to neutralize them. The excess of oxidative species can cause "fatal" damage to lipids within the cell membranes, cellular proteins and nucleic acids, as well as to the constituents of the extracellular matrix, such as collagens, proteoglycans, *etc*. [29].

Oxidative and/or nitrosative stress has been implicated in various pathological conditions involving several diseases, which fall into two groups:

(i) diseases characterized by "inflammatory oxidative conditions" and enhanced activity of either NAD(P)H oxidase (leading to atherosclerosis and chronic inflammation) or xanthine oxidase-induced formation of oxidants (implicated in ischemia and reperfusion injury),

(ii) diseases characterized by the implication of pro-oxidants that shift the thiol/disulphide redox equilibrium and cause impairment of glucose tolerance - the so-called "mitochondrial oxidative stress" conditions (leading to cancer and diabetes mellitus) [3].

Table 3. Main ROS and RNS

Radical		Non-radical	
hydroxyl	$^{\bullet}OH$	peroxynitrite anion	$ONOO^{-}$
superoxide anion radical	$O_2^{\bullet -}$	hypochloric acid	HOCl
nitric oxide	$^{\bullet}NO$	hydrogene peroxide	H_2O_2
thyil	$-RS^{\bullet}$	singlet oxygen	$^{1}\Delta_g$ ($^{-1}O_2$)
alkoxyl	$RO^{\bullet}$	ozone	O_3
peroxyl	$ROO^{\bullet}$	nitrosyl cation	NO^{+}
		nitroxyl anion	NO^{-}
		nitryl chloride	NO_2Cl

Oxidants

In a broader sense, oxidation concerns the reaction of any substance with molecules of oxygen, the primary oxidant. In chemistry, however, the term "oxidant" is used for all species able to render one or more (unpaired) electrons.

In a simplified way, oxidants can be classified as free-radical and non-radical species (cf. Table 3; adapted from [30]). They are often classified as reactive oxygen species (ROS) and reactive nitrogen species (RNS). Although the latter, similarly to ROS, contain oxygen atom(s) – e.g., NO^{+}, NO^{-}, NO_2Cl – the RNS usually participate at nitrosylation reactions.

Oxygen Metabolism – Source of Energy

Several oxidant species are produced at the processes occurring in animal cells, including human ones, during metabolism of oxygen, when these cells generate energy. Although the substrate (O_2) is – by a cascade of enzymatically driven reactions – reduced within subcellular organelles, mitochondria, to a completely harmless substance, the waste product – water, a fraction of generated ROS may escape from the enzymatically controlled processes:

$$O_2 + 1e^{-} \rightarrow O_2^{\bullet -} \quad (9)$$

$$O_2^{\bullet -} + 1e^{-} + 2H^{+} \rightarrow H_2O_2 \quad (10)$$

$$H_2O_2 + 1e^{-} + H^{+} \rightarrow {}^{\bullet}OH + H_2O \quad (11)$$

$$^{\bullet}OH + 1e^{-} + H^{+} \rightarrow H_2O \quad (12)$$

net reaction

$$O_2 + 4e^{-} + 4H^{+} \rightarrow 2H_2O \quad (13)$$

Table 4. Standard reduction potential (and half-life) for some dioxygen species in water, pH 7, 25 °C[a]

Species	(reaction)	E^{θ} [V]	$t_{1/2}$ [s]
O_2	(9)	-0.33[b]	reactive
$O_2^{\bullet-}$	(10)	+0.89	10^{-6}
H_2O_2	(11)	+0.38	long living
$^{\bullet}OH$	(12)	+2.31	10^{-9}

[a]Adapted from [31].
[b]The greater the positive E^{θ} value, the greater is generally the species reactivity, i.e., the ability to catch an electron [cf. reactions (9)–(12)].

As indicated by the reaction steps (9), (10), and (11), oxidants, namely $O_2^{\bullet-}$, H_2O_2, and $^{\bullet}OH$ are intermediate products of the enzymatically controlled cascade. Their reactivity and presumable site of action can be assessed by physico-chemical parameters, such as standard reduction potential (E^{θ}) and half-life ($t_{1/2}$) of the given species (cf. Table 4).

With regard to the high (positive) value of E^{θ} and to the short half-life values, escape of $^{\bullet}OH$ and $O_2^{\bullet-}$ from the sphere immediately surrounding mitochondrion can be virtually excluded. Yet the neutral molecule H_2O_2 is considered to be movable one, which can escape as from the "body" of the mitochondrion as well as from the cell body itself. It is comprehensible that in some tissues the actual H_2O_2 concentrations may reach 100 µM or more as e.g., in human and other animal aqueous and vitreous humors. The hydroperoxide levels at or below 20–50 µM seem, however, to have limited cytotoxicity to many cell types [32].

Oxygen Metabolism – A Defence Mechanism against Viral/bacterial Invaders

Along with the above four-electron reaction (13), several specialized cells – or more precisely their specific (sub)cellular structures – are able to reduce O_2 molecules producing the superoxide anion radical, which in aqueous (acidic) milieu can form the reactive perhydroxyl radical ($^{\bullet}O_2H$).

Nitric oxide, called also nitrogen monoxide ($^{\bullet}NO$), a (bioactive) free radical, is produced in various cells/tissues by NO-synthase (NOS) enzymes. The three distinct NOS isoforms are P_{450}-related hemoproteins that during L-arginine oxidation to L-citrulline produce $^{\bullet}NO$. Two of the permanently present enzymes that participate in the regulation of the blood vessel tonus are termed constitutive NOS (cNOS), while the third one is called an inducible NOS (iNOS). The level of $^{\bullet}NO$ produced by iNOS increases markedly during inflammation, a process accompanied with abundant production of the superoxide anion radical.

The two radical intermediates – $O_2^{\bullet-}$/$^{\bullet}O_2H$ and $^{\bullet}NO$ – serve as precursors of various ROS and RNS, including hydrogen peroxide, peroxynitrite/peroxynitrous acid, hypochlorous acid, *etc*. On respiring air, human beings by utilizing one mole of O_2 ingest 6.023×10^{23} molecules of oxygen, of which approximately 1–3 % is assigned to the generation of ROS/RNS that defend the organism against viral/bacterial invaders [15].

It has been noted that certain organ systems are predisposed to greater levels of oxidative stress and/or nitrosative stress. Those organ systems most susceptible to damage are the pulmonary system (exposed to high levels of oxygen), brain (exhibits intense metabolic activity), eye (constantly exposed to damaging UV light), circulatory system (victim to fluctuating oxygen and nitric oxide levels) and the reproductive systems (at risk from the intense metabolic activity of sperm cells) [30]. In some cases, however, the intermediate and/or the "final" reactive oxidative species may also damage cells/tissues of the human host. Imbalance between the extent of damage and self-repair of the functionally essential structures may result in a broader host tissue injury, eventually leading to a specific disease.

Because of the highly reactive nature of ROS/RNS, it is difficult to directly demonstrate their presence *in vivo*. It is considerably more practical to measure the "footprints" of ROS and RNS, such as their effects on various lipids, proteins, and nucleic acids [29].

INDIRECT ROS/RNS EVIDENCE

Most ROS/RNS have very short half-live times thus they cannot be directly detected in the organisms.. That is why, as reported also by Valko et al. [3], convincing evidence for the association of oxidative/nitrosative stress and acute and chronic diseases lies on validated biomarkers of these stresses. Table 5 summarizes most representative biomarkers of oxidative damage associated with several human diseases.

There are numerous further diseases whose pathology involves reactive oxidative/oxygen-derived species, i.e., ROS and/or RNS, at the onset and/or at later stages of the disease [33]. The magnitude and duration of the change in the concentrations of these species appear to belong among the main regulatory events (cf. Figure 5).

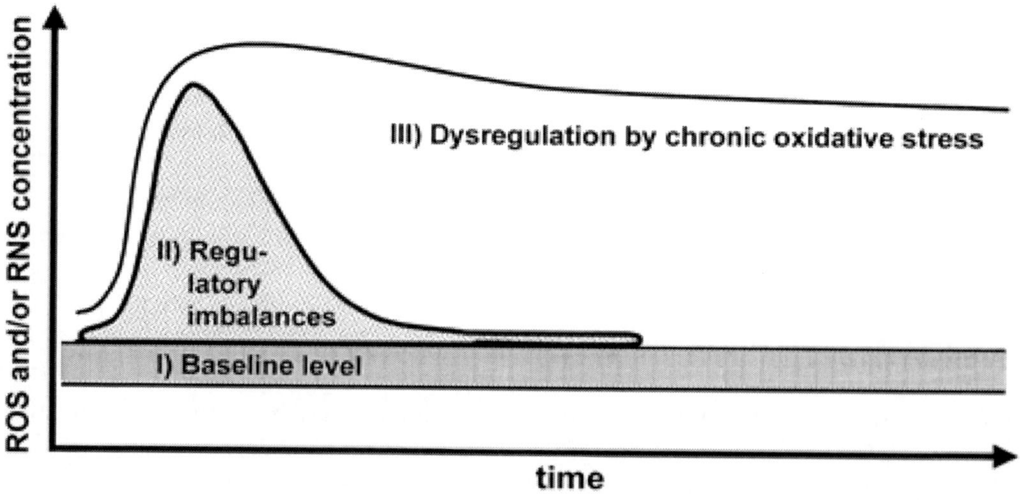

Figure 5. Regulatory events and their dysregulation depend on the magnitude and duration of the change in ROS and/or RNS concentration(s) (adapted from [34]).

Table 5. Biomarkers of oxidative damage associated with several chronic human diseases (adapted from [3])

Disease / *Biomarker*[a]	Alzheimer's disease	Atherosclerosis	Cancer	Cardiovascular disease	Diabetes mellitus	Parkinson's disease	Rheumatoid arthritis
8-OH-dG			+				
Acrolein		+		+			
AGE	+				+		
Carbonylated proteins						+	
F_2-isoprostanes	+	+		+	+		+
GSH/GSSG	+		+	+	+	+	+
HNE	+	+		+		+	
Iron level						+	
MDA	+	+	+		+		
NO_2-Tyr	+	+	+	+	+		
S-glutathiolated proteins					+		

[a]Abbreviations: 8-OH-dG, 8-hydroxy-20-deoxyguanosine; AGE, advanced glycation end products; GSH/GSSG, ratio of glutathione/oxidized glutathione; HNE, 4-hydroxy-2-nonenal; MDA, malondialdehyde; NO_2-Tyr, 3-nitro-tyrosine.

Table 6. Some characteristics registered within SF during inflammatory joint diseases[a]

Diagnosis	SF viscosity	Blood characteristics			H_2O_2 flux [μM/min]
		White cells/μL	% of PMNLs	PMNLs/μL	
Healthy	normal	<200	7	<14	<0.003
OA	decreased	600	13	48	0.017
RA	decreased	1900	66	1254	0.276

[a]Adapted from [37].

Abbreviations: PMNL, polymorphonuclear leukocyte; OA, osteoarthritis; RA, rheumatoid arthritis.

Today it is a widely accepted fact that ROS and RNS normally occur in living tissues at relatively low steady-state levels (cf. Figure 5, stage I "Baseline level"). The regulated increase in the production of superoxide anion radical or nitric oxide leads to a temporary imbalance, which forms the basis of redox regulation (stage II in Figure 5, "Regulatory imbalances"). The persistent production of abnormally large amounts of ROS or RNS, however, may lead to persistent changes in signal transduction and gene expression, which, in turn, may give rise to pathological conditions (as seen in Figure 5, stage III "Dysregulation by chronic oxidative stress") [34]. One of the classes of such diseases includes arthritic conditions – inflammatory diseases of joints. A substantial amount of evidence exists for an increased generation of oxidants in patients suffering from acute and chronic inflammatory joint diseases [36, 37] – see Table 6.

Regulatory Imbalances within a Synovial Joint

As schematically reported by Dröge [34], under physiological status, "Baseline level" (cf. Figure 5) of ROS and/or RNS concentration play an important role as regulatory mediators in signaling processes. In case of the composition of SF of healthy organisms, one may state two border concentrations of ROS (and RNS as well), which are primarily determined by the O_2 level within SF, or more precisely by the H_2O_2 level escaped from mitochondria of chondrocytes and from those of cells of the synovial membrane. A lower one exists at rest regimen of the joint and a higher H_2O_2 level at reoxygenation of the joint tissues during movement of the subject. The high-molar-mass HA however keeps most probably the joint ROS/RNS homeostasis between the two concentration values inside the "Baseline level" (see Figure 5, stage I).

On accepting the tenet that concentrations of H_2O_2 ranging around 50 μM (sometimes even up to 100 μM) are not toxic to any cells [32], the highest limit (cf. stage I, Figure 5) of the hydrogen peroxide level in SF, and thus in contact with both chondrocytes and synovial-membrane cells, is close to this concentration (<100 μM). The flux of H_2O_2 in the amount of less than 0.003 μM per minute does not change SF viscosity (cf. Table 6). In light of this observation one can propose that the ROS action, i.e., H_2O_2-degradative action on the high-molar-mass HA, is fully compensated by the *de novo* synthesis of megaDalton hyaluronans by the synoviocytes embedded within the synovial membrane of healthy human beings. Our detailed studies focusing on the H_2O_2-degradative action to HA macromolecules also showed that hydrogen peroxide up to hundreds of micromolar concentrations led to practically no cleavage/decay of high-molar-mass hyaluronan samples when the reaction system was “free” of any transition metal ions, namely those of iron and/or copper [M. Stankovská *et al.*, not published].

Let us now admit the situation of occurrence of temporary “Regulatory imbalances” (stage II in Figure 5), or more precisely the situation at which an acute inflammation is initiated within the synovial joint. On taking into account the data given in Table 6, the increase in ROS concentration, or more precisely the increase in H_2O_2 flux, appears to be functionally related to the rising number of PMNLs in the SF, presenting in the initial phase as Regulatory inbalance. This increase is however associated with the following events: i) infiltration of the increased number of white cells (PMNLS and/or macrophages) from the blood circulation into the SF, and ii) activation of these cells in the SF. Yet concerning the event given in ii), it has to be emphasized that at the time of infiltration movement of the white blood cells is impeded in the SF, due to its viscosity, which can be characterized as “normal” (cf. Table 6; see Figure 6) or high caused by the presence of high-molar-mass HA macromolecules. Moreover, it is a well known fact that especially high-molar-mass hyaluronans exert antiimflammatory action or more precisely, the long-sized HA chains quench the PMNLs and macrophages.

Thus one may admit that infiltration of an increased number of white cells into a millieu such as that of SF of healthy human beings need not immediately result in a rise of the ROS concentration or the H_2O_2 level enhancement, respectively. The demand of rapid/acute growth of ROS/RNS level within the joint during the stage II (cf. Figure 5, “Regulatory imbalances”) could not be met in this way. Resulting from our experimental findings, we may hereby offer/recommend our hypothesis/speculation in point of process sequencing which can very quickly, owing to their physiological status, bring about – for a temporary time period – the status possibly be defined as accute inflammation, or – by taking into account the Dröge scheme (cf. Figure 5 [34]) – the “Regulatory inbalances”.

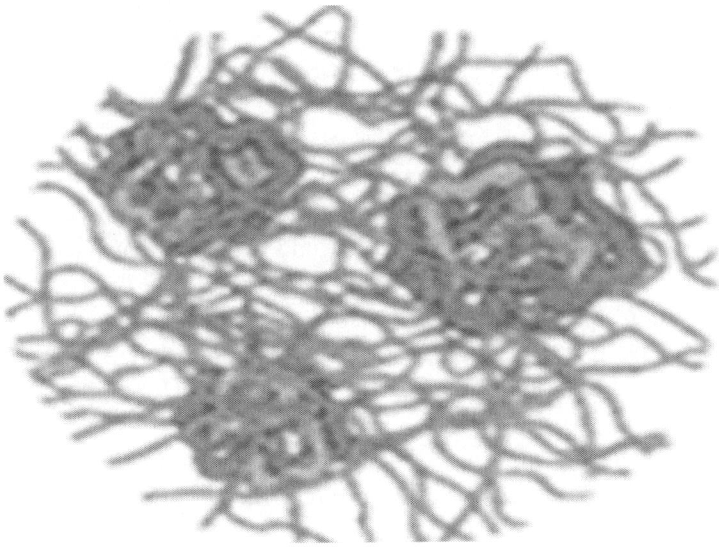

Figure 6. The movement of the white blood cells in the normal/highly viscous SF. The long-sized HA chains are sketched as strands.

Table 7. Comparison between acute and chronic inflammation (from [38])

Inflammation	Acute	Chronic
Causative agent	Pathogens, injured tissues	Persistent acute inflammation due to non-degradable pathogens, persistent foreign bodies, or autoimmune reactions
Major cells involved	Neutrophils, mononuclear cells (monocytes, macrophages)	Mononuclear cells (monocytes, macrophages, lymphocytes, plasma cells), fibroblasts
Primary mediators	Vasoactive amines, eicosanoids	IFN-γ and other cytokines, growth factors, reactive oxygen species, hydrolytic enzymes
onset	immediate	delayed
duration	few days	up to many months or years
outcomes	resolution, abscess formation, chronic inflammation	tissue destruction, fibrosis

INFLAMMATION

Inflammation generally means a complex biological response of tissues to harmful stimuli, such as infective pathogens, damaged cells, toxins, physical and/or chemical irritants. It is a protective attempt by the organism to remove injurious stimuli and to initiate the healing process for the tissue. Yet inflammation that runs unchecked can lead to various diseases (cf. Table 5), including those connected to synovial joints. Normally, however, inflammation is critically controlled and closely regulated by the body.

Inflammation can be classified as acute or chronic (Table 7). Acute inflammation is the initial response of the body to harmful stimuli and is achieved by the increased movement of PMNLs from the blood into the injured tissues. Then a cascade of biochemical events propagates and matures the (local) inflammatory response. Chronic inflammation usually leads to a progressive shift in the type of immune cells which are present at the site of inflammation and is characterized by destruction and often by (partial) healing of damaged tissues.

Acute inflammation – a short-term process appearing in a few minutes or hours – is usually characterized by five cardinal signs: rubor, calor, tumor, dolor, and *functio laesa*. However, the acute inflammation of an internal organ may not be manifested by the full set of signs.

Inflammation, and especially the acute one, is associated with elevated systemic levels of acute-phase proteins. These proteins prove beneficial in acute inflammation.

Acute-Phase Proteins

Acute-phase proteins are a class of proteins whose plasma concentrations increase (positive acute-phase proteins) or decrease (negative acute-phase proteins) in response to inflammation. This response is called the acute-phase reaction or acute-phase response. The acute-phase reactants are produced by the liver in response to specific stimulations. The following positive acute-phase proteins belong to the physiologically most prominent ones: C-reactive protein, α_1-antitrypsin and α_1-antichymotrypsin, fibrinogen, prothrombin, complement factors, ferritin, serum amyloid A, α_1-acid glycoprotein, ceruloplasmin, and haptoglobin. Others – negative acute-phase proteins such as albumin, transferrin – give negative feedback on the inflammatory response.

Ceruloplasmin

The concentration of ceruloplasmin, whose molar mass ($\approx$ 134 kDa) exceeds nearly twice that of albumin, increases markedly under certain circumstances – including those of acute inflammation. Since each ceruloplasmin macromolecule complexes/binds up to eight Cu(II)/Cu(I) ions of which two can liberate relatively easily [39], at the early stage of acute inflammation the actual copper level increases markedly. The consequence of higher ceruloplasmin concentration in blood plasma – accompanied with a rise in the concentration of copper ions – would mean a larger amount of this biogenic trace element that might cross the synovial membrane [16]. Yet, due to the gel-like consistency of SF, the copper ions entering into this specific environment start their redox action in the vicinity of the synovial membrane.

Weissberger's Oxidative System

The concentration of ascorbate in SF of healthy subjects reaches the values close to those established in blood serum, i.e., 40–140 μM [28]. Ascorbate, an "actor of physiologic HA catabolism in SF" with copper liberated from ceruloplasmin, creates easily the so-called Weissberger's oxidative system [40, 41] – ascorbate-Cu(I)-oxygen – generating H_2O_2 (cf. Scheme 4) [42-44]. Moreover, due to the simultaneous decomposition of hydrogen peroxide by the redox active copper ions, a large flux of hydroxyl radicals may occur [45].

As evident from the data listed in Table 1, iron and copper are the two prevailing redox active transition metals in SF. Although just only a minor fraction of their respective total levels equaling 5.2 μM and 4.3 μM is disposable for Weissberger's and/or Fenton-type reactions, it are the copper ions that better fulfill the requirement of acute (rapid) generation of ROS – particularly of $^{\bullet}OH$ radicals (cf. Figure 7).

Figure 7 illustrates the degradative action of ROS by monitoring the viscosity-time profiles of a HA solution into which – along with 100 μM ascorbate – a single transition metal was added [46]. As evident, a significant reduction of the solution dynamic viscosity (η), corresponding to the degradation of the high-molar-mass HA sample, clearly indicates a concentration-dependent manner for each metal (cf. left and right panels in Figure 7). While the character of the time dependence of η value upon the addition of $FeCl_2$ (5.0 μM) can be described as a gradual monotonous decline, the addition of $CuCl_2$ (5.0 μM) resulted in a literally "dramatic" drop of η value within a short time interval (30 min). A similar drop of η value and two-phase reaction kinetics are identifiable upon the addition of even a minute (0.1 μM) amount of $CuCl_2$ (see Figure 7, left panel). A possible explanation of this dissimilarity lies most probably in different reaction kinetics of the processes leading to generation of oxygen-derived reactive species in the system ascorbate *plus* $CuCl_2$ and in that comprising ascorbate *plus* $FeCl_2$.

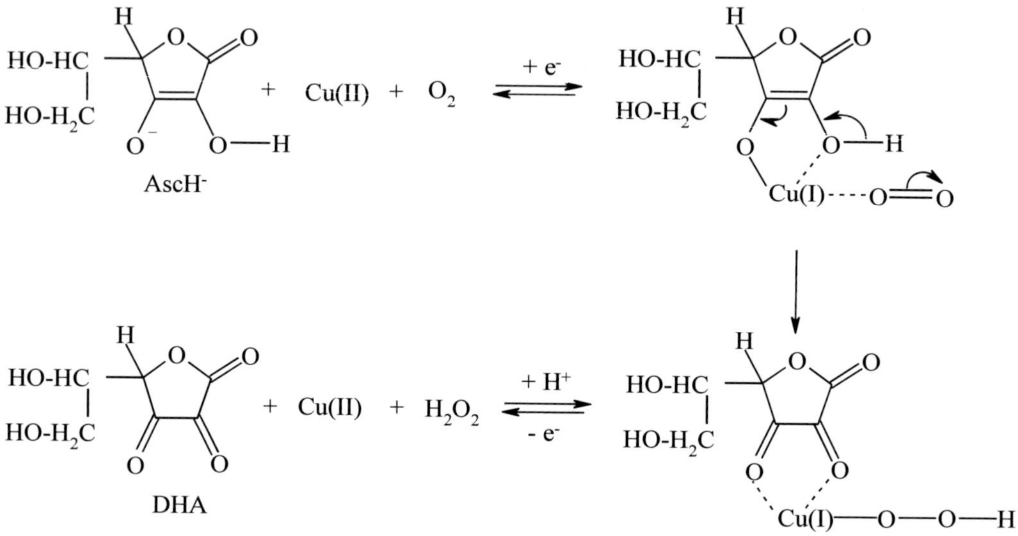

Scheme 4. Generation of H_2O_2 by Weissberger's system from ascorbate and Cu(II) under aerobic conditions (adapted from Fisher and Naughton [44]).

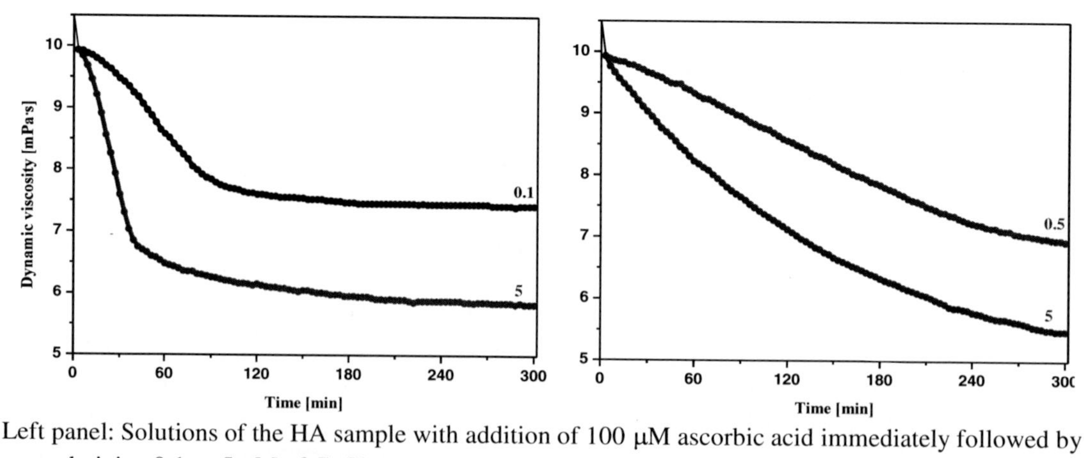

Left panel: Solutions of the HA sample with addition of 100 μM ascorbic acid immediately followed by admixing 0.1 or 5 μM of $CuCl_2$.

Right panel: Solutions of the HA sample with addition of 100 μM ascorbic acid immediately followed by admixing 0.5 or 5 μM of $FeCl_2$.

Figure 7. Time dependences of dynamic viscosity of solutions of a high-molar-mass HA sample.

As seen in Figure 7, the transition metal – either iron or copper – can play an active role in oxidative HA catabolism. However, the increase in Cu(II) concentration within the joint (and particularly in SF) could lead to an extremely rapid degradation of the native HA macromolecules. How efficiently the chemically generated •OH radicals are "scavenged" within this microenvironment by the locally disposable albumin as well as by the HA polymer fragments of lower molecular size, remains questionable. The oxidative process may escape the control mechanisms and damage/disrupt the synovial membrane. Moreover, the intermediate-sized HA-polymer fragments generated within this microenvironment could participate in the activation of "defender" cells. They may further intensify the inflammation state of the injured tissue(s) as the HA-polymer fragments can in turn augment the inflammatory responses. As reported by Jiang *et al.*, the HA fragments in the e.g. 2×10^5 Da range induce the expression of a number of inflammatory mediators in macrophages, including chemokines, cytokines, growth factors, proteases, and nitric oxide [47]. In this way, the oxidants generated by activated defender cells may enlarge the damage within the involved joint tissues such as the synovial membrane (cf. Figure 8). Such an increase in unmediated reactive radicals, generally termed oxidative stress, is an active area of research in a variety of diseases where copper may play an insidious role.

Moreover, reactive oxygen species appear to disrupt copper binding to ceruloplasmin, thereby releasing "free" copper ions, which in turn may promote oxidative pathology [39]. The damage can be manifested by visually localizable cardinal signs of inflammation – i.e., rubor, calor, tumor, dolor, and *functio laesa*, yet less distinct, repeated (micro-acute) inflammatory injures may lead to a disastrous outcome, e.g., an autoimmune disease such as rheumatoid arthritis.

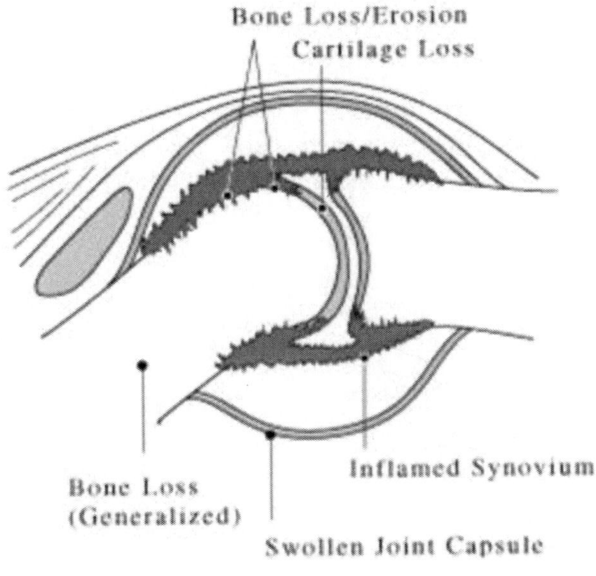

Figure 8. Damages within the inflamed joint tissues.

Relevancy and Function of Weissberger's Oxidative System at Acute Inflammation of the Joint

As demonstrated by the results depicted in Figure 7 (left panel) Weissberger's oxidative system is really a prompt/ultimate generator of hydrogen peroxide leading immediately to dramatic flux of $^{\bullet}OH$ radicals. Subsequently these radicals initiate a significant degradation of long-chain HA macromolecules, the process which diminishes markedly the dynamic viscosity of the hyaluronan solution. A similar HA degradative process can be anticipated in SF at the early stage of acute (synovial) joint inflammation. The lower SF viscosity may markedly promote the transition of defender cells from blood through the synovial membrane and further enhance the movement of these cells to the target synovial and periarticular tissues. These cells may simultaneously undergo activation in contact with/binding to biopolymer fragments resulted from ($^{\bullet}OH$) radical degradation of native high-molar-mass hyaluronans present in SF. The infiltrated defender cells thus may start their more or less specific action inside the intraarticular space.

Chronic Inflammation

In acute inflammation, if the injurious agent persists, chronic inflammation will ensue. This process marked by inflammation lasting many days, months or even years, may lead to the formation of a chronic wound. Chronic inflammation is characterized by the dominating presence of macrophages in the injured tissue. These cells are powerful defensive agents of the body, but the "toxins" they release – including ROS and/or RNS – are injurious to the organism's own tissues. Consequently, chronic inflammation is almost always accompanied by tissue destruction. Destructed tissues are recognized by the immunity system and, when "classified" by the body as foreign ones, a cascade of autoimmune reactions could start. Such reactions are well established in diseases such as rheumatoid arthritis, where – along with the (synovial) joints – several further tissues/organs, e.g., lungs, heart, and blood vessels, are permanently atacked, i.e., miss-recognized as foreign ones.

Medications Used to Treat Inflammatory Joint Diseases

There are many medications available to decrease joint pain, swelling, inflammation and to prevent or minimize the progression of the inflammatory disease. These medications include:

- Non-steroidal anti-inflammatory drugs (NSAIDs – such as acetylsalicylic acid/aspirin, ibuprofen or naproxen).
- Corticosteroids (such as prednisone).
- Anti-malarial medications (such as hydroxychloroquine).
- Other medications, including methotrexate, sulfasalazine, leflunomide, anti-TNF medications, cyclophosphamide, and mycophenolate.

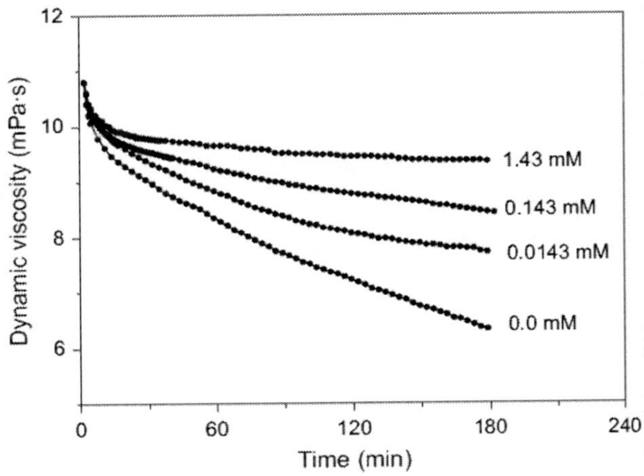

Concentration of acetylsalicylic acid added into the system before initiation of HA degradation in mM: 0.0, 0.0143, 0.143, and 1.43.

Figure 9. Effect of acetylsalicylic acid on HA degradation in the system 0.1 μM $CuCl_2$ + 100 μM ascorbic acid + 2 mM NaOCl.

As reported in the Section "Relevancy and function of Weissberger's oxidative system at acute inflammation of the joint", the early acute-phase of (synovial) joint inflammation should, most plausibly, be accompanied with generation of ROS (and RNS) – particularly with $^{\bullet}OH$ radicals. These, however, due to their extrememly high electronegativity (-2.31 V) should – in contact with any hydrogen atom containing compounds – entrap a proton ($^{\bullet}H$). By that process the $^{\bullet}OH$ radicals are partially or fully scavenged (cf. Figure 9). If the resulting radical generated from the given compound/medication is not able to initiate HA degradation, we speak of drug-scavenging, which could moderate the free radical process within the inflamed joint.

Figure 9 illustrates such an *in vitro* testing of the scavenging efficiency of acetylsalicylic acid/aspirin. As evident, this drug – based on its activity under aerobic conditions within the system HA-ascorbate-Cu^{2+}-NaOCl – can be classified as a potent scavenger of $^{\bullet}OH$ radicals [48].

CONCLUSION

With the current understanding that free radicals can act as cell signaling or "messenger" agents it is likely that they also play a role in normal cellular function as well as various disease etiologies. Researchers are now making rapid progress in understanding the role of oxidative stress and nitrosative stress in cardiovascular diseases such as atherosclerosis, ischemia/reperfusion injury, restenosis and hypertension; cancer; inflammatory diseases such as acute respiratory distress syndrome (ARDS), asthma, inflammatory bowel disease (IBD), dermal and ocular inflammation and arthritis; metabolic diseases such as diabetes; and diseases of the central nervous system (CNS) such as amyotrophic lateral sclerosis (ALS), Alzheimer's, Parkinson's, and stroke. The increased awareness of oxidative stress related to disease and the need to measure the delicate balance that exists between free radicals and the given systems in regulating them has given rise to a demand for new research tools.

ACKNOWLEDGMENT

The work was supported by the VEGA grant No. 2/0011/11 and the APVV grant No. 0351-10.

REFERENCES

[1] M. Valko, C. J. Rhodes, J. Moncol, M. Izakovic, M. Mazur, *Chem.-Biol. Interact.*, 160 (2006) 1–40.

[2] D. Harman, *J. Gerontol.*, 11 (1956) 298–300.

[3] M. Valko, D. Leibfritz, J. Moncol, M. T. D. Cronin, M. Mazur, J. Telser, *Int. J. Biochem. Cell Biol.*, 39 (2007) 44–84.

[4] http://www.niams.nih.gov/Health_Info/Rheumatic_Disease/graphics/joint_vert.gif.

[5] R. Servaty, J. Schiller, H. Binder, K. Arnold, *Int. J. Biol. Macromol.*, 28 (2001) 121–127.

[6] http://www.nature.com/ncprheum/journal/v2/n7/fig_tab/ncprheum0216_F1.html#figure-title (by F. H. Chen, K. T. Rousche, R. S. Tuan, *Nat. Clin. Prac. Rheumatol.*, 2 (2006) 373–382).

[7] B. M. Praest, H. Greiling, R. Kock, *Carbohydr. Res.*, 303 (1997) 153–157.

[8] K. M. N. Oates, W. E. Krause, R. H. Colby, *Mat. Res. Soc. Symp. Proc.*, 711 (2002) 53–58.

[9] E. Rovenská, *Rheumatologia*, 15 (2001) 57–64.

[10] R. O. Day, A. J. McLachlan, G. G. Graham, K. M. Williams, *Clin. Pharmacokin.*, 36 (1999) 191–210.

[11] G. Kogan, L. Šoltés, R. Stern, R. Mendichi, "Hyaluronic acid: A biopolymer with versatile physico-chemical and biological properties" in the book "Handbook of Polymer Research: Monomers, Oligomers, Polymers and Composites, Chapter 31", pp. 393–439, Eds. R. A. Pethrick, A. Ballada, G. E. Zaikov, Nova Science Publishers, New York 2007.

[12] V. R. Ryabina, S. E. Vasyukov, V. P. Panov, S. . Starodubtsev, *Khim.-Farm. Zh.* 21 (1987) 142–149.

[13] T. Hardingham, "Solution properties of hyaluronan" in the book "Chemistry and Biology of Hyaluronan", pp 1–19, Eds. H.G. Garg, C.A. Hales, Elsevier Press, Amsterdam 2004.

[14] J. S. Gibson, P. I. Milner, R. White, T. P. Fairfax, R. J. Wilkins, *Pflugers Arch.* 455 (2008) 563–573.

[15] Y.-Z. Fang, S. Yang, G. Wu, *Nutrition*, 18 (2002) 872–879.

[16] L. Šoltés, G. Kogan, "Impact of transition metals in the free-radical degradation of hyaluronan biopolymer" in the book "Kinetics & Thermodynamics for Chemistry & Biochemistry, Vol. 2", pp. 181–199, Eds. E.M. Pearce, G. E. Zaikov, G. Kirshenbaum, Nova Science Publishers, New York 2009; ibid *Polym. Res. J.*, 2 (2008) 269-288.

[17] P. I. Milner, R. J. Wilkins, J. S. Gibson, *Osteoarthr. Cartilage*, 15 (2007) 735–742.
[18] C. L. Hawkins, M. J. Davies, *Free Radic. Biol. Med.*, 21 (1996) 275–290.
[19] E. C. Kennett, M. J. Davies, *Free Radic. Biol. Med.*, 47 (2009) 389–400.
[20] C. L. Hawkins, M. J. Davies, *Biochem. Soc. Trans.*, 23 (1995) S248.
[21] G. Kogan, L. Šoltés, R. Stern, J. Schiller, R. Mendichi, "Hyaluronic Acid: Its Function and Degradation in In Vivo Systems", in the book "Studies in Natural Products Chemistry, Vol. 35 Bioactive Natural Products, Part D", pp. 789–882, Ed. by Atta-ur-Rahman, Elsevier, Amsterdam 2008.
[22] P. Myint, D. H. Deeble, P. C. Beaumont, S. M. Blake, G. O. Phyllips, *Biochim. Biophys. Acta,* 925 (1987) 194–202.
[23] D. Presti, J. E. Scott, *Cell Biochem. Func.*, 12 (1994) 281–288.
[24] M. Grootveld, E. B. Henderson, A. Farrell, D. R. Blake, H. G. Parkes, P. Haycock, *Biochem. J.*, 273 (1991) 459–467.
[25] W. Niedermeier, J. H. Griggs, *J. Chronic Dis.*, 23 (1971) 527v536.
[26] "Transport and Storage of Metal Ions in Biology", in the book "Biological Inorganic Chemistry: Structure and Reactivity" pp. 57v77, Eds. I. Bertini, H. B. Gray, E. I. Stiefel, J. S. Valentine, University Science Books, Sausalito California 2007.
[27] B. Halliwell, J. M. C. Gutteridge, *Methods Enzymol.*, 186 (1990) 1v85.
[28] S. F. Wong, B. Halliwell, R. Richmond, W. R. Skowroneck, *J. Inorg. Biochem.*, 14 (1981) 127v134.
[29] C. A. Hitchon, H. S. El-Gabalawy, *Arthritis Res. Ther.*, 6 (2004) 265–278.
[30] http://www.oxisresearch.com/library/oxidative_stress.shtml.
[31] http://books.google.sk/books?id=a52CE0vWRXQC&pg=PA321&lpg=PA321&dq=standard+reduction+potential+H2O2/%E2%80%A2OH&source=bl&ots=YJL8hdl4Wh&sig=ChpmRnhvt--GEuyfkbotmWkvCVg&hl=sk&ei=aTxcSvu9HoyF_AbQl6HwDA&sa=X&oi=book_result&ct=result&resnum=7.
[32] http://www.gaiaresearch.co.za/hydroperoxide.html.
[33] M. Martínez-Cayuela, *Biochimie*, 77 (1995) 147v161.
[34] W. Dröge, *Physiol. Rev.*, 82 (2002) 47v95.
[35] B. Halliwell, J. R. Hoult, D. R. Blake, *FASEB J.*, 2 (1988) 2867v2873.
[36] J. Schiller, B. Fuchs, J. Arnhold, K. Arnold, *Curr. Med. Chem.*, 10 (2003) 2123–2145.
[37] K. D. Brandt, "An Atlas of Osteoarthritis" The Parthenon Publishing Group Inc., 2001, NY, USA.
[38] http://en.wikipedia.org/wiki/inflammation.
[39] N. Shukla, J. Maher, J. Masters, G. D. Angelini, J. Y. Jeremy, *Atherosclerosis*, 187 (2006) 238–250.
[40] A. Weissberger, J. E. LuValle, D. S. Jr. Thomas, *J. Am. Chem. Soc.*, 65 (1943) 1934–1939.
[41] M. M. Khan, A. E. Martell, *J. Am. Chem. Soc.*, 89 (1967) 4176–4185.
[42] A. E. O. Fisher, D. P. Naughton, *Med. Hypotheses*, 61 (2003) 657–660.
[43] A. E. O. Fisher, D. P. Naughton, *Nutr. J.*, 3 (2004) 1–5.
[44] A. E. O. Fisher, D. P. Naughton, *Curr. Drug Deliv.*, 2 (2005) 261–268.
[45] L. Šoltés, M. Stankovská, V. Brezová, J. Schiller, J. Arnhold, G. Kogan, P. Gemeiner, *Carbohydr. Res.*, 341 (2006) 2826–2834.
[46] K. Valachová, G. Kogan, P. Gemeiner, L. Šoltés, "Hyaluronan degradation by ascorbate: Protective effects of manganese(II) chloride" in the book "Kinetics &

Thermodynamics for Chemistry & Biochemistry, Vol. 2", pp. 201–215, Eds. E.M. Pearce, G.E. Zaikov, G. Kirshenbaum, Nova Science Publishers, New York 2009; ibid *Polym. Res. J.*, 2 (2008) 269–288.

[47] D. Jiang, J. Liang, P. W. Noble, *Annu. Rev. Cell Dev. Biol.*, 23 (2007) 435–461.

[48] M. Stankovská, J. Arnhold, J. Rychlý, H. Spalteholz, P. Gemeiner, L. Šoltés, *Polym. Degrad. Stabil.*, 92 (2007) 644–652.

In: News in Chemistry, Biochemistry and Biotechnology ISBN: 978-1-63117-273-1
Editors: G. E. Zaikov, G. Nyszko, L. P. Krylova et al.

Chapter 2

SURFACE PROPERTIES OF POLYIMIDE COPOLYMERS

***Igor Novák*[1,*], *Peter Jurkovič*[2,†], *Jan Matyašovský*[2,‡], *Petr Sysel*[3], *Milena Špírková*[4] *and Ladislav Šoltés*[5,§]**

[1]Polymer Institute, Slovak Academy of Sciences, Bratislava, Slovakia
[2]VIPO, a.s., Partizanske
[3]Department of Polymers, Institute of Chemical Technology, Prague, Czech Republic
[4]Institute of Macromolecular Chemistry AS CR, Prague, Czech Republic
[5]Institute of Experimental Pharmacology of the Slovak Academy of Sciences, Bratislava, Slovakia

ABSTRACT

Several sorts of block polyimide based copolymers, namely poly(imide-co-siloxane) (PIS) block copolymers containing siloxane blocks in their polymer backbone have been investigated. In comparison with pure polyimides the PIS block copolymers possess some improvements, e.g., enhanced solubility, low moisture sorption, and their surface reaches the higher degree of hydrophobicity already at low content of polysiloxane in PIS copolymer. This kind of the block copolymers are used as high-performance adhesives and coatings. The surface as well as adhesive properties of PIS block copolymers depends on the content and length of siloxane blocks. The surface properties of PIS block copolymers are strongly influenced by enrichment of the surface with siloxane-based segments. Micro phase separation of PIS block copolymers occurs due to the dissimilarity between the chemical structures of siloxane, and imide blocks even at relatively low lengths of the blocks. The surface analysis of PIS block copolymers using various methods of investigation e.g., contact angle measurements, SEM, TEM, AFM, ATR-FTIR, and XPS, was performed, and the strength of the adhesive joint to more polar polymer was studied. The surface and adhesive properties are discussed in view of the varied composition of PIS block copolymers.

* E-mail: upolnovi@savba.sk.
† E-mail: pjurkovic@vipo.sk.
‡ E-mail: jmatyasovsky@vipo.sk.
§ E-mail: ladislav.soltes@savba.sk.

INTRODUCTION

Polyimides present an important class of polymers, necessary in microelectronics, printed circuits construction, and aerospace investigation, mainly because their high thermal stability and good dielectric properties [1-4]. Poly(imide-co-siloxane) (PIS) block copolymers containing siloxane blocks in their polymer backbone have been investigated [5, 6]. In comparison with pure polyimides the PIS block copolymers possess some improvements, e.g., enhanced solubility, low moisture sorption, and their surface reaches the higher degree of hydrophobicity already at low content of polysiloxane in PIS copolymer. This kind of the block copolymers are used as high-performance adhesives and coatings. The surface as well as adhesive properties of PIS block copolymers depends on the content and length of siloxane blocks. The surface properties of PIS block copolymers are strongly influenced by enrichment of the surface with siloxane segments [7]. Micro phase separation of PIS block copolymers occurs due to the dissimilarity between the chemical structures of both siloxane, and imide blocks.

EXPERIMENTAL

2-Aminoterminated ODPA-BIS P polyimides with controlled molecular weight were synthesized by solution imidization (first step in NMP at room temperature for 24 h, second step in NMP–BCB mixture at 180 °C). The number-average molecular weights of products were in the range M_n = 2000–18,000 g/mol (by ^{1}H NMR spectroscopy. The surface morphology (height image) and local surface heterogeneities (phase image) were measured by AFM. All measurements were performed under ambient conditions using a commercial atomic force microscope (NanoScopeTM Dimension IIIa, MultiMode Digital Instruments, USA) equipped with the PPP-NCLR tapping-mode probe (NanosensorsTM Switzerland; spring constant 39 N/m, resonant frequency 160 kHz). The surface energy of PIS block copolymer was determined via measurements of contact angles of a set of testing liquids (i.e., re-distilled water, ethylene glycol, formamide, methylene iodide, 1-bromo naphthalene) using SEE (Surface Energy Evaluation) system completed with a web camera (Masaryk University, Czech Republic) and necessary PC software. The drop of the testing liquid (V = 3 μl) was placed with a micropipette (0–5 μl, Biohit, Finland) on the polymer surface, and a contact angle of the testing liquid was measured. The peel strength of adhesive joint (P_{peel}) to polyacrylate was measured by 90° peeling of adhesive joint using universal testing machine Instron 4301 (Instron, England) with 100 N measuring cell. The adhesive joints for peel tests were fixed in aluminum peeling circle.

RESULTS AND DISCUSSION

The AFM measurements of the PIS copolymers are shown in Figure 1. AFM measurements of the surface topography (height image) and tip-sample interaction (phase image) of the samples containing 0–33 wt.% of siloxane monomer revealed differences in both characteristics. Only characteristic samples, i.e., 0, 10, 20, and 33 wt.% of siloxane are shown in the Figure 1; sample containing 30 wt.% of siloxane is very similar in height and phase images to the sample with 33 wt.% siloxane and thus it is not shown here.

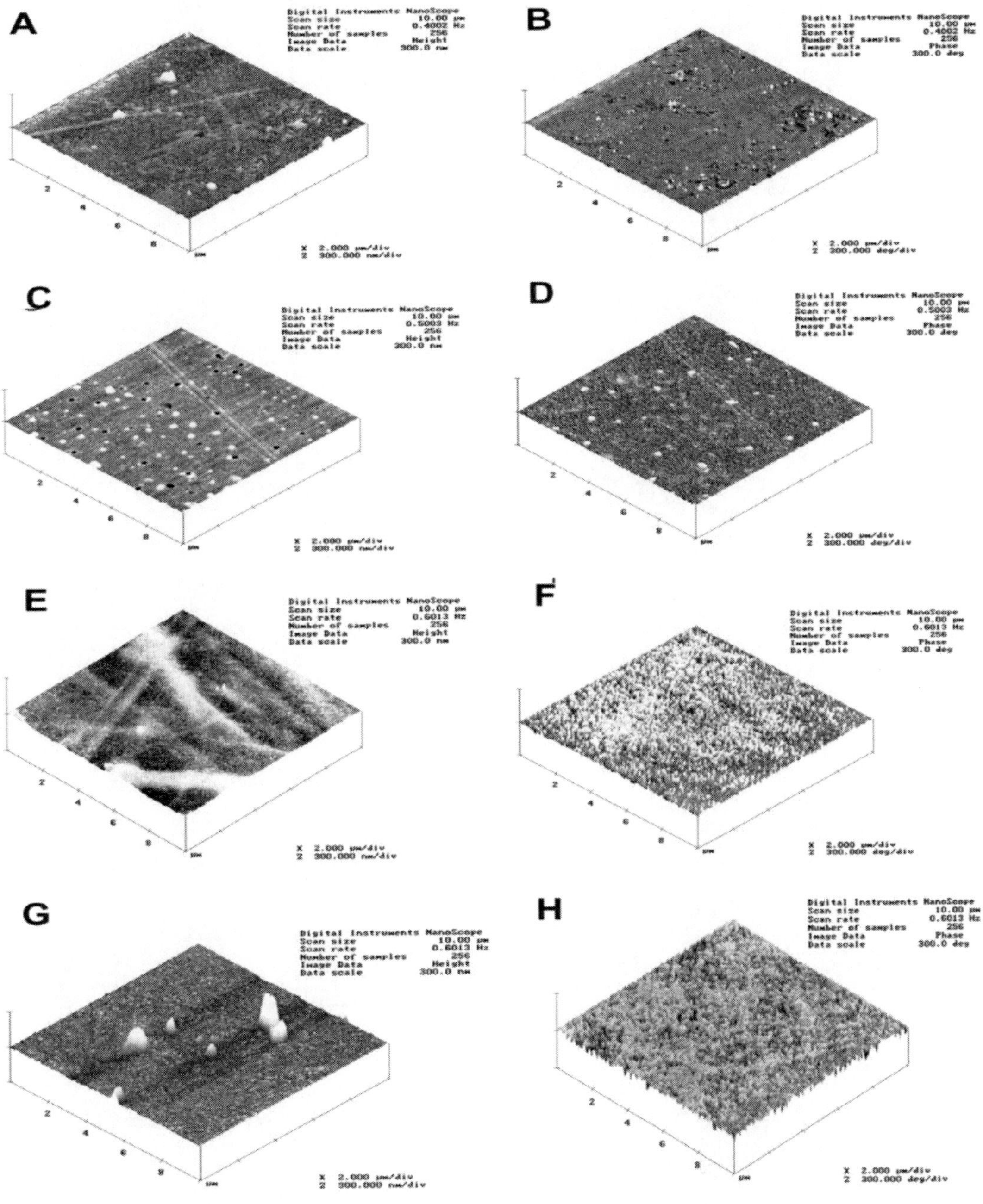

Figure 1. AFM images of PIS block copolymers films: pure polyimide (A, B), 10 wt.% of siloxane (C, D), 20 wt.% of siloxane (E, F), and 33 wt.% of siloxane (G, H) Height images (A, C, E, G), and phase (B, D, F, H) images, respectively.

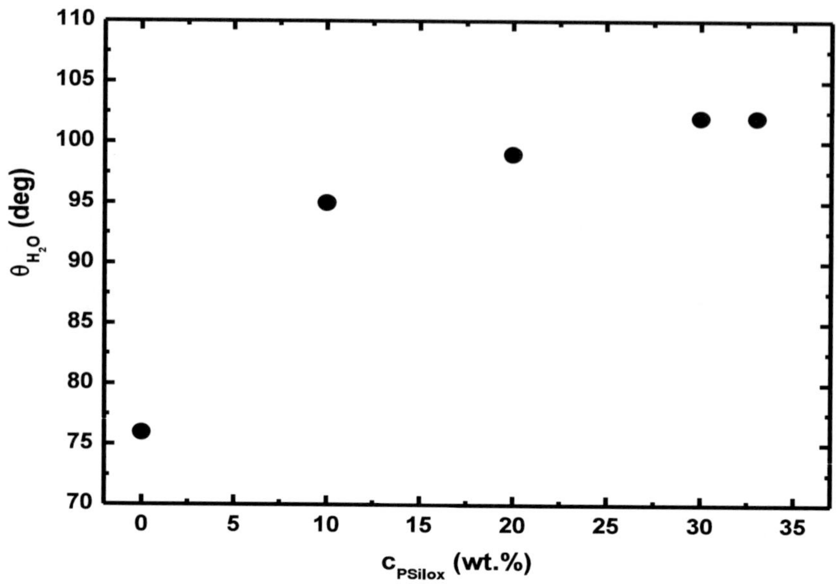

Figure 2. Contact angles of water vs. siloxane content in PIS block copolymer.

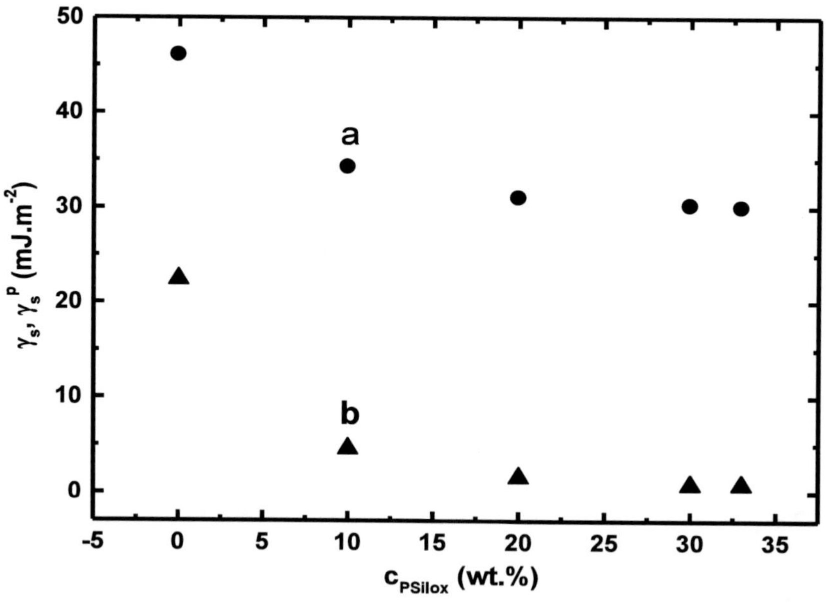

Figure 3. Surface energy, and its polar component of PSI block copolymer vs. siloxane content.

The comparison of height images: samples containing 20% (Figure 1E) and 30% (not shown here) have rugged and funicular surface relief. On the other hand, surfaces of pure polyimide (Figure 1A), 10% copolymer (Figure 1C) and 33% copolymer (Figure 1D) contain individual formations on the surfaces – ‘‘hills” of different size and height (tens–hundreds nm) and furthermore holes (tens of nm size) on 10% sample. Moreover, funicular formations are shadowed also in the Figure 1A and C. Comparison of phase images: Figure 1B vs. 1D, and 1F vs. 1H exhibit mutually similar relief. If compared the phase images with the relevant topography images, i.e., Figure 1A vs. C and Figure E vs. G, it is evident while height images are similar for first couple as well, significant differences for second couple exist.

Figure 2 shows the contact angles of re-distilled water deposited on PIS block copolymer surface vs. content of siloxane in copolymer. The contact angles of water by Figure 2 increased by growth of siloxane content and/or Si/N ratio in copolymer. The contact angles of PIS block copolymer increase from 76° for pure polyimide, to 95° for 10% of siloxane in copolymer up to 102° for 30% of siloxane in copolymer. Micro phase separation in PIS block copolymer occurs even at relatively low block lengths due to dissimilarity between the chemical structures of the siloxane, and imide blocks.

The dependencies of the surface energy, and its polar component of PIS block copolymer determined by OWRK (Owens-Wendt-Rabel-Kaelble) method [7] vs. content of siloxane in copolymer are shown in Figure 3. The surface energy of PIS block copolymer decreases significantly with the concentration of siloxane from 46.0 $mJ.m^{-2}$ (pure polyimide) to 34.2 $mJ.m^{-2}$ (10 % of siloxane), and to 30.2 $mJ.m^{-2}$ (30 % of siloxane). The polar component of the surface energy reached the value 22.4 $mJ.m^{-2}$ [pure polyimide], which decreases with content of siloxane in PIS copolymer to 4.6 $mJ.m^{-2}$ (10 % of siloxane) and 0.8 $mJ.m^{-2}$ (30 % of siloxane) The surface energy of pure polyimide is 46 $mJ.m^{-2}$, while the value of the surface energy of poly (dimethyl siloxane) is only 20.9 $mJ.m^{-2}$. At room temperature the siloxane molecules are above their glass temperature, their segments are capable to migrate to the polymeric surface, so making it more hydrophobic. The surface of the PSI copolymer films should be covered with polysiloxane segments having their thickness in molecular order.

Figure 4 shows the dependence of the peel strength of adhesive joint PSI block copolymer to epoxy vs. content of siloxane. It is seen that the peel strength of adhesive joint PIS copolymer-epoxy decreases with growth in siloxane content in the whole concentration range.

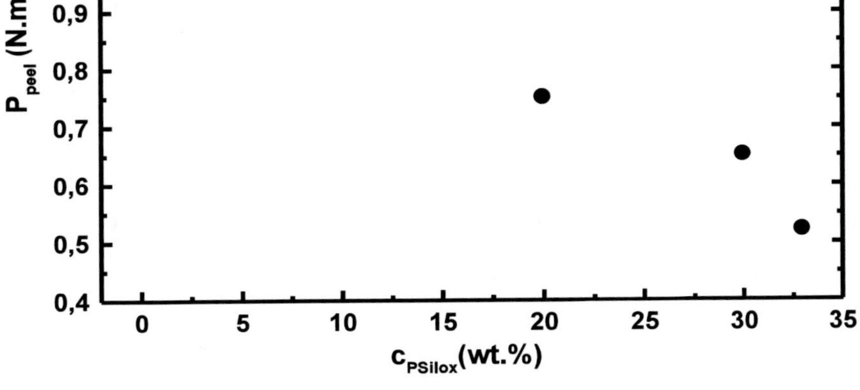

Figure 4. Peel strength of adhesive joint PSI block copolymer-epoxy vs. concentration of siloxane.

The fact that the strength of the adhesive joints decreases with increase in siloxane content reflects the increases hydrophobicity of the polymeric surface. The peel strength of adhesive joint to epoxy adhesive diminished from 1.2 MPa (pure polyimide), to 1.05 MPa (10 % of siloxane), and to 0.65 MPa (30 % of siloxane). This decrease of peel strength of adhesive joint is relatively steady for all investigated content of siloxane in block copolymer. Comparing polyimide with PSI block copolymer containing 30 % of siloxane shows that the peel strength of adhesive joint to epoxy decreased more than two times. The presence of siloxane in PSI block copolymer caused the more hydrophobic surface of copolymer (surface energy of copolymer containing 10 % of siloxane was 34.2 $mJ.m^{-2}$).

CONCLUSION

Ther morphology of PIS block copolymer has been changed due segregation of siloxane segments; constitution of polyimide continuous phase in copolymer was affirmed. A significant increase of roughness of PSI copolymer surface, if the content of siloxane is growing, was observed. The values of contact angles of water extremely increased by rising of siloxane content in PSI block copolymer and at higher composition were levelled off. The content of siloxane in copolymer increased, the surface energy, and its polar component of PSI copolymer diminished, the dispersive component of the surface energy on opposite increased, and if the content of siloxane in PIS copolymer rises up, strength of adhesive joint to epoxy decreased almost linearly.

ACKNOWLEDGMENT

This publication was prepared as part of the project „Application of Knowledge-based Methods in Designing Manufacturing Systems and Materials“ co-funded by the Ministry of Education, Science, Research and Sport of the Slovak Republic within the granted stimuli for research and development from the State Budget of the Slovak Republic pursuant to Stimuli for Research and Development Act No. 185/2009 Coll. and the amendment of Income Tax Act No. 595/2003 Coll. in the wording of subsequent regulations in the wording of Act. No. 40/2011 Coll., and by project of Ministry of Education of the Slovak Republic and Slovak Academy of Science VEGA, project No.2/0199/14.

REFERENCES

[1] F. Niklaus, P.Enoksson, E. Kälvesten, G. Stemme, *J. Micromech. Microeng*. 2001, 11, 100.

[2] M. B Chan-Park, S. S. Tan, *Int. J. Adhes. Adhesives* 2002, 22, 471.

[3] Engel, J Chen, C Liu, *J. Micromech. Microengn*. 2003, 13, 359.

[4] P. Sysel, R. Hobzová, V. Šindelář, J. Brus, *Polymer* 2001, 42, 10079.

[5] P. Sysel, D. Oupicky, *Polym. Intern*. 1996, 40, 275.

[6] J. E. McGrath, D. L. Dunson, S. J. Mecham, J. L. Hedrick, *Adv. Polym. Sci.* 1999, 140, 61.
[7] I. Novák, M. Števiar, I. Chodák, I. Krupa, T. Nedelčev, M. Špírková, M. M. Chehimi, J. Mosnáček, A. Kleinová: *Polym. Adv. Technol.* 2007, 18, 97.

In: News in Chemistry, Biochemistry and Biotechnology ISBN: 978-1-63117-273-1
Editors: G. E. Zaikov, G. Nyszko, L. P. Krylova et al.

Chapter 3

ANTIBACTERIAL POLYVINYLCHLORIDE PRE-TREATED BY BARRIER PLASMA

Igor Novák[1,*], Anton Popelka[1,6,+], Ján Matyašovský[2,†], Peter Jurkovič[2,‡], Marián Lehocký[3], Alenka Vesel[4], Ladislav Šoltés[5,§] and Ahmad Asadinezhad[7,ˇ]

[1]Polymer Institute, Slovak Academy of Sciences, Bratislava, Slovakia
[2]VIPO, a.s., Partizánske, Slovakia
[3]Tomas Bata University in Zlín, Zlín, Czech Republic
[4]Department of Surface Engineering, Plasma Laboratory, Jožef Stefan Institute, Ljubljana, Slovenia
[5]Institute of Experimental Pharmacology of the Slovak Academy of Sciences, Bratislava, Slovakia
[6]Center for Advanced Materials, Quatar University, Doha, Qatar
[7]Isfahan University of Technology, Isfahan, Iran

ABSTRACT

A multistep physicochemical approach making use of plasma technology combined with wet chemistry has fueled considerable interest in delivery of surface-active anti-adherence materials. In the first step of the approach, concerning an inherent lack of befitting functional groups on pristine substrate, plasma treatment at low temperature and atmospheric pressure has been substantiated to be productive in yielding reactive entities on the surface [1, 5]. The highlights the functionality of the adopted multistep physicochemical approach to bind polysaccharide species onto the medical-grade PVC surface. DCSBD plasma is capable of raising roughness, surface free energy, and

* Email: upolnovi@savba.sk.
† Email: jmatyasovsky@vipo.sk.
‡ Email: pjurkovic@vipo.sk.
+Email: Anton.Popelka@savba.sk
§Email: ladislav.soltes@savba.sk.
ˇ Email: asadinezhad@cc.iut.ac.ir.

introducing oxygen-containing functionalities anchored onto the surface. A structured poly(acrylic acid) brush of high graft density is synthesized using surface-initiated approach to further improve hydrophilicity and develop a stable brush-like assembly to yield a platform for biomolecular binding. In vitro bacterial adhesion and biofilm formation assays indicate incapability of single chitosan layer in hindering the adhesion of Staphylococcus aureus bacterial strain. Chitosan could retard Escherichia coli adhesion and plasma treated and graft copolymerized samples are found effective to diminish the adherence degree of Escherichia Coli.

INTRODUCTION

A new modification method using plasma technology combined with wet chemistry represents an efficient way in delivery of surface-active anti-adherence materials [1-4]. The atmospheric pressure electric discharge plasma has been substantiated to be productive in yielding reactive entities on the surface [5,6]. However, the need for treatment duration to a few seconds remains a pressing obstacle to extensive applications of this type of plasma [7]. A novel technology coined as diffuse coplanar surface barrier discharge (DCSBD) has been developed [8], which enables the generation of a uniform plasma layer under atmospheric pressure with a high surface power density in the very close contact of modified polymer.

EXPERIMENTAL

Materials: PVC pellets, extrusion medical-grade RB1/T3M of 1.25 $g \cdot cm^{-3}$ density, were obtained from ModenPlast (Italy) and used as received. Pectin from apple, (BioChemika, with esterification of 70-75%), acrylic acid (AA) (99.0%, anhydrous), and *N*-(3-dimethyl aminopropyl)-*N'*-ethyl carbodiimide hydrochloride (EDAC, 98.0%) were supplied by Fluka (USA). Chitosan from crab shells with medium molecular weight and deacetylation degree of 75-85%.

Plasma modification was implemented in static conditions by DCSBD plasma technology (figure 1) of laboratory scale with air as the gaseous medium at atmospheric pressure and room temperature. A schematic profile of the plasma system is given in Scheme 1. It basically comprises a series of parallel metallic electrodes inset inside a ceramic dielectric located in a glass chamber which allows the carrier gases to flow. All samples were treated on both sides with plasma power of 200 W for 15 sec.

For *grafting by AA* PVC substrates were immersed into spacer solutions containing 10 vol.% AA aq. solution. The reaction was allowed to proceed for 24 h at 30 °C. PAA grafted PVC samples were immersed into EDAC aq. solution at 4 °C for 6 h in order to activate the carboxyl groups on the surface. The highly active key intermediate, O-acylisourea, is produced having potential to react with reducing agents. Subsequently, they were transferred to chitosan and kept there for 24 h at 30 °C.

Sample 1 – pristine PVC, sample 2 –PVC treated by DCSBD plasma, sample 3 – PVC treated by plasma and grafted by AA, sample 4 – PVC treated by plasma, AA and chitosan, sample 5 – PVC treated by plasma AA, chitosan and pectin.

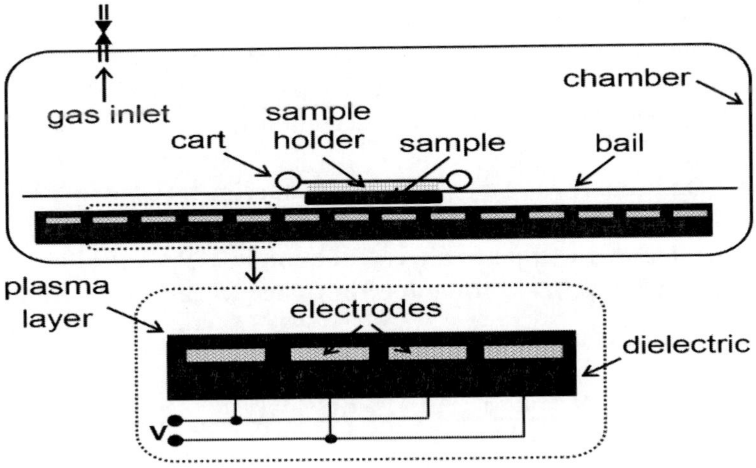

Scheme 1. Scheme of DSBD plasma source.

Scanning electron microscopy (SEM) was carried out on VEGA II LMU (TESCAN) operating in the high vacuum/secondary electron imaging mode at an accelerating voltage of 5-20 kV.

Bacterial adhesion and biofilm experiments were performed using gram-positive (*S. Aureus* 3953) and gram-negative (*E. Coli* 3954) bacteria. The circular shape specimens (d ≈ 8mm) were cut from the pristine and modified PVC samples before further investigation. After 24 hours incubation at 37 °C under continuous shaking at 100 rpm. The bacteria adhered on the surface of the specimens were removed by vigorous shaking of the test tube at 2000 rpm for 30 sec and quantified by serial dilutions and spread plate technique.

RESULT AND DISCUSSION

Surface Energy

Table 1 includes the contact angle values of deionized water (θ_w) recorded on different samples. Each sample has been designated by a number from 1 to 5 whose notation is inserted in the title of Table 1. Based on the given data, sample 1 exhibits a hydrophobic characteristic which after being treated by plasma, an evident change in θ_w arises and hydrophilicity ascends as anticipated. This trend continues as to sample 3 on which polyacrylic acid (PAA) chains are grafted where more hydrophilic propensity is shown inferred from θ_w value. The elevated hydrophilicity upon multistep modifications is assumed to come from the inclusion of superficial hydrophilic entities. The hydrophilicity then decreases as polysaccharides are coated onto the surface, though is well higher than that of sample 1, as the inherent hydrophilicity of chitosan is beyond doubt. Furthermore, sample 5 exhibits higher wettability than sample 4 implying a more effective binding of chitosan onto the surface, as remarked in other efforts as well. The hydrophilicity then decreases as polysaccharides are coated onto the surface, though is well higher than that of sample 1, as the inherent hydrophilicity of chitosan is beyond doubt. Furthermore, sample 5 exhibits higher wettability than sample 4 implying a more effective binding of chitosan onto the surface, as remarked in other efforts as well. To further explore the physicochemical parameters of the examined surfaces, an extensively used theory, Lifshitz-van der Waals/acid-base (LW/AB), has been exploited for free surface energy evaluation whose outputs with reference to diiodomethane, ethylene glycol, and deionized water as wetting liquids are supplied in Table 1. Sample 1 exhibits a basic character ($\gamma^{-} > \gamma^{+}$) as proposed by the data, even though acidity or basicity of neat PVC is yet controversial.

Table 1. Contact angle analysis results of different specimens using deionized water (w), ethylene glycol (E), diiodomethane (D), a nd formamide (F) as wetting agents. Sample 1: pristine/control; Sample 2: plasma treated: Sample 3: PAA grafted; Sample 4: chitosan coated; Sample 5: chitosan/pectin coated (mean±standard deviation)

Specimen	θ_w (°)	θ_E (°)	θ_D (°)	θ_F (°)	$\gamma^{-}_{LW/AB}$ (mJ/m²)	$\gamma^{+}_{LW/AB}$ (mJ/m²)	$\gamma^{AB}_{LW/AB}$ (mJ/m²)	$\gamma^{LW}_{LW/AB}$ (mJ/m²)	$\gamma^{tot}_{LW/AB}$ (mJ/m²)	$\gamma^{a)}_{Wu}$ (mJ/m²)	$\gamma^{b)}_{KN}$ (mJ/m²)	$\gamma^{c)}_{LN}$ (mJ/m²)
Sample 1	85.9 (±2.5)	60.5 (±3.0)	43.5 (±3.5)	64.2 (±6.0)	5.1	0.0	1.0	37.8	38.8	37.8	33.3	33.6
Sample 2	64.9 (±3.0)	49.4 (±4.0)	36.2 (±5.5)	51.0 (±6.0)	24.9	0.5	6.7	41.5	48.2	41.5	40.4	40.7
Sample 3	46.5 (±4.0)	51.3 (±5.5)	38.0 (±5.0)	47.7 (±4.5)	62.9	2.7	26.1	40.6	66.7	51.9	43.1	43.4
Sample 4	63.7 (±5.5)	43.4 (±3.0)	28.2 (±2.5)	44.9 (±5.0)	22.2	0.3	4.9	45.0	49.9	45.0	42.8	43.0
Sample 5	50.5 (±3.5)	40.0 (±2.5)	31.5 (±4.5)	31.0 (±3.5)	42.2	0.6	10.5	43.6	54.1	50.0	46.4	46.6

a) Surface free energy value according to Wu equation of state [33]; b) Surface free energy value according to Kwok-Neumann model [33]; c) Surface free energy value according to Li-Neumann model [33].

This increase is principally assisted by the polar (acid-base) component (γ^{AB}), rather than the apolar one (γ^{LW}), implying an incorporation of superficial polar oxygen-containing entities thanks to the air plasma treatment. A significant rise in γ^{tot} and γ^{AB} values is noticed for sample 3, in comparison with samples 1 and 2, indicative of the presence of carboxyl-containing units on the surface. As for samples 4 and 5, a reduction in γ^{AB} and γ^{tot} values is observed compared to sample 3, however, their γ^{tot} values rise above that of sample 1. The minimum values of θ_E and θ_F are found for sample 5 which reflect that the surface is seemingly coated by alcoholic and amine containing moieties which in fact points to the more efficient binding of chitosan when compared to sample 4.

Surface Morphology

The surface topography of samples 1-5 investigated by SEM as a common surface qualitative technique are presented in Figure 2. Sample 1 shows a level and uniform morphology which goes through a significant alteration ensuing the plasma treatment taking on an etched pattern with an unevenly shaped texture. The generated morphology is favorable for next coupling processes due to an enhanced surface area and roughness. The developed pattern on sample 2 is indeed, an outcome of the competing functionalization and ablation phenomena which brings on a reorganization of the surface microstructure.

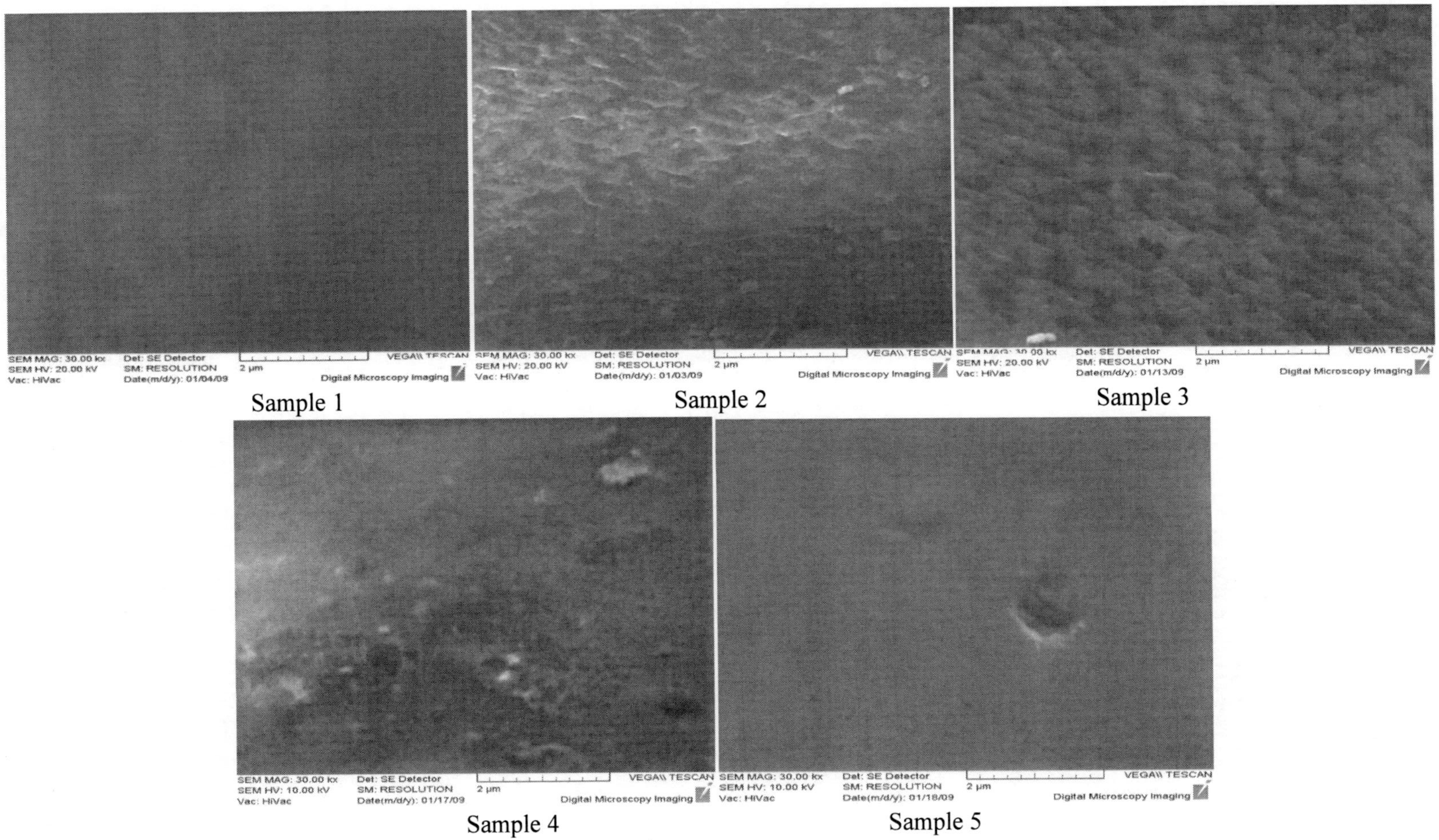

Figure 1. SEM micrographs of samples 1-5 taken at $3x10^4$ magnification.

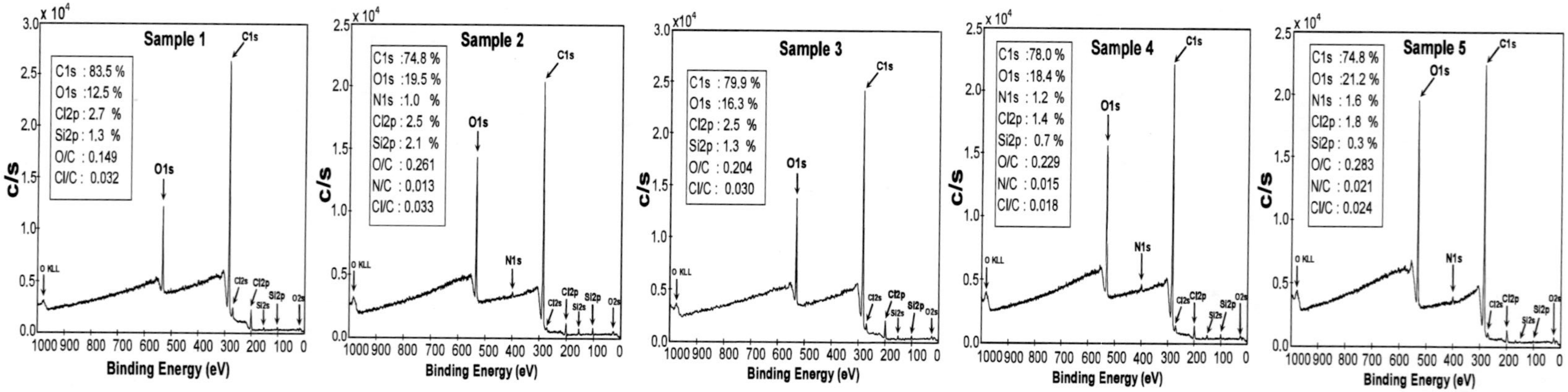

Figure 2. XPS survey-scan spectra of samples 1-5 along with atomic compositions.

The incident of the ablation is validated by gravimetric analysis where a weight loss of 4 $\mu g \cdot cm^{-2}$ has been observed due to the plasma treatment for 15 sec implying an approximate etching rate of 2 nm/s in terms of the used PVC grade density. Based on the sample 3 micrograph, PAA chains develop superficial domains of submicron dimension and brush-like features are then recognizable on the surface. As the grafting moves forward, clustering takes place because of the domains size growth. An additional compelling factor in controlling the surface microstructure is the grafting mechanism which is actually initiated by generated surface radicals.

Surface Chemistry

XPS Analysis

XPS, with a probe depth measuring around 5 nm, has been put to use to more thoroughly monitor the bearings of the surface modifications by picking up a quantitative perception into the surface elemental composition. The recorded survey spectra along with the corresponding surface atomic compositions and ratios of samples 1-5 are all provided in Figure 4. Carbon (C), oxygen (O), chlorine (Cl), and silicon (Si) elements are found on the sample 1 surface whose composition and elemental ratios are presented in the legend of the respective graph. The Cl2p atomic content is substantially lower than the amount found for a neat PVC containing no additives which refers to the existence of several additives and also X-ray degradation. The same rationale accounts for the considerable amount of O1s detected in sample 1 which is not a typical element in standard PVC.

Upon binding chitosan on the surface (sample 4), pronounced changes appear in the surface chemistry, as O1s content and O/C fraction increase and also N1s signal emerges, while Cl2p and Si2p bands abate due to the surface coverage by polysaccharide species. This trend yet continues for sample 5 as higher O1s and N1s as well as O/C and N/C atomic rations are detectable compared to sample 4 giving support to the notion that chitosan can be more stably, i.e., in higher quantity, attached onto the surface when layered along with pectin. In other words, use of pectin can promote the quality of chitosan binding.

Bacterial Adhesion and Biofilm Assay

The most crucial step of the biofilm formation is bacterial adhesion considered as a sophisticated topic in biointerface science whose plenty of aspects have not yet been well conceived. As a matter of fact, adhesion phenomenon is an interplay of myriad factors. Figure 5 shows the histograms of bacterial adhesion extent for samples 1-5 after 24 h incubation. As Regards the adherence degree of *S. aureus* onto the samples 2-4, no reduction is evident in the number of viable adhered colonies, compared to sample 1, signifying an inability of the modifications in hampering the *S. aureus* adhesion to the surface. From sample 1 to 3, both hydrophilicity and roughness rise, as remarked earlier, and then decrease in the case of samples 4 and 5. The adhesion degrees vary with a similar trend as well. Considering sample 5, it is inferred that chitosan/pectin assembly imparts biocidal effects against *S. aureus.* Chitosan single layer and chitosan/pectin multilayer restrain the adherence degree by 50% and 20%, respectively.

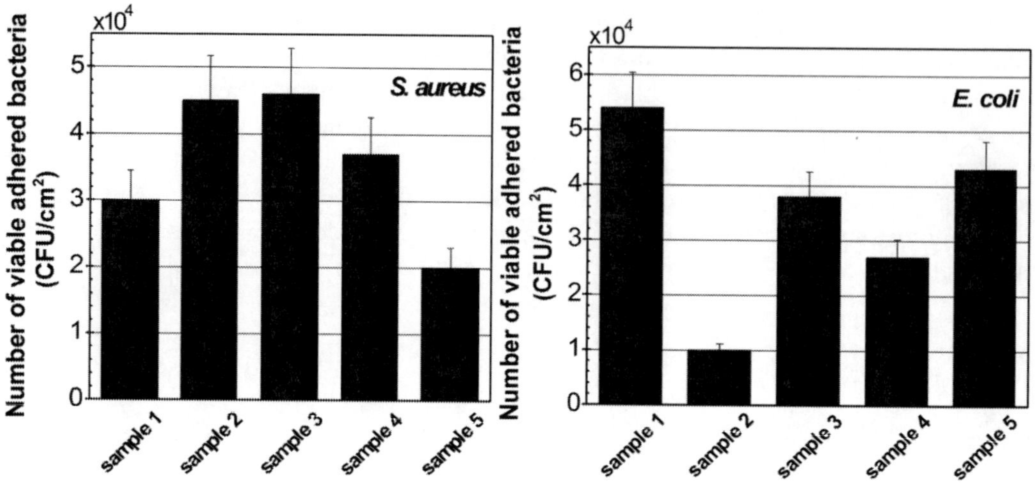

Figure 3. Histograms of bacterial adhesion degree for samples 1-5 after 24 h incubation against two microorganisms.

Chitosan/pectin multilayer is found to be effective against both gram-positive and gram-negative strains which can be translated as a higher quality of chitosan coating when it is applied along with pectin.

Conclusion

DCSBD plasma is capable of raising roughness, surface free energy, and introducing oxygen-containing functionalities anchored onto the PVC surface. A structured PAA brush of high graft density is synthesized using surface-initiated approach to further improve hydrophilicity and develop a stable brush-like assembly to yield a platform for biomolecular binding. In vitro bacterial adhesion and biofilm formation assays indicate incapability of single chitosan layer in hindering the adhesion of *S. aureus* bacterial strain, while up to 30% reduction is achieved by chitosan/pectin layered assembly. On the other hand, chitosan and chitosan/pectin multilayer could retard *E. coli* adhesion by 50% and 20%, respectively. Furthermore, plasma treated and graft copolymerized samples are also found effective to diminish the adherence degree of *E. coli*.

Acknowledgments

This paper was processed in the frame of the APVV project No. APVV-351-10 as the result of author's research at significant help of APVV agency Slovakia, and by project of Ministry of Education of the Slovak Republic and Slovak Academy of Science VEGA, project No.2/0199/14.

REFERENCES

[1] T. Desmet, R. Morent, N. D. Geyter, C. Leys, E. Schacht, P. Dubruel, *Biomacromolecules* 2009, 10, 2351.

[2] G. Speranza, G. Gottardi, C. Pederzolli, L. Lunelli, R. Canteri, L. Pasquardini, E. Carli, A. Lui, D. Maniglio, M. Brugnara, M. Anderle, *Biomaterials* 2004, 25, 2029.

[3] K. Triandafillu, D. J. Balazs, B. O. Aronsson, P. Descouts, P. T. Quo, C. van Delden, H. J. Mathieu, H. Harms, *Biomaterials* 2003, 24, 1507.

[4] E. R. Kenawy, S. D. Worley, R. Broughton, *Biomacromolecules* 2007, 8,1359.

[5] F. S. Denes, S. Manolache, *Prog. Polym. Sci.* 2004, 29, 815.

[6] P. K. Chu, J. Y. Chen, L. P. Wang, N. Huang, *Mat. Sci. Eng.* 2002, R 36, 143.

[7] M. Černák, L. Černáková, I. Hudec, D. Kováčik, A. Zahoranová, *Eur. Phys. J. Appl. Phys.* 2009, 47, 22806p1.

[8] M. Černák, J. Ráhel, D. Kováčik, M. Šimor, A. Brablec, P. Slavíček, *Contrib. Plasma Phys.* 2004, 44, 492.

In: News in Chemistry, Biochemistry and Biotechnology ISBN: 978-1-63117-273-1
Editors: G. E. Zaikov, G. Nyszko, L. P. Krylova et al.

Chapter 4

NEW TYPES OF NANOCOMPOSITES BASED ON ETHYLENE COPOLYMERS

Igor Novák,[1,*] Peter Jurkovič,[2] Ján Matyašovský[2] and Ladislav Šoltés[3]

[1]Polymer Institute of the Slovak Academy of Sciences, Bratislava, Slovakia
[2]VIPO, a.s., Partizánske, Slovakia
[3]Institute of Experimental Pharmacology of the Slovak Academy of Sciences, Bratislava, Slovakia

ABSTRACT

The paper deals with adhesive and mechanical properties study of nanocomposites based on ethylene-acrylic acid copolymer during aluminium bonding. The main objective was to describe the changes of co-polymer properties during increasing of the nanofiller's concentration. Based on executed experiments it was found out, that the properties of tested nanocomposite system were mostly improved depending on the contents of the nanofiller in the system. The optimum concentration of nanofiller Aerosil 130 SLP in the composite was 2.5 weight % for cohesive mechanical properties of the system and 3.5 weight % for adhesive ones. Thermal properties of the composite system showed their maximum within concentration of 4.5 weight % of nanofiller.

Keywords: Composite, hot-melt adhesives, nanofillers, EAA co-polymers

INTRODUCTION

When compared with other types of composites, thermoplastics have some advantages. They are solvent-free and non-toxic (in most cases); they are characterized by short time of creation of adhesive bond respectively foil; they are applicable at low temperatures; they ensure high adhesion to different material and high impact strength of the joint; they ensure

* upolnovi@savba.sk.

suitable initial strength of adhesive joints; they have good storage stability; they are proper for gluing automation and increasing labour productivity; no undesirable moisture is brought into the materials – it means that there is not necessary the long-term storage of products in conditioned environment.

Nowadays, adhesives based on EAA (ethylene – acrylic acid) copolymers, EVA (ethylene – vinyl acetate) copolymers, thermoplastic polymers, polyamide, polyesters, polyethylene, and cellulose [1-6] belong to most often used composites. By addition of a proper type of filler, mentioned properties can be even improved. The aim of this contribution is to evaluate the influence of nanofiller on the properties of EAA copolymer.

EXPERIMENTAL PART

As a polymer, EAA copolymer MICHEM Adhesive 20 EAA, with the ratio of 20 % wt. of acrylic acid and the ratio of 80 % wt. of ethylene, was used. Characteristic properties of the product are:

- appearance: slight turbidity, almost transparent polymer,
- density: 1.3 $g.cm^{-3}$,
- melt flow: 1.8 $g.10min^{-1}$,
- content of volatiles: less than 0.1 wt. %.

Aerosil 130 SLP (Degussa comp.) was used as filler into nanocomposite system. Aerosil is a flame-patterned silica oxide with an average particle size from 40 to 50 nm. Picture 1 shows the microscopic image of used filler. As we can see, the structure of the filler is spherical with a minimal difference in particles size and non-porous/solid surface.

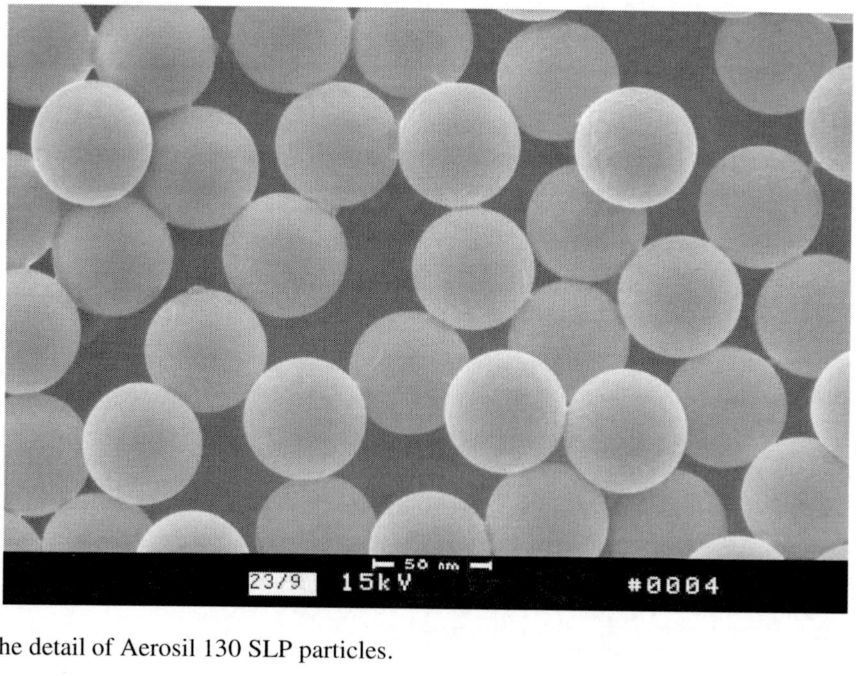

Figure 1. The detail of Aerosil 130 SLP particles.

For preparation of nanocomposite system, EAA copolymer was used as the base for copolymer matrix; which was blended with the filler in concentrations 0, 0.5, 1, 1.5, 2.5, 3.5, and 4.5 % wt. To mix the mixture, we used Plastograf Brabender PLE 331 heated by silicone oil in fully filled tank W-50-h (volume 50 cm^3). The temperature at mixing of nanocomposite was adjusted to 180 °C by a thermostat containing tempered silicone medium. Mixing was at 35 rpm^{-1} for 10 minutes at predetermined temperature. Considering the properties of individual components, it was preferable to use a triangular blade.

At measurement of adhesive characteristics, the aluminium sheet with thickness of 2 mm and chemical composition listed in table 1 was used.

Table 1. Chemical composition of adherends

Elements	Al	Cu	Fe	Mg	Mn	Ni	Si	Zn
Content (wt. %)	99,5	0,0025	0,32	0,002	0,0035	0,013	0,12	0,007

To measure the peeling strength of adhesive joint, the aluminium foil AlMgSi 0,5 with thickness of 0.1 mm was used.

Before gluing, the surface of adherents was grinded with 120 grit sandpaper and then scratches were aligned with 1000 grit sandpaper. Afterwards, the surface was cleaned of grease and other dirtiness with a mixture of benzin and toluene (volume ratio 1:1). To ensure a constant spacing between bonded adherents and an equal thickness of adhesive, two distant wires with diameter ϕ 0.15 mm were placed parallel on the bottom board.

The surface of aluminium foil used in the peeling test was only ungreased with a mixture of benzin and toluene. To measure cohesive characteristics, it was necessary to make test blades according to Figure 2.

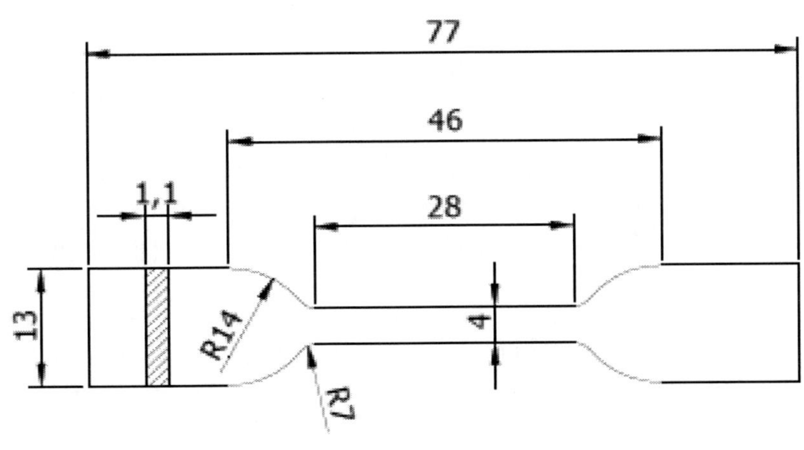

Figure 2. Specimen for testing of tensile strength.

To make them, first boards from filled and unfilled systems (dimensions of 74 x 100 x 1.1) were prepared in a shape in hydraulic press at 180 °C, pressure 250 kPa, for 5 minutes. After cooling of them in a mechanical press, test blades were scissored.

For preparing the samples for testing of adhesive properties (Figure 3), thin layer of hot-melt adhesive was inserted between two cleaned and ungreased aluminium boards with distant wires ϕ 0.15 mm. Lap joint was foil-wrapped into teflon foil and the whole sample was fixed with aluminium foil and put between press plates tempered at 180 °C. At pressure of 100 kPa during 5 minutes, lap joint was formed. The specimens for peeling test were made similarly.

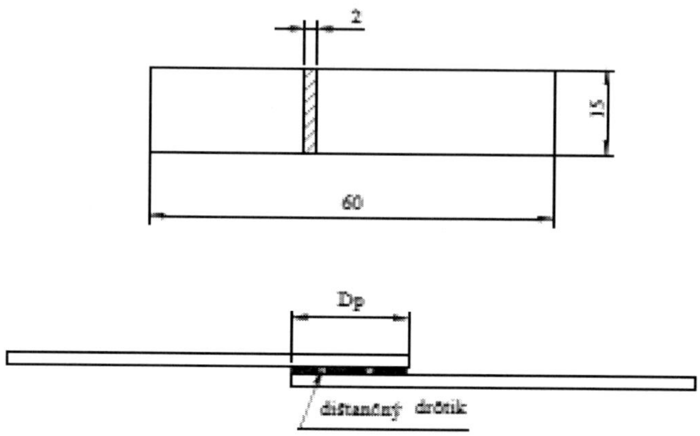

Figure 3. Lap adhesive joint .

Methods of testing included mechanical tests (measurement of cohesive properties and hardness), adhesive tests (measurement of shear strength of adhesive joint at loading by tension [2], measurement of strength of adhesive joint at peeling [3], measurement of surface properties, thermo gravimetric analysis, and measurement of thermal properties.

Measurement of cohesive characteristics included the loading the test blade by tensile (Figure 1) at rate of separation of the jaws 50 mm.min^{-1} with machine Instron 4301 (Instron, England), when following characteristics were evaluated: maximal tensile strength (MPa), maximal elongation (%), elongation at rupture (%), tensile strength at rupture (MPa), Young module of elasticity (MPa), yield strength (MPa), and elongation at yield.

Measurement of hardness in °ShD was done according to ASTM D 2122-2. Equipment D Scale Durometer PTC 307 – L designed for plastics and react-plastics was used.

To measure adhesive characteristics, the test machine Instron 4301 was used (rate of separation of the jaws 50 mm.min^{-1}). Following characteristics were evaluated: shear strength (MPa), relative elongation (%), Young module of elasticity (MPa), and energy of destruction of adhesive joint (J).

At peel test, the tested specimen was fixed in testing machine Instron 4301. Board A1 was fixed in the low jaw and aluminium foil was fixed in the upper movable jaw. Rate of separation was slower, only 10 mm.min^{-1}. The values evaluated were: strength of the joint at maximal loading (MPa), average peel power (N), and average tear tension (N.mm^{-1}). Besides, also thermo-graphic analysis was done with a thermogravimeter TG-1 (Perkin Elmer, USA).

Results and Discussion

The Figure 4 presents the dependence of maximal tensile strength (R_{max}) and tensile strength at breaking (R_r) on the content of filler in composite adhesive. From measured results follow, that with increasing content of nanoparticles of filler in EAA, the maximal tensile strength of composite is non-linearly increased. It can be assumed, that further filling will increase the value of maximal tensile strength, but only for certain concentration. At this concentration, EAA composite will be saturated with Aerosil 130 SLP, what causes insufficient wetting of surface of filler particles and following lowering of max. tensile strength.

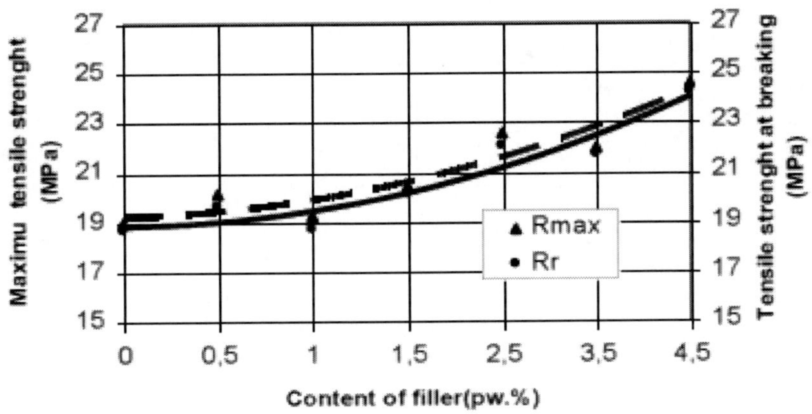

Figure 4. Dependence of max. tensile strength (R_{max}), tensile strength at breaking (R_r) on the content of filler.

The dependence of adhesive shear strength of joint on the content of filler is on the Figure 5. Considering the high specific surface of nanoparticle filler (130 $m^2.g^{-1}$), intense change of investigated parameter occurs already at low concentrations of filler. Increased dispersion of measured values can be justified by the possible presence of non-homogeneous in the composite system, as well as the deteriorative wetting of the aluminium substrate in the growth of filler content. Substantially is worsened the spreading of copolymer melt adhesive on the glued surface due to an increase melt viscosity of hot melt glue, which deteriorates the surface wetting.

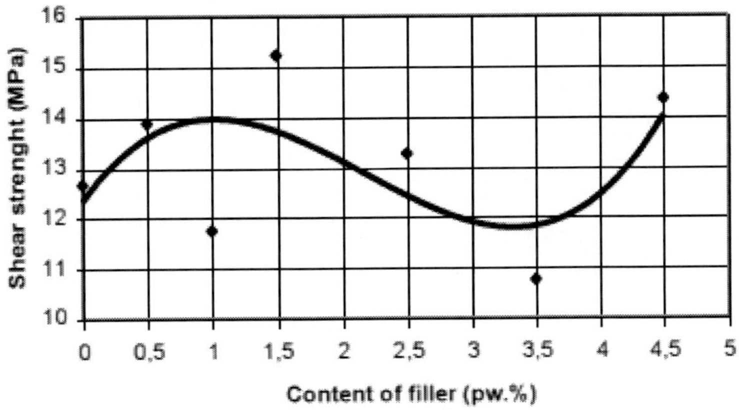

Figure 5. Dependence of adhesion joint shear strength on the filler content.

Character of dependence of average peeling stress is parabolic with the maximum at the content of filler 3,5 weight % (Figure 6). Also in this case, measured values show higher variance, similarly as at measurement of adhesive shear strength of joint.

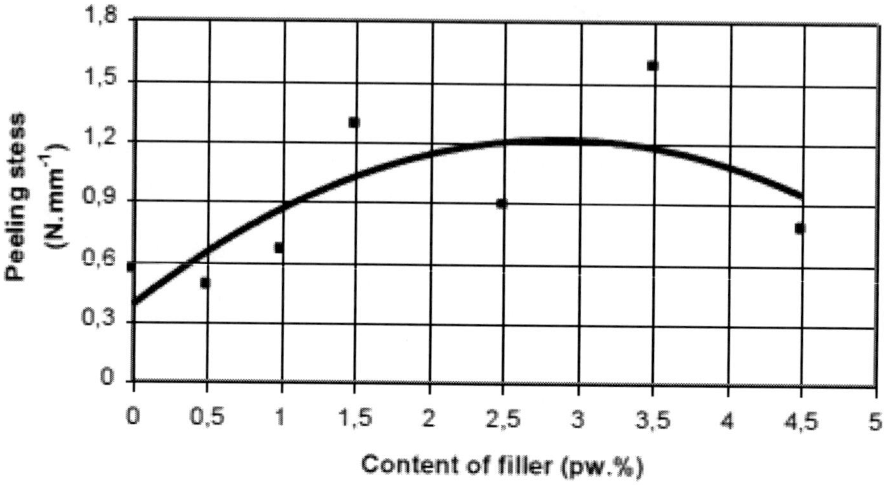

Figure 6. Dependence of peeling stress on adhesive concentration.

Thermo gravimetric analysis confirmed, that temperature of 10 % weight loss and temperature of sudden weight loss (Figure7) had after initial decrease increasing tendency. With the increase of filler particles, the temperature of loss 10% weight is increasing from 360 °C to 385 °C, which represents a rise up to 8 %. The reason is higher absorption of heat with Aerosil 130 SLP. Temperatures of sudden loss reach lower values (342 °C up to 374 °C) in comparison with the temperature of loss 10 % weight.

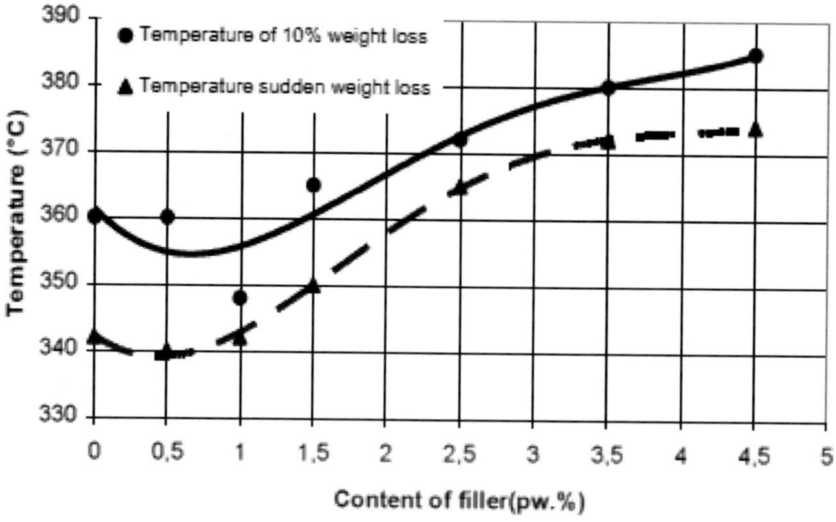

Figure 7. The dependence of temperature of 10% weight loss and temperature of sudden weight loss on the content of Aerosil 130 SLP.

CONCLUSION

On the base of realised experiments it can be concluded, that nanoparticle filler Aerosil 130 SLP influences individual properties of filled EAA system differently. The filler has positive impact to improve the cohesion and adhesion strength, heat resistance, peeling tension and surface properties of the system. On the other hand, reduces the relative extension, factors of heat and thermal conductivity and specific volume heat capacity. The cohesive mechanical parameters of the system can be stated as an optimal concentration of nanofiller Aerosil 130 SLP 2.5 wt. %, the adhesion properties of 3.5 wt. %. Nanoparticles composite systems showed the highest heat resistance in filler concentration from 3.5 to 4.5 wt. %. For practical application of filled EAA nanocomposite systems is therefore necessary to know how to use, environment, application temperature and method of stress and accordingly select the optimal concentration nanofiller.

ACKNOWLEDGMENT

This contribution is the result of the project implementation: „Research of the Application Potential of Renewable and Recycled Materials and Information Technologies in the Rubber Industry" (project code ITMS: 26220220173) supported by the Research & Development Operational Programme funded by the ERDF, and project of Ministry of Education of the Slovak Republic and Slovak Academy of Science VEGA, project No.2/0199/14.

REFERENCES

[1] Novák, I., Pollák, V., 2002: Adhezíva využívané na lepenie v automobilovom priemysle. In: Chemagazín, No.4, ročník XII, p. 4-5.

[2] STN EN 1465: 2000. Stanovenie pevnosti v šmyku preplátovaného lepeného spoja pri namáhaní v ťahu.

[3] STN EN 28510-2: 2000. Skúška odlupovania lepeného spoja skúšobného telesa z ohybného a tuhého adherendu. Časť 2: Odlupovanie pod uhlom 180°.

[4] Novák, I., Florián, Š., Pollák, V., Žigo, O., 2010: Tlakovo-citlivé elektricky vodivé adhezíva. In: Chemagazín, No. 4, ročník XX, p. 22-23.

[5] Kinloch, A. J., 1994: Adhesion and Adhesives. Chapman and Hall, UK.

[6] Lu, D., Wong, C. P., 2000: Intern. J. Adhesion and Adhesives, 20, p.189.

In: News in Chemistry, Biochemistry and Biotechnology ISBN: 978-1-63117-273-1
Editors: G. E. Zaikov, G. Nyszko, L. P. Krylova et al.

Chapter 5

INTERACTION OF HYBRID ANTIOXIDANTS: ICHPHANS WITH AN ERYTHROCYTE MEMBRANE

E. Yu. Parshina[1*]***, L. Ya. Gendel***[2] ***and A. B. Rubin***[1]
[1]Faculty of Biology, Lomonosov Moscow State University, Moscow, Russia
[2]Emanuel Institute of Biochemical Physics, Russian Academy of Sciences, Moscow, Russia

ABSTRACT

Morphological transformation of erythrocytes and structural changes in the erythrocyte membrane have been revealed by scanning electron microscopy and spin-probe technique. These effects were caused by the incorporation of ichphans, new generation drugs combining antioxidant and anticholinesterase effects, into the erythrocyte membrane and their distribution in the intramembrane space. Different distribution and modulatory effect of the derivatives with different hydrophobic properties have been shown. The derivatives with 8 and 10 carbon atoms in the aliphatic substituent were the most efficient modifiers of the membrane structure and morphology of erythrocytes.

Keywords: Hybrid compounds, synthetic antioxidants, anticholinesterase drugs, membrane transport, erythrocyte morphology, membrane microviscosity, spin-probes method

INTRODUCTION

The search for new efficient drugs is an urgent problem of physicochemical biology and pharmacology. The development of drugs with combined effect, i.e., compounds including fragments with different types of biological activity, is one of the search trends. Such drugs include hybrid antioxidants ichphans synthesized at the Emanuel Institute of Biochemical Physics (Russian Academy of Sciences, Moscow) (Nikiforov et al., 2003). Their structure

* e-mail: parshinae@gmail.com.

includes two fragments providing for antioxidant and anticholinesterase activities, respectively, as well as a side alkyl substituent with a different number of carbon atoms in the aliphatic chain, which provides for hydrophobic properties of these drugs. It was proposed that such chemical structure allows ichphans to be considered as promising drugs against Alzheimer's disease (Braginskaya et al., 1996; Molotchkina et al., 2002).

High antioxidant activity and the inhibitory effect on acetylcholinesterase from human erythrocytes have been shown for ichphans as well as different efficiency of their derivatives with different hydrophobic properties (Braginskaya et al., 1996; Ozerova, 2000; Molotchkina et al., 2002; Nikiforov et al., 2003, Burlakova et al., 2008).

The functional activity of erythrocytes largely depend on their shape, which changes after the exposure to biologically active compounds, pathological processes in the body, and other factors (Sheetz and Singer, 1974; Bessis,1974; Gendel and Kruglyakova, 1986; Luneva et al., 2002). The changes in the erythrocyte shape induced by the membrane transport of certain ichphans have been demonstrated (Parshina et al., 2011, Parshina et al., 2012).

In this work continued scanning electron microscopy investigation of the effect of ichphan derivatives with different hydrophobic properties on the erythrocyte morphology and used the spin-probe technique to study their effect on the erythrocyte membrane structure.

Materials and Methods

In this work, we used ichphan derivatives with different aliphatic substituents at the quaternary nitrogen atom (Nikiforov et al., 2003), which were synthesized at the Emanuel Institute of Biochemical Physics (Russian Academy of Sciences, Moscow) (table).

Experiments were carried out on erythrocytes from outbred albino rats (with heparin as an anticoagulant). Erythrocytes were isolated as described elsewhere (Luneva et al., 2002). In the experiments, erythrocytes were suspended in buffer A (145 mM NaCl, 5 mM KCl, 4 mM Na_2HPO_4, 1 mM NaH_2PO_4, 1 mM $MgSO_4$, 1 mM $CaCl_2$, 10 mM glucose, pH 7.4, t = 4°C). The cell concentration in the suspension was 1 x 10^7 cells per microliter. The isolated erythrocyte mass was stored at 4°C and used within 8 h.

The effect of ichphans on the erythrocyte morphology was studied by incubating the cells with each ichphan derivative for different time periods (from 2 to 120 min) at 18 ± 2°C. Ichphans were used as ethanol solutions and the ethanol concentration in samples did not exceed 0.8% by volume. Samples incubated under similar conditions in the absence of the tested compound served as control.

The samples were fixed in 1% glutaraldehyde (Serva, Germany) in buffer A for 3 h, after which the cells were washed according to published recommendations (Kozinets and Simovart, 1984). The cell monolayer was applied onto a slide, air-dried, coated with a platinum-palladium layer in an EICO IB-3 ion coater (Japan), and examined under a CamScan scanning electron microscope (United Kingdom).

The effect of ichphans on the erythrocyte membrane structure was studied by the method of spin-probes, 5-and 16-doxyl stearates (hereafter, probes I and II, respectively).

The probes were added to erythrocyte suspension as ethanol solutions to the final probe concentration of 1 x 10^{-4} M. After a 7-min incubation, a tested compound was added and the samples were immediately prepared to record electron spin resonance (ESR) spectra. The total ethanol concentration in samples did not exceed 1.2% by volume. Samples incubated under similar conditions in the absence of tested compounds served as control.

The structure-sensitive parameters included the distance (in G) between the extremes in probe I ESR spectrum in the low- and high-field regions $2A'_{\|}$ and the correlation time τ equal to the 90° rotation of probe II (Antsiferova et al., 1977). τ was evaluated using equation:

$$\tau = 6.65\Delta H_{+1}\left(\sqrt{\frac{I_{+1}}{I_{-1}}} - 1\right) \times 10^{-10}\mathrm{s},$$

where I_{+1} and I_{-1} are the amplitudes of the low-field and high-field components of the ESR spectrum, respectively, and ΔH_{+1} is the width of the low-field components, G.

An ESR spectrometer RE-1307 (Russia) was used. Samples were incubated and ESR spectra were recorded at 18 ± 2°C.

The obtained data were statistically analyzed using Student's -test.

Results and Discussion

Table presents the structural formulas of ichphan derivatives studied in this work. The left part of the structure of all derivatives is a shielded phenol, which is responsible for their antioxidant properties. The right part is a fragment responsible for the anticholinesterase activity as well as an alkyl substituent bound to the quaternary nitrogen atom, the aliphatic chain of which contains from 1 to 16 carbon atoms. The longer is the substituent aliphatic chain, the more hydrophobic is the derivative. The quaternary nitrogen atom in the structure of ichphans imparts a positive charge and organic cation properties to them.

Structural formulas and designations of ichphan derivatives

Structural formula	X	R	Designation
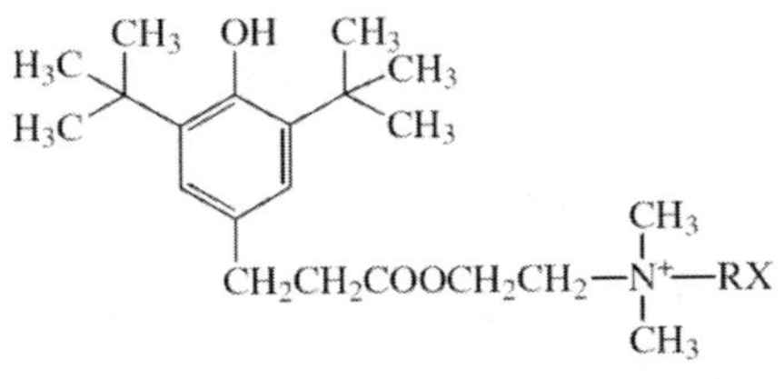	J^-	$-CH_3$	I(C-1)
	Br^-	$-C_8H_{17}$	I(C-8)
		$-C_{10}H_{21}$	I(C-10)
		$-C_{12}H_{25}$	I(C-12)
		$-C_{16}H_{33}$	I(C-16)

Notes: R, carbohydrate radical; X, halogen anion.

Normally, the bulk of erythrocytes in mammals have the shape of discocytes, while other morphological forms (echinocytes, stomatocytes, etc.) are minor (Bessis, 1974; Kozinets and Simovart, 1984). Figure 1 shows the typical kinetic curves reflecting the changes in the concentration of discocytes, echinocytes, and stomatocytes in erythrocyte suspension after the addition of ichphan derivatives at the concentration of 1 x 10^{-4} M.

One can see that the incubation of erythrocyte suspension with ichphans changed the cell shape. The concentration of discocytes considerably decreased (Figure 1a) and the concentration of echinocytes increased (Figure 1b) relative to control. Later, an inverse process was observed: the number of echinocytes decreased, the number of discocytes increased, and stomatocytes appeared and increased in number (Figure 1c). The obtained data indicate that the efficiency of the transforming effect and the pattern of its kinetic curves substantially different for the membrane transport of the derivatives with different structure and hydrophobicity of the side aliphatic substituent.

The least hydrophobic compound I(C-1) had only a marginal echinocytogenic effect. The most pronounced increase in the concentration of echinocytes at the early incubation stages and generation of the greatest stomatocytes numbers after long-term incubation were observed for the derivatives containing 8, 10, and 12 carbon atoms in the carbohydrate chain of the site substituent. I(C-8) showed the maximum modifying effect (Figs. 1a, 1c).

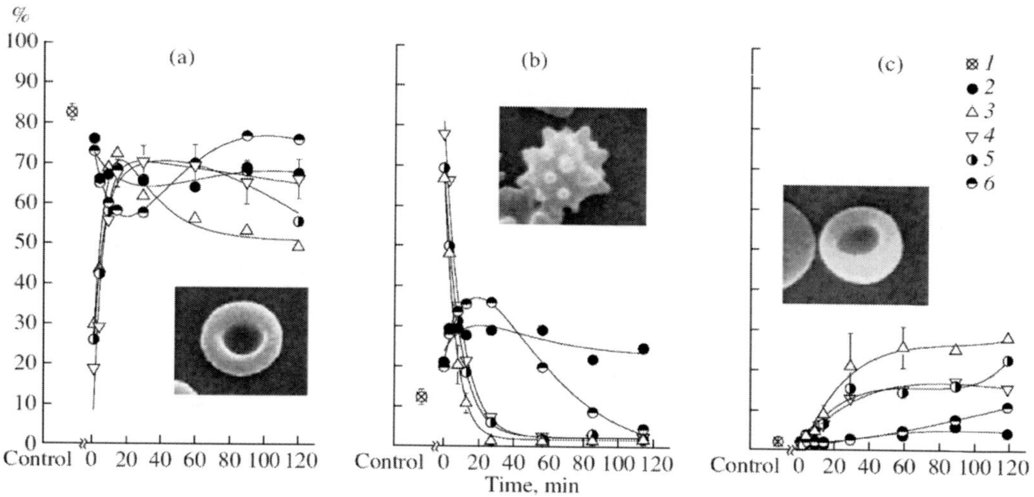

Figure 1. Kinetic curves of the concentrations (%) and electron micrographs (1000x) of discocytes (a), echinocytes (b), and stomatocytes (c) in erythrocyte suspension under the influence of ichphans (1 x 10-4 M) with different length of the aliphatic chain of the hydrophobic substituent; 1, control; 2, I(C-1); 3, I(C-8); 4, I(C-10); 5, I(C-12); 6, I(C-16).

I(C-16) with the largest and most hydrophobic side substituent differed from other ichphan derivatives by both lower echinocytogenic effect and the kinetics of morphological transformations: the induced increase and decrease in echinocytes concentration were observed later compared to other derivatives.

The time-related changes in the erythrocyte morphology during the incubation with ichphans can be explained in terms of the coupled bilayer hypothesis (Sheetz and Singer, 1974) as follows. The interpolation of these compounds into the outer monolayer of the erythrocyte membrane is the initial stage of their membrane transport, which increases the outer monolayer area relative to the inner one and gives rise to echinocytes. Later, electrostatic forces mediate the penetration of positively charged ichphan molecules into the inner membrane monolayer, which is negatively charged due to the prevalence of phosphatidylserine in it (Sheetz and Singer, 1974). In this case, the area disbalance is first smoothed out and cells become discoid as the compensatory effect is realized. Later, as ichphan molecules are accumulated in the inner membrane monolayer, the area disbalance increases the area of this monolayer, which gives rise to stomatocytes. As stated previously (Parshina et al., 2011), only a fraction of cells in erythrocyte suspension is involved in the shape transformation under the conditions used in the experiment. Apparently, ichphans integrate into the membranes of these cells. The proportion of cells transformed into stomatocytes is relatively low compared to those transformed into echinocytes. In this context, one can propose that, under equilibrium conditions, the distribution of ichphans in the membrane of a considerable fraction of cells causes the compensatory effect.

The pronounced modifying effect of ichphans on the erythrocyte morphology allowed us to propose that the intercalation of ichphan molecules into the erythrocyte membrane and their membrane transport should affect the structural state of the intramembrane space. The experiments involving the spin probe technique have confirmed this proposal.

Complementary data on the structural state of different parts of the lipid bilayer were obtained using spin probes I and II, the radical fragments of which lie in the surface and deep parts of the membrane, respectively.

The impact of ichphans on the bilayer structure was described by the $2A'_{\parallel}$ and τ parameters. A decrease in $2A'_{\parallel}$ and τ reflects a microviscosity decrease in the regions of spin probe distribution in the intramembrane space. Figure 2 shows typical curves reflecting relative changes in $2A'_{\parallel}$ for probe I ($\Delta 2A'_{\parallel} = 2A'_{\parallel\ experiment} - 2A'_{\parallel\ control})100/2A'_{\parallel\ control}$) and in τ for probe II ($\Delta\tau = (\tau_{experiment} - \tau_{control})100/\tau_{control}$) during erythrocyte incubation with ichphan derivatives. In most cases, ichphans decreased the mean $2A'_{\parallel}$ and τ values. This indicates a decrease in the microviscosity of both surface and deep regions in the intramembrane space induced by these substances.

The amplitude of the induced changes in $2A'_{\parallel}$ and τ was not high similar other biologically active compounds (Kury and McConnell, 1975; Gendel et al.,1997; Parshina et al., 2012).

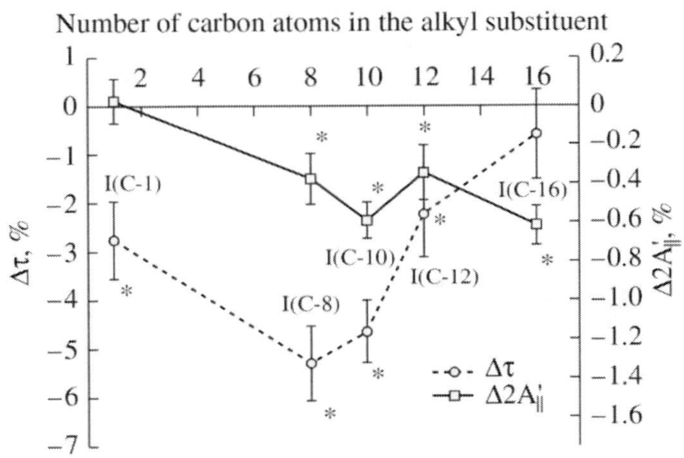

Figure 2. Changes in the mean distance between the $2A'_{\parallel}$ extremes in probe I (5-doxyl stearate) ESR spectrum and the correlation time τ of probe II (16-doxyl stearate) after erythrocyte suspension incubation with ichphan derivatives (5×10^{-4} M) for 30 min, %. Note: * $p < 0.05$.

The most pronounced decrease in these parameters was induced by I(C-8) and I(C-10). I(C-1) had no notable effect on the microviscosity of the membrane surface region but decreased the microviscosity of the deep membrane regions. In contrast, I(C-16) containing the largest side substituent largely modified the structure of the membrane surface regions.

The obtained data indicate that the membrane transport of ichphan derivatives with different structure of the side substituent and different hydrophobic properties modifies the structure of different regions in the intramembrane space, which likely correspond to their localization sites. The differences in the efficiency of the modifying effect can be due to a limited holding capacity of the erythrocyte membrane for the arriving substance, accordingly, larger homologs can reach lower concentrations in the membrane and the overall effect is lower. The impact of the side substituent structure and hydrophobicity on the distribution of ichphan derivatives in the erythrocyte membrane and its holding capacity for the interpolating agent has been previously demonstrated for homologs of spin-labeled nonelectrolytes (Gendel and Kruglyakova, 1986; Luneva et al., 2002). In this work, the introduction of substituents with different hydrophobic properties into the molecule of organic cation also had a considerable impact on their distribution in the membrane.

Thus, morphological transformation of erythrocytes and structural changes in the erythrocyte membrane induced have been revealed by scanning electron microscopy and spin-probe technique. These effects were caused by the incorporation of ichphan derivatives with different hydrophobic properties into the erythrocyte membrane and their distribution in the intramembrane space.

Different distribution and modulatory effect of the derivatives have been shown. A complex pattern of the relationship between the efficiency of ichphans and their hydrophobicity has been demonstrated. Among the studied compounds, the derivatives I(C-8) and I(C-10) were the most efficient modifiers of the membrane structure and morphology of erythrocytes in the studied. According to published data, these substances have the most pronounced antioxidant and anticholinesterase activities (Braginskaya et al., 1996; Ozerova, 2000; Molotchkina et al., 2002; Nikiforov et al., 2003, Burlakova et al., 2008). The obtained data suggest that the activity of membrane-bound acetylcholinesterase can be modulated by the changes in the surface architectonics and microviscosity of the erythrocyte membrane. The pattern of the morphological changes in erythrocytes induced by the membrane transport of ichphans indicates that the antioxidant effect of compounds used in this work can be realized in both monolayers of the erythrocyte membrane.

REFERENCES

Antsiferova, L.I., Vasserman, A.M., Ivanova, A.N., et al., *Atlas spektrov elektronnogo paramagnitnogo rezonansa spinovykh metok i zondov* (Atlas of Electron Paramagnetic Resonance Spectra of Spin Tags and Probes), Moscow: Nauka, 1977.

Bessis, M., *Corpuscles: Atlas of Red Blood Cell Shape*, New York: Springer, 1974.

Braginskaya, F.I., Molochkina, E.M., Zorina, O.M., et al., New Synthetic Bioantioxidants—Acetylcholinesterase (AChE) Inhibitors in *Alzheimer Disease: From Molecular Biology to Therapy,* Becker, R. and Giacobini, E., Eds., 1996, pp. 337–342.

Burlakova E.B., Molochkina E.M., Nikiforov G.A. Hybrid antioxidants. Oxidation Communications Journal,2008,Vol. 31, No. 4, pp. 739-757 .

Gendel, L.Ya. and Kruglyakova, K.E., Structural and Functional Interactions of Physiologically Active Compounds with Biomembranes, in *Metod spinovykh metok i zondov* (The Method of Spin Labels and Probes), Moscow: Nauka, 1986, pp. 163–194.

Gendel, L.Ya., Yakovleva, N.E., Lelekova, T.V., et al., The Effect of Thyroliberin on the Structural Characteristics of Rat Erythrocytes, *Izv. Akad. Nauk, Ser. Biol.,* 1997, no. 1, pp. 103–106.

Kozinets, G.I. and Simovart, Yu.A., *Poverkhnostnaya arkhitektonika kletok perifericheskoi krovi v norme i pri zabolevaniyakh sistemy krovi* (Surface Architectonics of Peripheral Blood Cells in Normalcy and in Pathologies of Blood System), Tallinn: Valgus, 1984.

Kury, P.G. and McConnell, H.M., Regulation of Membrane Flexibility in Human Erythrocytes, *Biochemistry*, 1975, vol. 14, no. 13, pp. 2798–2803.

Luneva, O.G., Gendel, L.Ya., and Kruglyakova, K.E., Features of Organic Nonelectrolyte Binding to the Erythrocyte Membrane, *Biofizika*, 2002, vol. 47, no. 1, pp. 38–44.

Nikiforov, G.A., Belostotskaya, I.S., Vol'eva, V.B., Komissarova, N.L., and Gorbunov, D.B., *Nauchn. vestn. Tyumenskoi meditsinskoi akademii, spetsial'nyi vypusk "Biooksidanty"* (Sci. Herald of Tyumen Medical Academy: Special Issue "Biooxidants"), 2003, pp. 50-51.

Molotchkina, E.M., Ozerova, I.B., Burlakova, E.B., *Free Radical Biology and Medicine,* 2002, vol. 33, Issue 2S1, no. 610, pp. S229-S230.

Ozerova, I.B. New Antioxidants – Screened Phenols – as Modulators of Acetylcholine Esterase Activity in vivo and in vitro, Cand. Sci. (Biol.) Dissertation, Moscow: *Emanuel Inst. Biochem. Phys.,Russ. Acad. Sci.*,2000.

Sheetz, M.P. and Singer, S.J., Biological Membranes as Bilayer Couple. A Molecular Mechanism of Drag - Erythrocyte Interaction, *Proc. Natl. Acad. Sci.USA,1974,vol.71,pp4457-4461.*

Parshina E. Yu., Gendel L. Ya., Rubin A. B. Effect of new hybrid antioxidants—Ichphans—on the surface architectonics of erythrocytes. In “Progress in Study of Chemical and Biochemical Reactions. *Kinetics and Mechanism 2011, pp.71-78***.**

Parshina E.Yu., Gendel L.Ya., Rubin A.B., “Influence of hydrophobic properties of ichphans antioxidants on their membranotropic activity,” *Pharmaceutical Chemistry Journal, 2012,vol. 46, no. 2, pp. 82–85.*

In: News in Chemistry, Biochemistry and Biotechnology ISBN: 978-1-63117-273-1
Editors: G. E. Zaikov, G. Nyszko, L. P. Krylova et al.

Chapter 6

ANTIFUNGAL ACTIVITY OF AMINATED CHITOSAN AGAINST THREE DIFFERENT FUNGI SPECIES

T. M. Tamer[1], M. M. Sabet[1], E. A. Soliman[1], A. I. Hashem[2] and M. S. Mohy Eldin[1]

[1]Polymer Materials Research Department, Advanced Technologies and New Materials Research Institute (ATNMRI), City of Scientific Research and Technological Applications (SRTA- City), Alexandria, Egypt

[2]Organic Chemistry Department, Faculty of Science, Ain-Shams University, Cairo, Egypt

ABSTRACT

The antifungal activity of aminated chitosan against three different fungal species *Aspergillus Niger, Alternaria Alternata and Fusarium Moniliforme* was measured and evaluated. Aminated chitosan was produced by chemically amination of chitosan via introducing further amino groups to the back bone of chitin using parabenzoquinone (pBQ) as activation agent and ethylene di amine (EDA) as amino group source. The aminated chitin was further deacetylated to obtain finally chemically modified chitosan with higher content of amine groups. The success of grafting process has been confirmed using FT-IR, TGA, DSC and SEM. It was found that the antifungal activity of the modified chitosan is better than the native one, and increases by increasing external amine groups against given fungal species. Modification improves solubility of polymer along different acidic pH but still un soluble in neutral and alkaline pH.

Keywords: antifungal, aminated chitosan, *Aspergillus Niger, Alternaria Alternata* and *Fusarium Moniliforme*

INTRODUCTION

Chitin and Chitosan

Chitin is the second most abnormal polysaccharide in nature, second only to cellulose; and is primary present in the exoskeletons of crustaceans such as crab, shrimp, and lobster, etc.). In addition to crustaceans, it is also found in various insects, worms, fungi and mushrooms, in varying proportions from species and from region Table (1).

Chitin has the same backbone as cellulose, but it has an acetamide group on the C-2 position instead of a hydroxyl group and its molecular weight, purity and crystal morphology are dependent on its source (Salmon and Hudson, 1997). Chitosan is the N-deacetylated derivative of chitin; and so it is a linear polysaccharide consisting of β-(1-4) 2 amino-2-deoxy-D glucopyranose as shown in Figure (1).

Table 1. Approximate chitin content in various living species (Mrunal R. Thatte, 2004)

Species	Weight % chitin by dry weight body
Fungi	5-20
Worms	20-38
Squids octopus	3-20
Scorpions	30
Spiders	38
Cockroaches	35
Water Beetle	37
Silkworm	44
Hermit Crab	69
Edible Crab	70

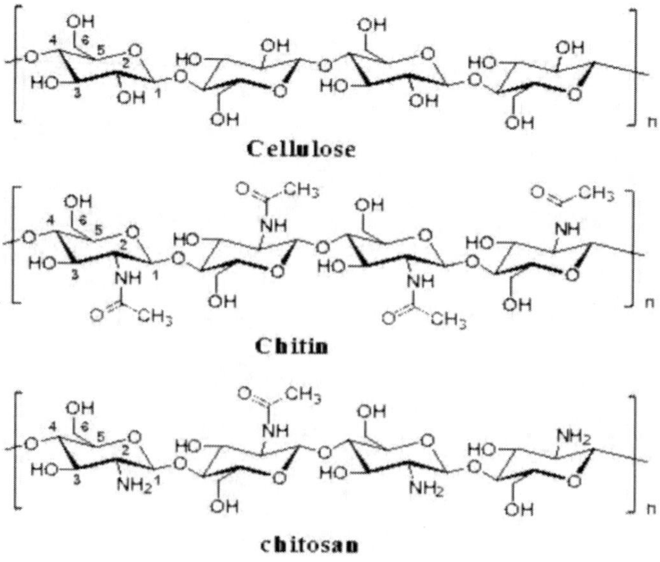

Figure 1. Chemical structures of cellulose, chitin and chitosan.

Isolation of Chitin and Synthesis of Chitosan

Within its natural resource of commercial interest, chitin exists not as a stand alone biopolymer, but rather in conglomeration with other biomaterials, mainly proteins, lipids, and inorganic salts. The isolation process of chitin starts at sea-food industry (figure 2), (Brine, 1984). Shells from crab, shrimp….etc are first crushed into fine powder to help make a greater surface area available for the heterogeneous processes to follow. An initial treatment of the shell with 5 % sodium hydroxide dissolves various proteins, leaving behind chitin. Then treatment with 30% hydrochloric acid hydrolyzes lipids and calcium salts (mainly as CaCO3) and other mineral inorganic constituents. Chitin thus obtained can be hydrolyzed using 50% sodium hydroxide at high temperature (100-150 °C) to provide chitosan, Figure 2.

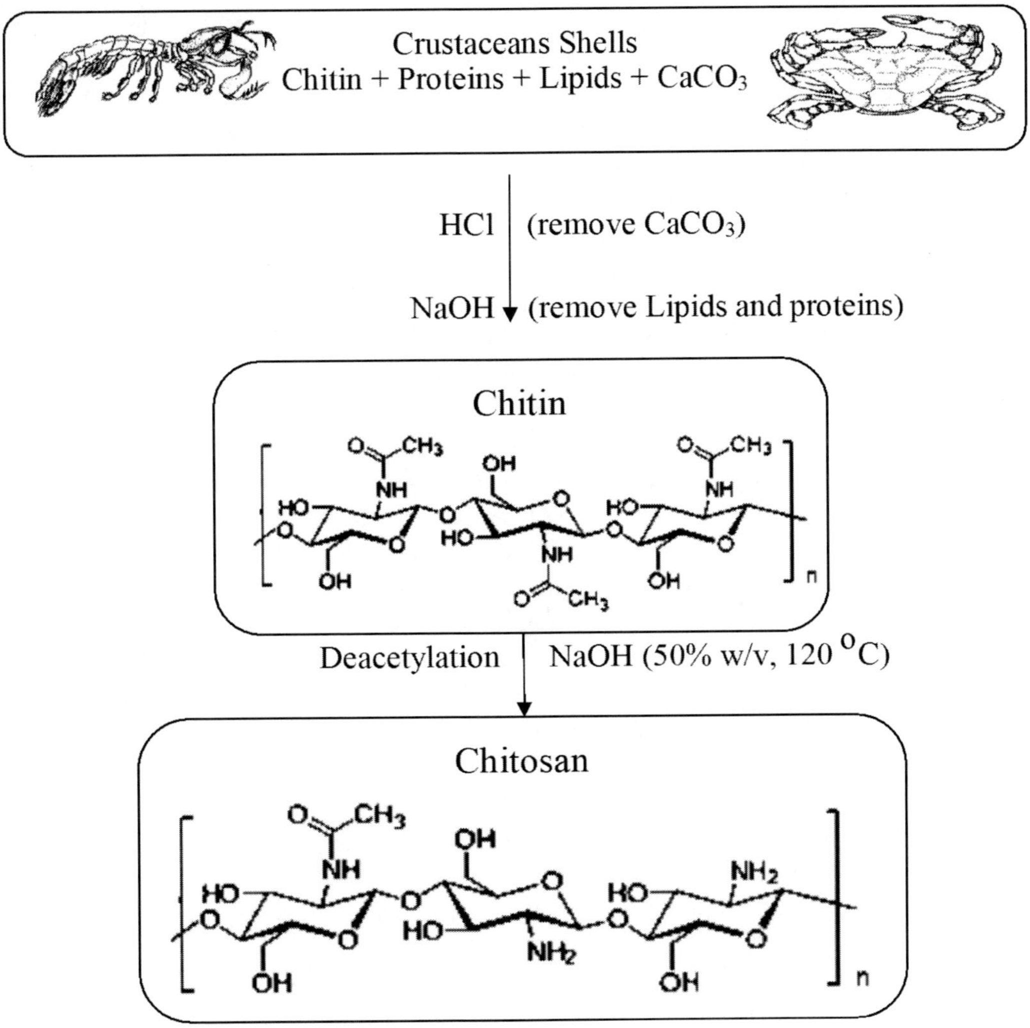

Figure 2. Schematic extraction of chitin and preparation of chitosan.

Antifungal Activity of Chitosan

Chitin and chitosan have been investigated as an antimicrobial material against a wide range of target organisms like algae, bacteria, yeasts and fungi in experiments involving *in vivo* and *in vitro* interactions with chitosan in different forms (solutions, films and composites). Early research describing the antimicrobial potential of chitin, chitosan, and their derivatives dated from the 1980-1990s (Chen, C. S et al. 1998, Hadwiger, L. A et al. 1981, Papineau, A. M et al. 1991, Shahidi, F et al. 1999, Sudarshan, N. R. et al. 1992, Young, D. H et al. 1982).

Mechanism of the Antifungal Activity

Several different mechanisms for antifungal inhibition by chitosan have been proposed and recorded in the literature, but the exact mechanism is still unknown.

In general, it is known that the mode of chitosan action on phytopathogens fungi could development in an extra level (plasma membrane) and intracellular level (penetration of chitosan on fungal cell) (Guo et al., 2008; Palma- Guerrero et al., 2008).

Several studies suggest that positively charged chitosan interact with the negatively charged residues at the cell surface of fungi, which causes extensive cell surface alterations and alters cell wall permeability; therefore, this interaction causes the leakage of intracellular electrolytes and proteinaceous material of the cell (Guo et al., 2008). El Ghaouth et al. demonstrated that chitosan provoked the leakage of amino acids and proteins of the *Rhizopus stolonifer* cell (El Ghaouth et al., 1992). Similar results was obtained on three isolates on *R. stolonifer* grew in minimum medium, in that study there were an increased release of compounds at 260 and 280 nm with chitosan of different molecular weight (Guerra-Sánchez et al., 2009). In other studies, potassium ion leakage was demonstrated by effect of chitosan on fungal cell, being more pronounced for the first 5 min (Singh et al., 2008; García-Rincón et al., 2010). In general, it is known that chitosan treatment causes changes in the membrane integrity of spores, modifications in pH media and the proteins release. This effect was different depending on the isolate, kind of chitosan and used concentration (Hernández-Lauzardo et al., 2012).

On the other hand, the membrane integrity of *P. expansum* and *B. cinerea* spores was affected by chitosan. *P. expansum* was more sensible than *B. cinerea*; and the effect was related with the fungal species (Liu et al., 2007). In other studies, chitosan affected the membrane integrity on *S. sapinea* allowing the out flow of cell components (Singh et al., 2008). Besides, chitosan could be affecting the plasma membrane properties. It was demonstrated that this polymer caused a decrease in the H+-ATPase activity on plasma membrane of *R. stolonifer*; this effect could provoke the accumulation of protons inside the cell, which would result in the inhibition of the chemiosotic driven transport that allows the H^{+}/K^{+} exchange (García-Rincón et al., 2010).

Resent researches suggest that the plasma membrane forms a barrier to chitosan in chitosan-resistant but not chitosan-sensitive fungi. Additionally, it was reported that the plasma membranes of chitosan-sensitive fungi had more polyunsaturated fatty acids than chitosan-resistant fungi, suggesting that the permeabilization by chitosan may be dependent on membrane fluidity (Palma- Guerrero et al., 2010, Hernández-Lauzardo et al. 2011).

For low molecular weight and oligomer chitosan, the ability of molecules penetration will be increased that varying inhibition mechanism. Few reports demonstrated that chitosan could penetrate the fungal cell. Recent studies of chitosan-fungal cell interactions showed that the polymer penetrates the cell and cause intracellular affectations. Microscopic observation reported that chitosan oligomers diffuse inside hyphae interfering on the enzymes activity responsible for the fungus growth (Eweis, M et al. 2006). It was found that chitosan by an energy-dependent process quickly penetrated the conidia of *F. oxysporum* (less than 15 min) and caused ultra structural alterations (disorganized cytoplasm, retraction of the plasma membrane and loss of intracellular content) in the treated spores (Palma- Guerrero et al., 2008). However, is evident that a chitosan tracer is needed to evaluate the capture and dissemination within the cell.

Previous report showed that oligochitosan penetrated the fungal cell and caused disruption on endomembrane system of *Phytophthora capsici*, such as, distortion and disruption of most vacuoles, thickening of plasmalemma and appearance of unique tubular materials (Xu et al., 2007a). Additionally, other studies in this plant pathogenic fungus with oligochitosan marked confirmed that, the polymer penetrated the membrane and binds to nucleic acids (Xu et al., 2007b).

Factors Affect on Chitosan Antifungal Activity

The extent of the antifungal action of chitosan is influenced by intrinsic and extrinsic factors such as MW, pH, species of fungi,... etc. According to several authors, the antimicrobial activity of chitosan is directly proportional to the DD of chitosan (Tanigawa and co workers, 1992; Hirano and Nagao, 1989). The increase in DD means an increased number of amino groups on chitosan. As a result, chitosan has an increased number of protonated amino groups in an acidic condition and dissolves in solution completely, which leads to an increased chance of interaction between chitosan and negatively charged cell walls of microorganisms in solution.

The essential role of free amine groups in the antimicrobial mechanism of chitosan attract the attention of scientist to produces several derivatives of chitosan with higher amine contents. In this work, evaluation of antifungal activity of aminated chitosan was tested against three different fungal species *Aspergillus Niger, Alternaria Alternata and Fusarium Moniliforme.*

Chitosan Amination

Aminated chitosan was prepared as our previous work (Mohy Eldin et al., 2012), Briefly, Aminated chitosan derivatives were prepared through three steps. In the first step, 4 g of chitin was dispersed in 50 mL of PBQ–distilled water solution at a known pH and temperature and was stirred for 6 h. The PBQ-conjugated chitin was separated and washed with distilled water to remove unreacted PBQ. In the second step, PBQ-conjugated chitin was dispersed in 50 mL of EDA-distilled water solution of definite temperature and was stirred for 6 h. The aminated modified chitin was separated and washed with distilled water to remove unreacted EDA. In the last step, aminated modified chitin was deacetylated according to the method of Rigby and Wolfarn (Rigby G,1936) The aminated modified chitin derivative was treated with 50% aqueous solution of NaOH at 120–150 °C for 6 h. The obtained aminated chitosan derivatives were separated and washed with distilled water to remove unreacted NaOH.

Polymer Solubility

Chitosan solubility, biodegradability and biological reactivity depend on the amount of protonated amino groups in the polymeric chain, therefore on free amine groups along polymer backbone. The amino groups (pKa from 6.2 to 7.0) are completely protonated in acids with pKa smaller than 6.2 making chitosan soluble. Chitosan is insoluble in water, organic solvents and aqueous bases and it is soluble after stirring in acids such as acetic, nitric, hydrochloric, perchloric and phosphoric (Guibal, 2004; Klug et al., 1998; Kubota et al., 2000; Kurita, 2006; Anthonsen & Smidsroed, 1995; Rinaudo, 2006; Sankararamakrishnan & Sanghi, 2006). It was demonstrated that intra- and inter-molecular hydrogen bonds play a significant role in forming chitosan's crystalline domains, and appear to provide the main factor limiting its aqueous solubility (it is soluble in water at pH < 6). Protonation of amine group in acidic environment form polycationic form that distorted crystal structure and provide solution stabilities. Solubility of chitosan polymer is one of very important factor for its biological applications. It increases the chance of polymer-microorganism interaction.

Solubility of aminated chitosan comparing to chitosan itself was studied and presented in figure 4. A solubility test was performed by dissolving a weighted sample in 2% acetic acid and was stirred at room temperature for a one hour, and then the sample was filtrated, dried, and weighed. The solubility was determined by using the following equation:

Solubility % = [1- insoluble part/ total weight sample] x 100.

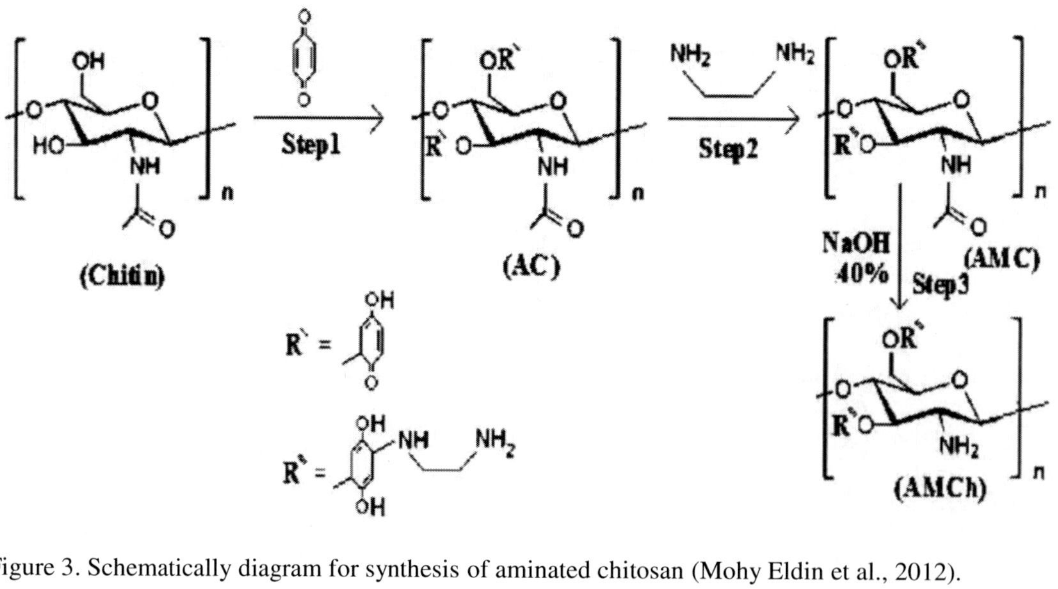

Figure 3. Schematically diagram for synthesis of aminated chitosan (Mohy Eldin et al., 2012).

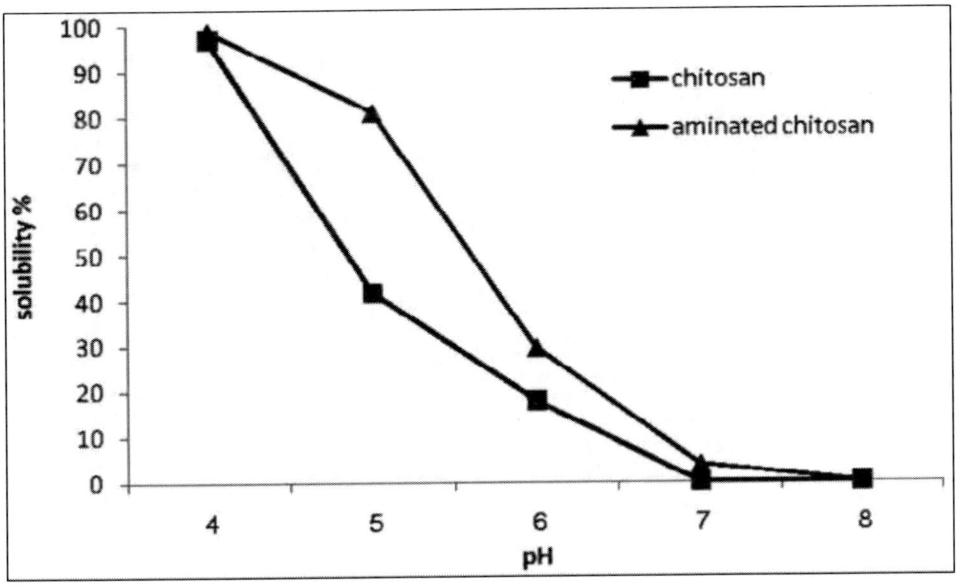

Figure 4. Solubility percent of chitosan and aminated chitosan at different pH. (Mohy Eldin et al., 2012).

Aminated Chitosan Characterization

FT-IR

FT-IR of chitosan and aminated chitosan was done using FTIR-8400S SHIMDZU, Japan. Figure 5 show stretching vibration band at 3430 cm^{-1} that attributed stretching vibration of NH_2 and OH groups. Aminated chitosan exhibit more fine at this region that may be attributed increase amine content in the modified polymer. Bands at 2970 cm^{-1} represent to (C-H stretching on methyl) and 2935 cm^{-1} for (C-H stretching in methylene). The bands at 1654 cm^{-1} correspond to stretching of carbonyl group (C=O) of primary amide (amide I). The band at 1633 cm^{-1} corresponds to deformation vibration of $-NH_2$ in plane. The band at 1568 cm^{-1} corresponds to deformation vibration of groups –NH– of amines. The bands at 1427, 1388 and 1159 cm^{-1} correspond to deformation vibration of C–N and the band at 1055 cm^{-1} corresponds to asymmetric stretching of C–N–C.

Thermal Gravimetric Analysis

Thermo Gravimetric analysis (TGA) of chitosan and aminated chitosan were carried out using TGA-50 SHIMADZU Japan. Figure 6; illustrate the thermal degradation of chitosan and aminated chitosan under Nitrogen atmosphere. First depuration from ambient temperature to about 150 °C was attributed to elevation of moisture content in polymer. Increasing of the amount of moisture from 7.2% to about 11.15% was attributed to increase hydrophilic nature of modified polymer by amination process. Table 2; illustrate the most important depuration in thermal degradation behavior of chitosan and aminated chitosan.

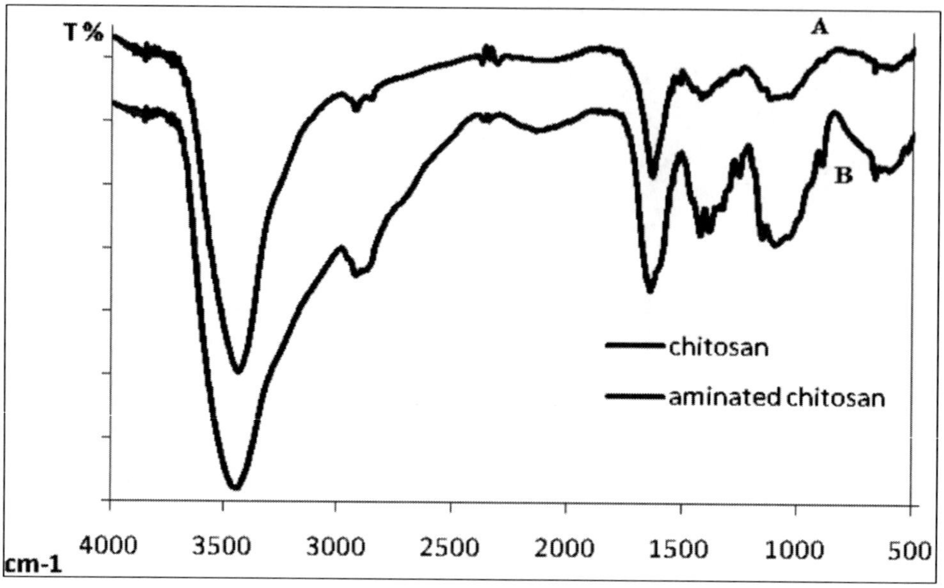

Figure 5. FT-IR of Chitosan A and aminated Chitosan B.

Table 2. Thermal gravimetric depuration steps of Chitosan and aminated chitosan

	Ambient-150 °C	220-350 °C	350-750	T50
Chitosan	7.2%	35.33%	12.16%	309.4 °C
Aminated chitosan	11.15%	36.35%	49.23%	347.22 °C

According to Pawlak and Mucha, the main depression of chitosan TGA curve ranged from 220 to 350 was a result of oxidative decomposition of the chitosan backbone. In this stage first depression was resulted from destruction of amine groups to form crosslinked fragment and the second decomposition step, which appears at high temperature, may result from the thermal degradation of a new crosslinked material formed by thermal crosslinking reactions occurring in the first stage of degradation process (Pawlak. and Mucha, (2003). Results in table 2 show decrease the thermal stability of chitosan as grafted with external amine groups. That may be attributed to role of amine group in enhancing thermal degradation process.

Differential Scanning Calorimetry

Differential scanning calorimetric analyses of chitosan and aminated chitosan are illustrated by Figure 7. The first endothermic peak that starting from 50 to 120 °C can be ascribed to the loss of moisture content. Polysaccharides usually have a strong affinity for water, and in solid state these macromolecules may have disordered structures which can be easily hydrated. As is known, the hydration properties of these polysaccharides depend on the primary and supramolecular structures (Kacurakova M. et al., 1998; Phillipsv G.O. et al., 1996).

The second thermal event may be related to the decomposition of Glucose amine (GlcN) units with correspondent exothermic peak at 295 °C (Guinesi L.S. and. Cavalheiro E.T.G, 2006, Kittur F.S. et al., 2002). Increase the intensity that attributed to increase decomposition process amine (GcN) unites by Increase of amine content.

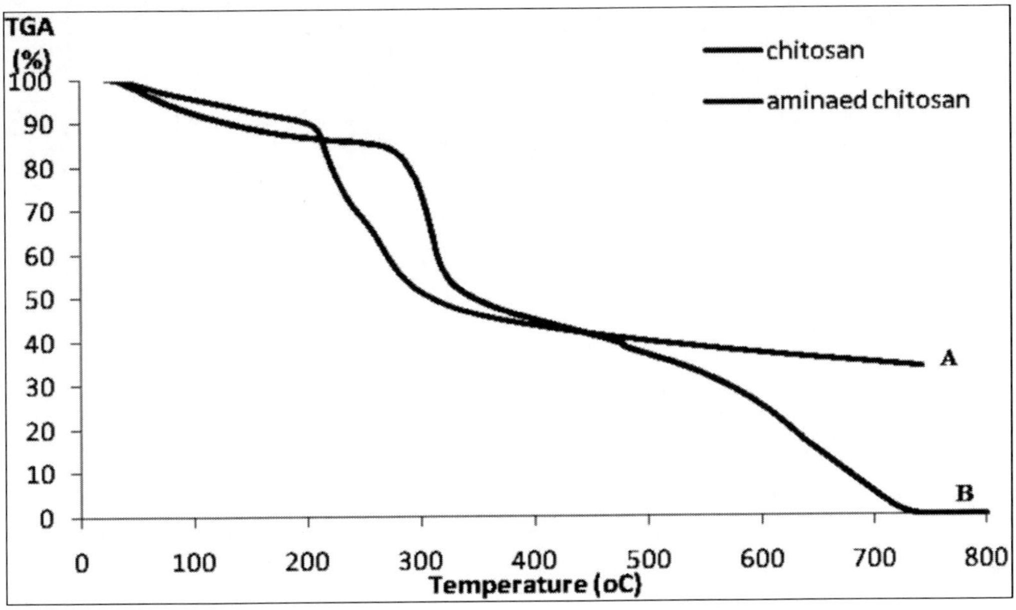

Figure 6. TGA analysis of (a) chitosan (b) aminated chitosan.

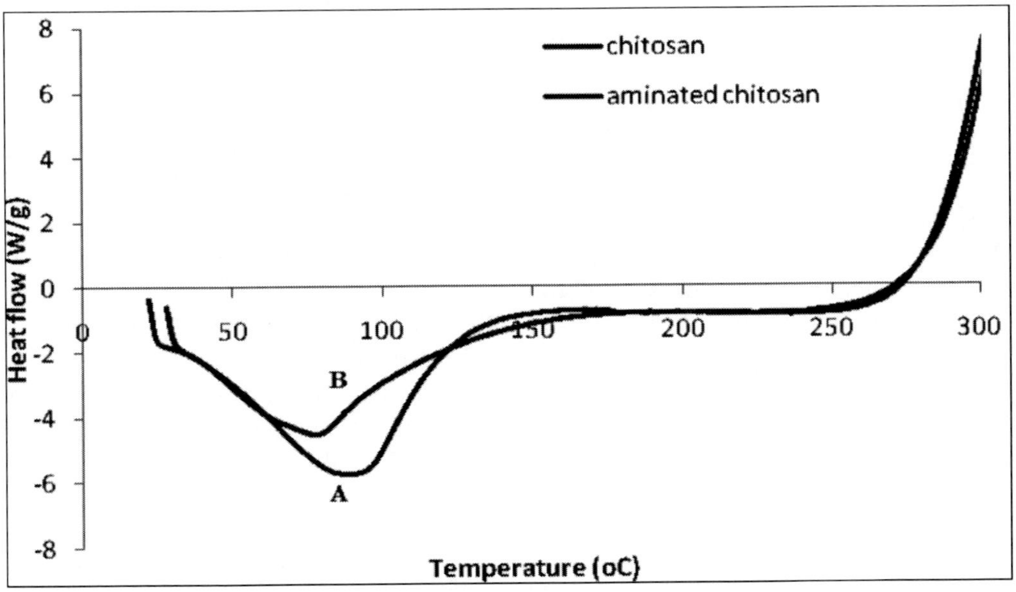

Figure 7. DSC analysis of chitosan (A) aminated chitosan(B).

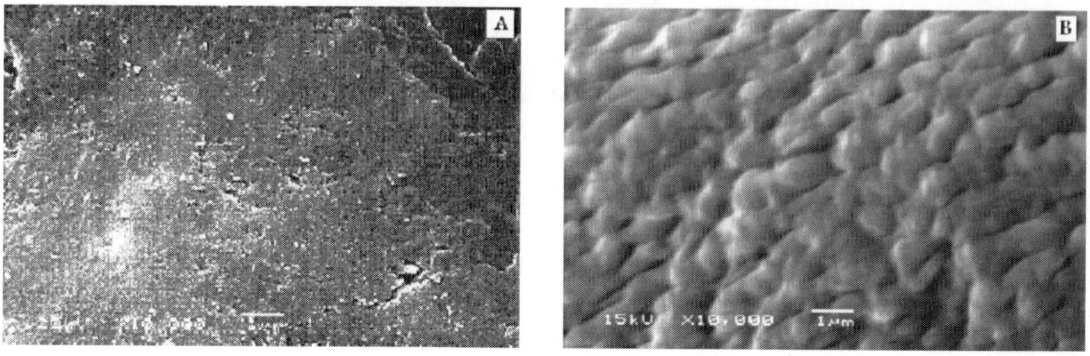

Figure 8. SEM graph of Chitosan A, and aminated Chitosan B (Mohy Eldin et al., 2012).

Scan Electron Microscope

Surface morphological analysis of chitosan and their aminated derivatives were performed using Scanning Electron Microscope (SEM).

The SEM graphs, Figures 8; show that the surfaces of chitosan become rougher upon amination that attributed to modification process. The increase in surface roughness is usually accompanied with increase in the surface area leading to enhanced adhesion with the microorganism's cell walls.

EVALUATION OF THE ANTIFUNGAL ACTIVITY OF AMINATED CHITOSAN

Antifungal activity of chitosan and aminated chitosan was tested against three different fungi species *Aspergillus Niger, Alternaria Alternata and Fusarium Moniliforme.*

Fusarium species are frequently reported as the causative agent in opportunistic infections in human (Godoy P et al., 2004). *A. niger* is the most common causative agent encountered in food contamination cases. Although it is not a common human pathogen, in high concentration, it may cause Aspergillosis (Sebti I et al., 2005). In the other hand Alternaria produces a number of toxins as pathogenicity factors, among them alternariol and alternariol monomethylether are major ones, since these are produced by most *Alternaria* species in large quantities (Heisler et al., 1980). Toxins of *Alternaria* have been detected as natural contaminants of plants like tomato fruit and tomato products (Stack et al., 1985), apples (Stinson et al., 1981) and olives (Visconti et al., 1986).

In this study, The mycelial disks (7 mm in diameter) from two-week-old cultures of the fungi were placed in the centre of Petri dishes (90 mm in diameter) with 10 ml solid PDA or PSA medium layered with 800μL of chitosan or aminated chitosan solution (2%), then incubated at 25°C. The mycelia growth was determined by measuring colony diameter daily and antifungal index was calculated as following equation

$$\text{Antifungal index } (\%) = (1 - D_a/D_b) \times 100$$

Where D_a is the diameter of the growth zone in the experimental dish (cm) and D_b is the diameter of the growth zone in the control dish (cm).

Figures (9-11), show daily growth of different fungi species. Aminated chitosan show always lower growth than that of chitosan in all selected fungi species. This could be explained by the fact that the negatively charged plasma membrane is the main target site of polycation (Singh T et al., 2008). Therefore, the polycationic aminated chitosan will interact more effectively with the fungus compared with free form of chitosan itself and disrupt the membrane integrity (Qi L et al., 2004).

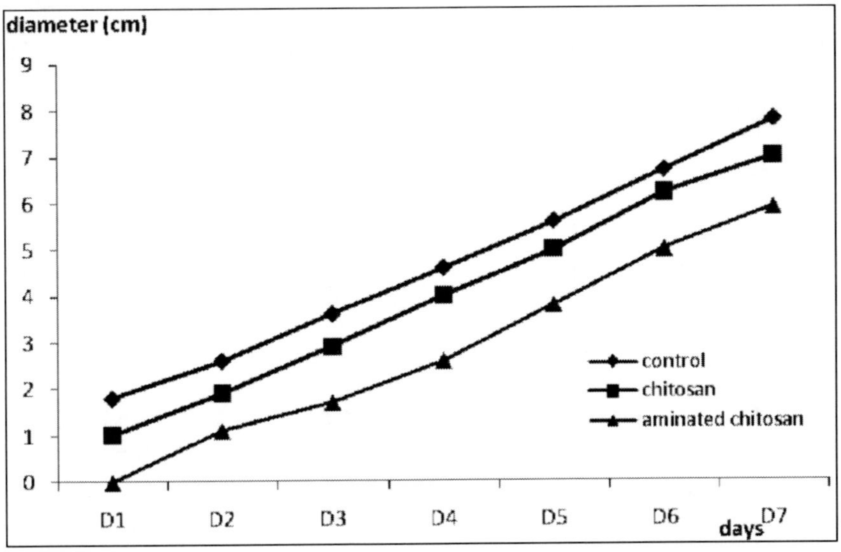

Figure 9. Antifungal activity of chitosan and aminated chitosan against *Aspergillus Niger*.

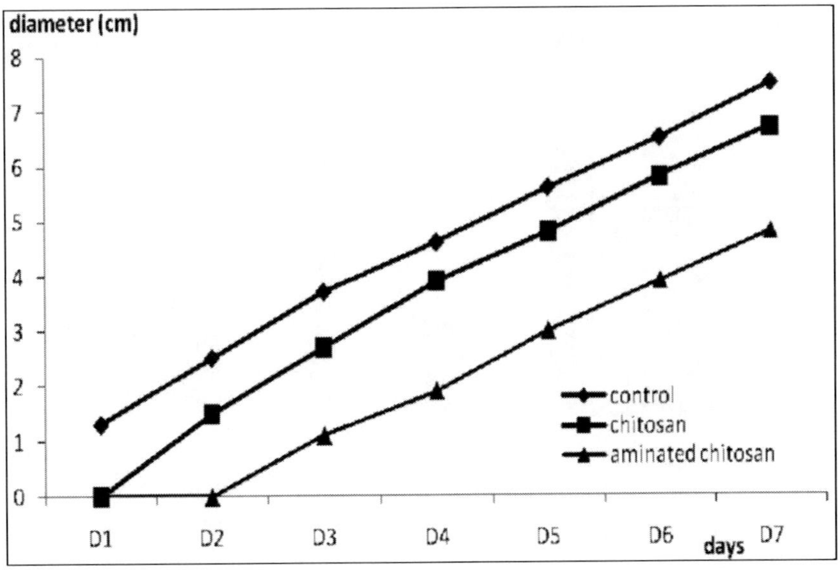

Figure 10. Antifungal activity of chitosan and aminated chitosan against *Alternaria Alternata*.

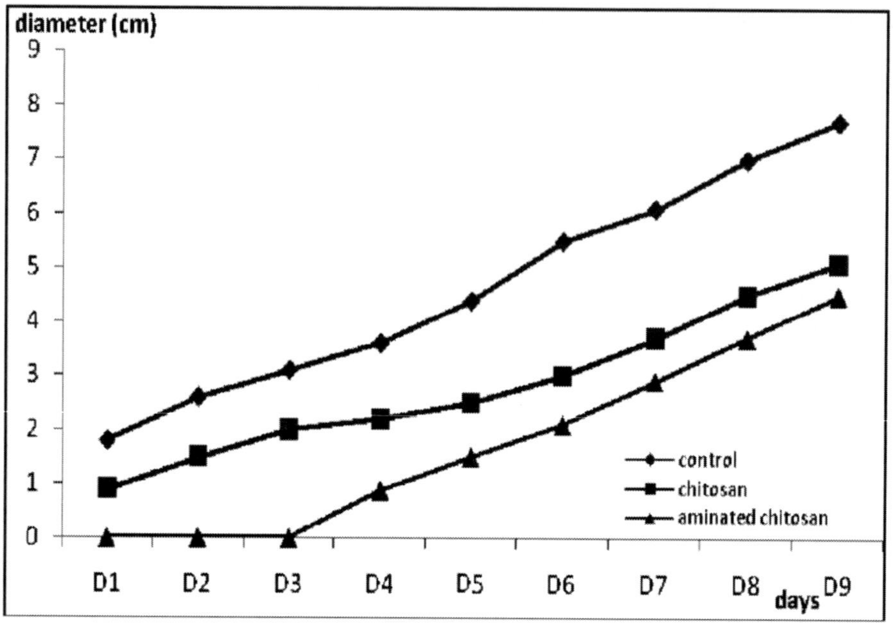

Figure 11. Antifungal activity of chitosan and aminated chitosan against *Fusarium Moniliforme.*

Figure 12 show antifungal index of chitosan and aminated chitosan against different fungi species. It's clear that promotion of antifungal activity by modifications, study show also increase the activity of polymer solutions against *Fusarium M* species rather than *Alternaria A* and *Aspergillus N* that may be attributed to its internal structure of cell wall membrane.

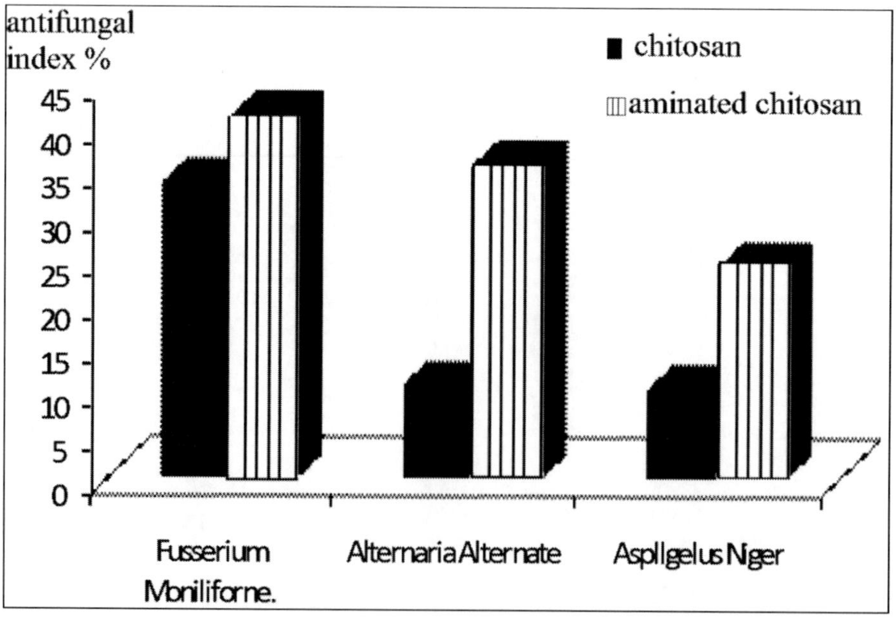

Figure 12. antifungal index of chitosan and aminated chitosan against *Aspergillus Niger*, *Alternaria Alternata* and *Fusarium Moniliforme.*

DISCUSSION

Mycotoxins can be characterized as secondary metabolites of various toxigenic fungi. Mycotoxins occur in a wide variety of foods and feeds and have been implicated in a range of human and animal diseases (Coker, 1997). Exposure to mycotoxins can produce both acute and chronic toxicities ranging from death to deleterious effects upon, for example, the central nervous, cardiovascular and pulmonary systems. Their general teratogenicity, cancerogenicity and their toxicological properties constitute a high human and animal health risk. The mycotoxins also attract attention because of the significant economic losses associated with their impact on human health and animal productivity.

Chitosan's fungal inhibition was observed on different development stages such as mycelial growth, sporulation, spore viability and germination, and the production of fungal virulence factors.

It has been commonly recognized that antifungal activity of chitosan depends on its molecular weight, deacetylation degree, pH of chitosan solution and, of course, the target organism. Mechanisms proposed for the antifungal activity of chitosan focused mainly on its effect on fungal cell wall (Allan C.R., and Hadwiger L.A, 1979) and cell membrane (Zakrzewska A., et al., 2005)

The antifungal activity of chitosan has been reported and developed in several studies both in vitro and in vivo, although chitosan activity against fungus has been shown to be less efficient as compared with its activity against bacteria (Tsai et al., 2000). The inhibitory efficiency of chitosan has been related to chitosan properties such as its molecular weight, deacetylation degree, pH of chitosan solution and, of course, the target organism.. In others works, researchers reported that the level of inhibition of fungi was also highly correlated with chitosan concentration, indicating that chitosan performance is related to the application of an appropriate rate. On the other hand, results from Bautista-Banos et al. (2006) and Guo-Jane et al. (2006), showed important differences among them. Nevertheless, all these studies indicated that the polycationic nature of chitosan is the key to its antifungal properties and that the length of the polymer chain enhances that activity. An additional explanation includes the possible effect that chitosan might have on the synthesis of certain fungal enzymes (El Ghaouth et al., 1992). Recent studies have shown that not only chitosan is effective in stopping the growth of the pathogen, but it also induces marked morphological changes, structural alterations and molecular disorganization of the fungal cells (Bautista-Banos et al., 2006). The positive charge of the chitosan is due to the protonisation of its functional amino group. This group reacts with the negatively charged cell walls of macromolecules, causing a dramatic increase in the level of the permeability of cell membrane, causing disruptions that lead to cell death (Sebti et al., 2005).

In the presented results, it can show increase of anrifungal activity of *Fusarium Moniliforme and Alternaria Alternata more than that in Aspergillus Niger*. This variation of antifungal activity may be attribute to nature and consists of fungal cell wall. *A. niger* was found to be highly resistant to both chitosan and aminated chitosan. Fungi that have chitosan as one of the components in the cell wall are more resistant to externally amended chitosan. This fact could therefore explain the high resistance of *A. niger* as it contains 10% of chitin in its cell wall (Klis F. M., et al., 2007).

References

Allan, C. R. & Hadwiger, L. A. (1979). "The fungicidal effect of chitosan on fungi of varying cell wall composition". *Exp. Myco.*, vol. *3*, 285–287.

Análise das isotermas de adsorção de Cu(II), Cd(II), Ni(II) e Zn(II) pela N-(3, 4-dihidroxibenzil) quitosana empregando o método da regressão não linear. *Quimica Nova*, Vol. *21*, 410–413.

Anthonsen, M. W. & Smidsroed, O. (1995). Hydrogen ion titration of chitosans with varying degrees of N-acetylation by monitoring induced 1H NMR chemical shifts. *Carbohydrate Polymers*, *26*, 303–305.

Bautista-Ban˜os, S., Herna´ndez-Lauzardo, A. N., Vela´zquez-del Valle, M. G., Herna´ndez-Lo´pez, M., Ait Barka, E., Bosquez-Molina, E., et al. (2006). Chitosan as a potential natural compound to control pre and postharvest diseases of horticultural commodities. *Crop Protection Review*, *25*, 108–118.

Brine, C. J. (1984). in chitin, chitosan and Related Enzymes, J.P.Zikakis, *Academic ress Inc Chpt.*,17-23,

Chen, C. S., Liau, W. Y. & Tsai, G. J. - J. (1998). *Food Prot.*, *61*, 1124-1128.

Coker, R. D. (1997). Mycotoxins and their control: Constraints and opportunities. *NRI Bulletin*, *73*. Natural Resources Institute, Central Avenue, Chatham Maritime, Chatham, Kent, ME4 4TB.

El Ghaouth, A., Arul, J., Asselin, A. & Benhamou, N. (1992 a). Antifungal activity of chitosan on post-harvest pathogens: induction of morphological and cytological alterations in Rhizopus stolonifer. *Mycol. Res.*, *96*(9), 769-779.

El Ghaouth, A., Arul, J., Asselin, A. & Benhamou, N. (1992). Antifungal activity of chitosan on post-harvest pathogens: induction of morphological and cytological alterations in Rhizopus stolonifer. *Mycology Research*, *96*, 769–779.

Eweis, M., Elkholy, S. S. & Elsabee, M. Z. - (2006). *Int. J. Biological Macrom.*, *38*, 1-8.

García-Rincón, J., Vega-Pérez, J., Guerra-Sánchez, M. G., Hernández- Lauzardo, A. N., Peña-Díaz, A. & Velázquez-Del, Valle M. G. (2010). Effect of chitosan on growth and plasma membrane properties of Rhizopus stolonifer (Ehrenb.: Fr.) Vuill. *Pestic. Biochem. Phys.*, *97*(3): 275-278.

Godoy. P., Nunes, F., Silva, V., Tomimori-Yamashita, J., Zaror, L. & Fischman, O. (2004). "Onychomycosis caused by *Fusarium solani* and *Fusarium oxysporum* in S˜ao Paulo, Brazil," *Mycopathologia*, vol. *157*, no. 3, 287–290.

Guerra-Sánchez, M. G., Vega-Pérez, J., Velázquez-del, Valle M. G. & Hernández-Lauzardo, A. N. (2009). Antifungal activity and release of compounds on Rhizopus stolonifer (Ehrenb.:Fr.) Vuill. by effect of chitosan with different molecular weights. *Pestic. Biochem. Phys.*, *93*(1), 18-22.

Guibal, E. (2004). Interactions of metal ions with chitosan-based sorbents: *A review. Separation and Purification Technology*, Vol. *38*, 43–74.

Guinesi, L. S. & Cavalheiro, E. T. G. (2006). *The use of DSC curves to determine the acetylation degree of chitin/chitosan samples Thermochimica Acta*, *444*, 128–133.

Guo, Z., Xing, R., Liu, S., Zhong, Z., Ji, X., Wang, L. & Li, P. (2008). The influence of molecular weight of quaternized chitosan on antifungal activity. *Carbohydr. Polym.*, *71*(4), 694-697.

Guo-Jane, Tsai, M. -T., Lee, J. -M. & Zhong, M. -Z. (2006). Effects of chitosan and a low-Molecular-Weight chitosan on Bacillus cereus and application in the preservation of cooked rice. *Journal of Food Protection*, *69*(9), 2168–2175.

Hadwiger, L. A., Kendra, D. G., Fristensky, B. W. & Wagoner, W. (1981). Chitosan both activated genes in plants and inhibits RNA synthesis in fungi, in: "Chitin in nature and technology". Muzzarelli, R. A. A., Jeuniaux, C. & Gooday, G. W. (Eds.), Plenum, New York.

Heisler, E. G., Siciliano, J., Stinson, E. E. & Osman, S. F. (1980). High performance liquid chromatographic determination of major mycotoxins produced by *Alternaria* molds. *J. Chromatogr.*, *194*, 89–94.

Hernández-Lauzardo, A. N., Guerra-Sánchez, M. G., Hernández-Rodríguez, A., Heydrich-Pérez, M., Vega-Pérez, J. & Velázquez-del, Valle M. G. (2010a). Assessment of the effect of chitosan of different molecular weights in controlling Rhizopus rots in tomato fruits. *Arch Phytopathology Plant Protect* (2012), *45* (1), 33-41.

Hernández-Lauzardo, Miguel Gerardo, Velázquez-del Valle, María Guadalupe & Guerra-Sánchez. (2011). Current status of action mode and effect of chitosan against phytopathogens fungi. *Afr. J. Microbiol. Res.*, Vol. *5*(25), 4243-4247.

Hirano, S., Nagao, N. & Agric. (1989), *Biol, Chem.*, (*53*), 3065- 66.

Kacurakova, M., Belton, P. S., Hirsch, J. & Ebringerova, A. (1998). *Physiology and Molecular plant Pathology*, *35*, 215-230.

Kittur, F. S., Prashanth, H., Sankar, K. U. & Tharanathan, R. N. (2002). characterization of chitin, chitosan and their carboxymethyl derivatives by different scanning calorimetry. *Carbohydr. Polym.*, *49*, 185.

Klis, F. M., Ram, A. F. J. & De Groot, P. W. J. (2007). "A molecular and genomic view of the fungal cell wall," in Biology of the Fungal Cell, R. J. Howard and N. A. R. Gow, Eds., 97–112, Springer, Berlin, Germany.

Klug, M., Sanches, M. N. M., Laranjeira, M. C. M., Favere, V. T. & Rodrigues, C. A. (1998).

Kubota, N., Tastumoto, N., Sano, T. & Toya, K. (2000). A simple preparation of half Nacetylated chitosan highly soluble in water and aqueous organic solvents. *Carbohydrate Research*, Vol. *324*, 268–274.

Kurita, K. (2006). Chitin and chitosan: Functional biopolymers from marine crustaceans. *Marine Biotechnology*, Vol. *8*, 203–226.

Liu, J., Tian, S., Meng, X. & Xu, Y. (2007). Effects of chitosan on control of postharvest diseases and physiological responses of tomato fruit. *Postharvest Biol. Technol.*, *44*(3), 300-306.

Mohy Eldin, M. S., Soliman, E. A., Hashem, A. I. & Tamer, T. M. (2012). Antimicrobial Activity of Novel Aminated Chitosan Derivatives for Biomedical Applications. *Advances in Polymer Technology*, Vol. *31*, No. 4, 414–428.

Mrunal R. Thatte (2004). Ph.D thesis. Synthesis and antibacterial assessment of water-soluble hydrophobic chitosan derivatives bearing quaternary ammonium functionality. Louisiana State University

Palma-Guerrero, J., Lopez-Jimenez, J. A., Pérez-Berná, A. J., Huang, I. –C., Jansson, H. –B., Salinas, J., Villalaín, J., Read, N. D. & Lopez-Llorca, L. V. (2010). Membrane fluidity determines sensitivity of filamentous fungi to chitosan. *Mol. Microbiol.*, *75*(4), 1021-1032.

Palma-Guerrero, J., Jansson, H., Salinas, J. & Lopez-Llorca, J. V. (2008). Effect of chitosan on hyphal growth and spore germination of plant pathogenic and biocontrol fungi. *J. Appl. Microbiol.*, *104*(2), 541- 553.

Palma-Guerrero, J., Jansson, H., Salinas, J. & Lopez-Llorca, J. V. (2008). Effect of chitosan on hyphal growth and spore germination of plant pathogenic and biocontrol fungi. *J. Appl. Microbiol.*, *104*(2), 541-553.

Papineau, A. M., Hoover, D. G., Knorr, D. & Farkas, D. F. (1991). *Food Biotechnol.*, *5*, 45-57.

Pawlak, A. & Mucha, M. (2003). Thermochim. *Acta*, 396 153.

Phillips, G. O., Takigami, S. & Takigami, M. (1996). *Food hydrocolloids*, *10*, 11-19.

Qi, L., Xu, Z., Jiang, X., Hu, C. & Zou, X. (2004). "Preparation and antibacterial activity of chitosan nanoparticles," *Carbohydrate Research*, vol. *339*, no. 16, 2693–2700.

Rigby, G. (1936), patent. "Substantially Undegraded Deacetylated Chitin and Processes for Producing the Same." USA 2,040,879. 19 May 1936.

Rinaudo, M. (2006). Chitin and chitosan: Properties and applications. *Progress in Polymer Science*, Vol. *31*, 603–632.

Salmon. S. & Hudson. S. M. (1997). *J. Macomol. Sci.*, M.C.C37, 199-276.

Sankararamakrishnan, N. & Sanghi, R. (2006). Preparation and characterization of a novel xanthated chitosan. *Carbohydrate Polymers*, Vol. *66*, 160–167.

Sebti, I., Martial-Gros, A., Carnet-Pantiez, A., Grelier, S. & Coma, V. "Chitosan polymer as bioactive coating and film against *Aspergillus niger* contamination," *Journal of Food Science*, vol. *70*, no. 2, M100–M104, 2005.

Sebti, I., Martial-Gros, A., Carnet-Pantiez, A., Grelier, S. & Coma, V. (2005). Chitosan polymer as bioactive coating and film against Aspergillus niger contamination. *Journal of Food Science*, *70*(2), 100–104.

Shahidi, F., Arachchi, J. & Jeon, Y. J. (1999). *Trends Food Sci. Technol.*, *10*, 37-51.

Singh, T., Vesentini, D., Singh, A. P. & Daniel, G. (2008). Effect of chitosan on physiological, morphological, and ultrastructural characteristics of wood-degrading fungi. *Int. Biodet. Biodeg.*, *62*(2), 116-124.

Singh, T., Vesentini, D., Singh, A. P. & Daniel, G. (2008). "Effect of chitosan on physiological, morphological, and ultrastructural characteristics of wood-degrading fungi," *International Biodeterioration and Biodegradation*, vol. *62*, no. 2, 116–124.

Stinson, E. E., Osman, S. F., Heisler, E. G., Siciliano, J. & Bills, D. D. (1981). Mycotoxin production in whole tomatoes, apples, oranges and lemons. *J. Agr. Food Chem.*, *29*, 790–792.

Sudarshan, N. R., Hoover, D. G. & Knorr, D. (1992). *Food Biotechnol.*, *6*, 257-272.

Tanigawa, T., Tanaka, Y., Sashiwa, H., Saimoto, H., Shigeasa, Y. J., Sandford, P. a. & Zikakis, J/P. (1992), (Eds.). Elsevier science Publishers Ltd., London, New York, 206-15.

Tsai, G. J., Wu, Z. Y. & Su, W. H. (2000). Antibacterial activity of a chitooligosaccharide mixture prepared by cellulase digestion of shrimp chitosan and its application to milk preservation. *Journal Food Protection*, *63*, 747–752.

Visconti, A., Logrieco, A. & Bottalico, A. (1986). Natural occurrence of *Alternaria* mycotoxins in olives—Their production and possible transfer into the oil. *Food Additives and Contaminants*, *3*, 323–330.

Xu, J., Zhao, X., Han, X. & Du, Y. (2007a). Antifungal activity of oligochitosan against Phytophtora capsici and other pathogenic fungi in vitro. *Pestic. Biochem. Phys.*, *87*(3), 220-228.

Xu, J., Zhao, X., Wang, X., Zhao, Z. & Du, Y. (2007b). Oligochitosan inhibits Phytophtora capsici by penetrating the cell membrane and putative binding to intracellular target. *Pestic. Biochem. Phys.*, *88*(2), 167-175.

Young, D. H., Kohle, H. & Kauss, H. (1982). *Plant Physiol.*, *70*, 1449-1454.

Zakrzewska, A., Boorsma, A., Brul, S., Hellingwerf, K. J. & Klis, F. M. (2005). "Transcriptional response of Saccharomyces cerevisiae to the plasma membrane-perturbing compound chitosan". *Eukaryot. Cell.*, vol. *4*, 703–715.

In: News in Chemistry, Biochemistry and Biotechnology ISBN: 978-1-63117-273-1
Editors: G. E. Zaikov, G. Nyszko, L. P. Krylova et al.

Chapter 7

COLLAGEN MODIFIED HARDENER FOR MELAMINE-FORMALDEHYDE ADHESIVE FOR INCREASING WATER RESISTANCE OF PLYWOOD

Ján Matyšovský[1*], Peter Jurkovič[1], Pavol Duchovič[1] and Igor Novák[2]

[1]Vipo a.s., Partizánske, Slovakia
[2]Polymer Institute, Slovak Academy of Sciences, Slovakia

ABSTRACT

One of the very important technological operations in woodworking industry is gluing. The aim of this work was preparation of hardener for melamine-formaldehyde (MEF) adhesives suitable for gluing of plywood. Commercial hardener was modified by biopolymers (waste animal polymers). Glued joints were expected to be classified as resistant to water in class 3.

In the experiments, two types of leather collagen hydrolysates *(VIPOTAR I* and *VIPOTAR II)* were applied into MEF adhesive. Leather collagen hydrolysates were obtained from waste produced by leather industry. Glued plywood specimens were preliminary conditioned by two different ways. Plywood glued with MEF adhesive with the modified hardener showed good strength properties when evaluated according to the standard STN EN 314-1, 2. Glued joints can be graded in class 3.

Keywords: Biopolymer, hardener, gluing, melamine adhesive, plywood, water resistance

INTRODUCTION

Great attention is paid to improvement of technology of gluing and development of new types of adhesives. The important effort is exploitation of available products that could

---

[*] Corresponding author: Email: jmatyasovsky@vipo.sk.

improve effectiveness of adhesive mixtures and reduce cost in production of adhesives. For the improvement of product quality (from the point of view of hygienic criteria), searching and using of raw materials reducing release of formaldehyde from glued joints is very important. Biopolymers could be such materials (e.g., waste from leather or food industry).

Melamine-formaldehyde (MEF) adhesives are thermo-reactive adhesives curing at neutral or acidic pH at higher temperatures (130-140 °C) usually at presence of hardeners. Laser scanning microscopy was used to investigate the distribution of adhesive in wood fibers. Cyr et al. (2007) researched penetration of melamine-urea-formaldehyde (MUF) adhesive at fiberboard (MDF) production. Atomic force microscopy (AFM) enabled to recreate the finest detail of fiber surface. Adhesive can penetrate into any layers of wood cell walls, uses its affinity to both water and wood polymers to penetrate through pores from surface to lumen.

Improvement the water resistance for challenge expositions, or modification of certain properties of joints can be achieved by a mixture of adhesives e.g., urea-formaldehyde (UF) with resorcinol, melamine or polyvinylacetate (PVAC). Problems of influence of melamine content in MUF adhesives on formaldehyde emission and cured resine structure was investigated by Tohmura et al. (2001). They used 6 MUF adhesives synthesized with different F/(M+U) and M/U molar ratios. The ^{13}C nuclear magnetic resonance (NMR) spectroscopy of cured MUF resins revealed that more methylol groups, dimethylene-ether, and branched methylene structures were present in the MUF resins with a higher F/(M+U) molar ratio, leading to increased bond strength and formaldehyde emmission. The lower formaldehyde emission from cured MUF adhesives with a higher M/U molar ratio may be ascribed to the stronger linkages between triazine carbons of melamine than those of urea carbons.

Dukarska and Lecka (2008) researched in preparation of adhesive mixture based on melamine adhesive for production of exterior plywood. Melamine-urea-phenol-formaldehyde (MUPF) and phenol-formaldehyde (PF) resins were filled by the waste from polyurethane (PUR) foam. Usage of adhesive mixtures based on MUF adhesive was searched by Jozwiak (2007). Fillers used were potato starch and rye flour. Obtained results showed that glued plywood met the standard for bond quality grade 3 and the mixture could be used for wood gluing at various levels of wood moisture content (6 – 21 %).

Cellulose and lignin, as the basic wood component, are able to interact with proteins. Experiments were carried out on the interaction with dried animal blood plasma and egg albumin (Polus-Ratajzak et al. 2003). Infrared FTIR spectroscopy was used to analyze chemical changes in cellulose and lignin during the reaction. Obtained spectra indicated on possible chemical reaction between the peptide chain and reactive groups associated with cellulose.

Shitij Chaba and Anil N. Netravali (2005) presented the research in modification of soy protein concentrate using glutaraldehyde and polyvinyl alcohol. The modified resin allow to process soy protein polymer without any plasticizer. The modified resin also showed increased tensile properties, improved thermal stability and reduced moisture resistance as compared to soy protein concentrate resin.

At present, the market has got an excess protein, especially protein hydrolysates from leather waste. Collagen belongs to the most important technical proteins, which enables more effective preparation of adhesive mixtures, Sedliačik (2008, 2009).

The aim of our research was to develop a hardener for MEF adhesive mixtures. The mixtures could be used for woof gluing in bond quality grade 3, according to the standard EN

314-1, 2. The adhesive joint of grade 3 is applicable at outdoor conditions – at unlimited climatic influences. Non modified commercial MEF adhesives provide glued joint in grade 2. The MEF hardener was modified by biopolymers of animal origin. Various waste biopolymers (leather waste) could be secondary used.

EXPERIMENTAL PART

The experiments were carried out with the adhesive (KRONOCOL SM 10) and the particular hardener (hardener - product of Duslo Šala). Required hardener addition is 3 %.

To prepare a modified hardener, biopolymers in the form of collagen substrates were used. Substrates were prepared by dechromation of chrome leather waste at two different temperatures and were specified as *activator VIPOTAR I* (prepared at 20 °C) and *activator VIPOTAR II* (at 30 °C). Substrates pH value was adapted to the value of 4,0. Solubility and hydrophobic improvement was assured by addition of lyotropic agent and hydrophobic agent (methylester of tannery fat MEKT). Commercial hardener was activated by addition of activators *VIPOTAR I* or *VIPOTAR II* in ratios 3,5 %. Adhesive mixtures were tested in 3-layer beech plywood. Pressing temperature was 130 °C, adhesive consumption 150 g.m^{-2}.

Shear strength was measured and evaluated using a tensile testing machine LaborTech 4.050 with 5 kN head. Glued joint quality was tested according to the standard STN EN 314-1. Bond quality was expressed as grade 1, 2 or 3. Requirements for joint quality at plywood are determined by the standard STN EN 314-2.

RESULTS AND DISCUSSION

To test the effectiveness of activator *VIPOTAR I*, influence of various concentrations of the activator on shear strength of prepared plywood was tested. If activator was added in hardener (in adhesive mixture), shear strength of the joint was increased. The improvement was observed only under specific activator concentration. The optimal addition was determined as 3,5 %. At higher ratio (5 %), the shear strength was lower when compared to ratios 3,5 % or 2,5 %.

Table 1. The shear strength of plywood specimens glued with the adhesive mixture with various amount of activator *VIPOTAR I*

sample	Activator addition in hardener [%]	Required standard value of shear strength [MPa]	Average shear strength [MPa]	Minimal measured shear strength [MPa]	Maximal measured shear strength [MPa]
reference	–	1,0	**1,1**	0,82	1,26
1	2,5	1,0	**1,6**	1,33	2,31
2	3,5	1,0	**1,9**	1,66	2,59
3	5,0	1,0	**1,3**	0,92	1,46

In table 1, mean values of shear strength evaluated according the method for grade 3, together with individual measured minimal and maximal values, are shown. Above mentioned tendency of shear strength is also evident in this detailed evaluation. Based on the experiments, further experiments were carried with addition of 3,5 %.

Table 2. The shear strength of preliminary conditioned plywood specimens (gluing in grade 2)

Sample /modifier/	Required standard value [MPa]	Average shear strength [MPa]	Standard deviation [MPa]	Coefficient of variation [%]	Minimal measured shear strength [MPa]	Maximal measured shear strength [MPa]	Number of samples
1- *VIPOTAR I*	1,0	**2,8**	0,20	7,3	2,4	3,2	12
2 – *VIPOTAR I*	1,0	**2,5**	0,23	9,2	2,2	2,9	12
3 –*VIPOTAR II*	1,0	**2,5**	0,28	11,1	2,0	3,0	12
4 –*VIPOTAR II*	1,0	**2,4**	0,26	10,9	2,1	3,0	12

Table 3. The shear strength of preliminary conditioned plywood specimens (gluing in grade 3)

Sample /modifier/	Required standard value [MPa]	Average shear strength [MPa]	Standard deviation [MPa]	Coefficient of variation [%]	Minimal measured shear strength [MPa]	Maximal measured shear strength [MPa]	Number of samples [ks]
1 – *VIPOTAR I*	1,0	**2,4**	0,29	11,7	1,7	2,9	15
2 – *VIPOTAR I*	1,0	**2,3**	0,37	16,1	1,6	2,8	13
3 – *VIPOTAR II*	1,0	**2,3**	0,22	9,7	1,8	2,8	15
4– *VIPOTAR II*	1,0	**1,9**	0,35	18,9	1,4	2,4	15

When preparing adhesive mixture for the experiments of water resistance, both activators *VIPOTAR I* and *VIPOTAR II* were used. Activators were added in the amount of 3,5 %. Resulting shear strength values for plywood conditioned for grade 2 are listed in table 2.

All mean values of shear strength for grade 2 markedly exceeded required standard value of 1,0 [MPa]; even all individual measured values were double than standard required value. The shear strength in comparison with the shear strength of the joint glued without the modifiers was significantly higher, more than doubled. Final shear strength values for plywood conditioned for grade 3 are listed in table 3.

Similarly as for grade 2, all mean shear strength values of grade 3 exceeded required standard value of 1,0 [MPa]. Moreover, all individual measured values were above the standard required value. Similarly as in the experiments for grade 2, the shear strength at grade 3 compared with the shear strength of the joint glued without the modifiers, was higher.

If we compare individual measured minimal and maximal values of shear strength for two various ways of conditioning of tested material, we can see that strength for grade 3 reached lower values when compared with grade 2. The same tendency was observed at mean values of shear strength. Such results can be expected, as preliminary conditioning for grading 3 is significantly more aggressive (longer total time of boiling interrupted with drying at higher temperature).

All tested adhesive mixtures and glued joints met the standard for grade 2 and grade 3, as well, and significantly exceeded the shear strength values of the reference sample.

Our findings confirmed the expected presumption; the strength and water resistance of adhesive bond is markedly influenced by the addition of a small amount of biopolymer (skin collagen). Commercial hardener was modified by activators *VIPOTAR I* and *VIPOTAR II* in the ratio 3,5 %. If we consider the ratio of activators in all volume of adhesive mixture, the concentration of them is very low; nevertheless, their impact on the resulting strength and water resistance of adhesive joints is so marked.

CONCLUSION

Our assumption that the addition of biopolymers in form of hydrolysates containing skin collagen can result in increased shear strength and increased water resistance, was confirmed. Collagen macromolecules dispersed in solution or in adhesive mixture have good adhesion to glued surface. In line with the results of other authors, we assume the right chemical reaction between the functional groups of protein and functional groups of the adhesive.

From the above results, it is visible that researched additives can become modifiers for adhesive mixtures based on MEF adhesives. MEF adhesives used in praxis are graded as adhesives class 2. Glued joints graded as class 2 are suitable in the environments with higher moisture (e. g. sheltered exterior, outdoor conditions – short-time climatic influences, indoor conditions with higher moisture when compared with grade 1). Both of the tested collagen substrates significantly increased the shear strength of glued joint, and enabled to grade the bond as 3. Adhesive bonds graded as 3 are applicable at outdoor conditions – at unlimited climatic influences.

ACKNOWLEDGEMENT

This publication was prepared as part of the project “Application of Knowledge-based Methods in Designing Manufacturing Systems and Materials” co-funded by the Ministry of Education, Science, Research and Sport of the Slovak Republic within the granted stimuli for research and development from the State Budget of the Slovak Republic pursuant to Stimuli for Research and Development Act No. 185/2009 Coll. and the amendment of Income Tax Act No. 595/2003 Coll. in the wording of subsequent regulations in the wording of Act. No. 40/2011 Coll.

REFERENCES

[1] Cyr, P. L., Riedl, B. & Wang, X. M. (2008). Investigation of Urea-Melamine-Formaldehyde (UMF) resin penetration in Medium-Density Fiberboard (MDF) by High Resolution Confocal Laser ScanningMicroscopy. In: *Holz als Roh-und Werkstoff, 66*, 129–134.

[2] Dukarska, D. & Lecka, J. (2008). Polyurethane foam scraps as MUPF and PF filler in the manufacture of exterior plywood. In: Annals of Warsaw University of Life Sciences - SGGW, Forestry and Wood Technology, Warszawa, No. *65*, 14–19.

[3] Jozwiak, M. (2007). Possibility of gluing veneers with high moisture content with the use modified MUF adhesives resin. In: Annals of Warsaw Agricultural University – SGGW. Forestry and Wood Technology, Warszawa, No. *61*.

[4] Polus-Ratajczak, I., Mazela, B. & Golinski P. (2003). The chemical interaction of animal origin proteins with cellulose and lignin in wood preservation. In: Annals of Warsaw Agricultural University - SGGW, Forestry and Wood Technology, Warszawa, No. *53*, 296–299.

[5] Sedliačik, J. & Sedliačiková, M. (2009). Innovation tendencies at application of adhesives in wood working industry. In: Annals of Warsaw University of Life Sciences – SGGW. Forestry and Wood Technology, Warszawa, No *69*, s. 262-266. ISSN 1898-5912.

[6] Sedliačik, J., Šmidriaková, M. & Jabloňski, M. (2008). Obniženie energetycznych wymagaň wytwarzania sklejek. Przemysl drzewny No.4, s. 24–26, ISSN 0373-9856.

[7] Shitij, Chaba. & Anil, N. Netravali. (2005). "Green" composites. Part 2: Characterization of flax yarn and glutaraldehyde/poly (vinyl alcohol) modified soy protein concentrate composites. In: *Journal of materials science*, *40*, 6275-6282.

[8] Stn EN 314-1: 2005. Preglejované dosky. Kvalita lepenia. Časť 1: Skúšobné metódy.

[9] Stn EN 314-2: 2005. Preglejované dosky. Kvalita lepenia. Časť 2: Požiadavky.

[10] Tohmura, S., Inoue, A. & Sahari, S. H. (2001). Influence of the melamine content in melamine-urea-formaldehyde resins on formaldehyde emission and cured resin structure. In *J. Wood Sci.*, *47*, 451- 457.

In: News in Chemistry, Biochemistry and Biotechnology ISBN: 978-1-63117-273-1
Editors: G. E. Zaikov, G. Nyszko, L. P. Krylova et al.

Chapter 8

POSSIBILITIES OF APPLICATION OF COLLAGEN COLOID FROM SECONDARY RAW MATERIALS AS A MODIFIER OF POLYCONDENSATION ADHESIVES

Ján Matyasovský[1*], Peter Jurkovič[1], Ján Sedliačik[2] and Igor Novák[3]
[1]VIPO, a.s., Partizánske, Slovakia
[2]Fac Wood Sciences and Technology, Technical University in Zvolen, Zvolen, Slovakia
[3]Polymer Institute, Slovak Academy of Sciences, Bratislava, Slovakia

ABSTRACT

This work presents the utilisation of collagen jelly as one of several possibilities of leather waste reprocessing. In the frame of experimental research, soluble collagen was used as a modifier for poly condensation adhesives composition. Based on the results it can be said, that collagen has a significant influence on basic properties of urea-formaldehyde (UF) and phenol-formaldehyde (PF) adhesives and also on mechanical and physical properties of glued joints.

INTRODUCTION

During the leather processing up to 25% mass of input raw material comes to chromium tanned waste. As presence of such large amount of waste presents economic and mainly ecologic problem for leather tanning industry, big effort is given to development of technologies for processing respectively disposal of chromium tanned waste in the world. Technologies based on different principles are result of this effort, which enable to separate chromium from collagen. Application of these procedures in industry and also the scale of evaluation of chromium waste depend also on effective application of obtained products.

Importance of lowering of formaldehyde (fd) emission from hardened UF adhesives and lowering of the price of PF adhesives was solved in the project of the 5th frame

* Corresponding author: Email: jmatyasovsky@vipo.sk.

programme European Commission with the name: "*Radical Environmentally Sustainable Tannery Operation by Resource Management*" (RESTORM). VIPO Partizanske – Slovak republic with UTB Zlin – Czech republic, CSIC Barcelona – Spain and UNC Northampton – Great Britain solved together described Work Packages.

The aim of the part of WP 6.4 was development of ecologic technology of dechromation of chromium shavings without oxidation Cr^{3+} to Cr^{6+}, with remained fibril structure with evaluation of the influence of pH, temperature and number of dechromation bathes on amount of removed chromium. Further, the influence of these parameters on looses of collagen by hydrolysis was followed and the degree of collagen destruction was evaluated by determination of iso-electric point and following of its increased solubility in the dependence on temperature of dechromation.

The aim of WP1 was to develop more valuable polycondensation adhesives with improved ecologic parameters by modification with natural non-toxic, biologically easy decomposable biopolymers. The influence of the amount of biopolymer in adhesive mixture was tested on lifetime, gel time, viscosity, lowering of formaldehyde emission and mechanical and physical parameters of plywood.

Experimental Part

Used material:

- chromium shavings,
- chemicals and materials for modification of adhesive mixtures:
    - resin from Chemko a.s., Strazske production:
      Diakol M1 – water solution of urea-formaldehyde polycondensate determined for production of board materials, used under heat in combination with hardener.
      look: milky liquid,
      dry content matter: 65 % weight,
      pH: 7,4 – 8,5,
      gel time: 60 – 80 s,
      content of free formaldehyde: max. 0,35 mg fd/g.
    - ***hardener R-60*** – water solution of ammonium nitrate, treated with formic acid to pH 4 – 5, concentration 57 – 60% weight,
    - technical flour,
    - collagen jelly,
    - beech veneer of thickness 1,8 mm.

Results and Discussion

Following results were reached by experimental trials:

1. Development of environmentally friendly technology of chromium shavings dechromation.

2. Modification of commercially produced polycondensation UF adhesive with the application of collagen jelly prepared from chromium shavings.
   - Technology of separation of soluble $Cr_2(SO_4)_3$ after alkali processing v $Ca(OH)_2$ was proposed for interruption of the bound Cr^{3+} — OOC – collagen – dechromation.
   - In experimental work, there was followed the influence of acid concentration – pH of water solution, influence of temperature and number of dechromation bathes on the degree of dechromation and looses of collagen by increasing of its solubility.
   - Collagen jelly was prepared for application into UF adhesives.

In experimental work, there was followed:

1. Amount of released Cr^{3+} after three dechromation bathes from chromium shavings at the value of pH 1,5 and temperatures 20 °C, 25 °C and 30 °C (samples were taken off every 30 min) is described in figure1 and the view on dechromed shavings is on the figure 2.

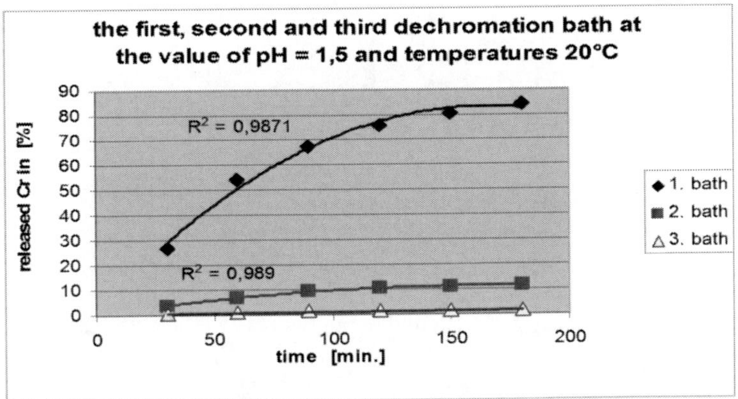

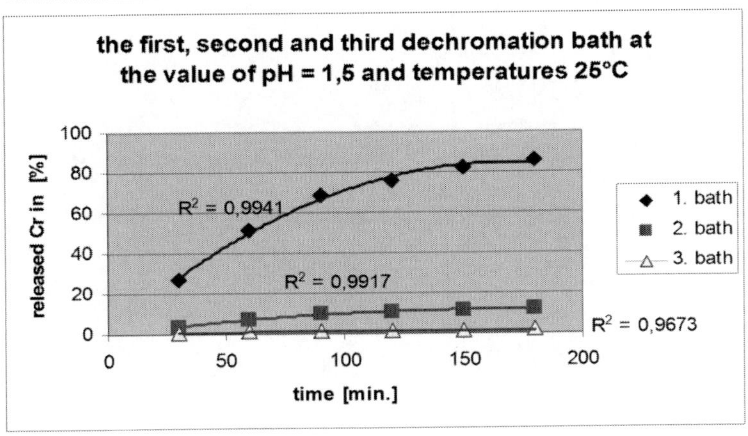

Figure 1. (Continued)

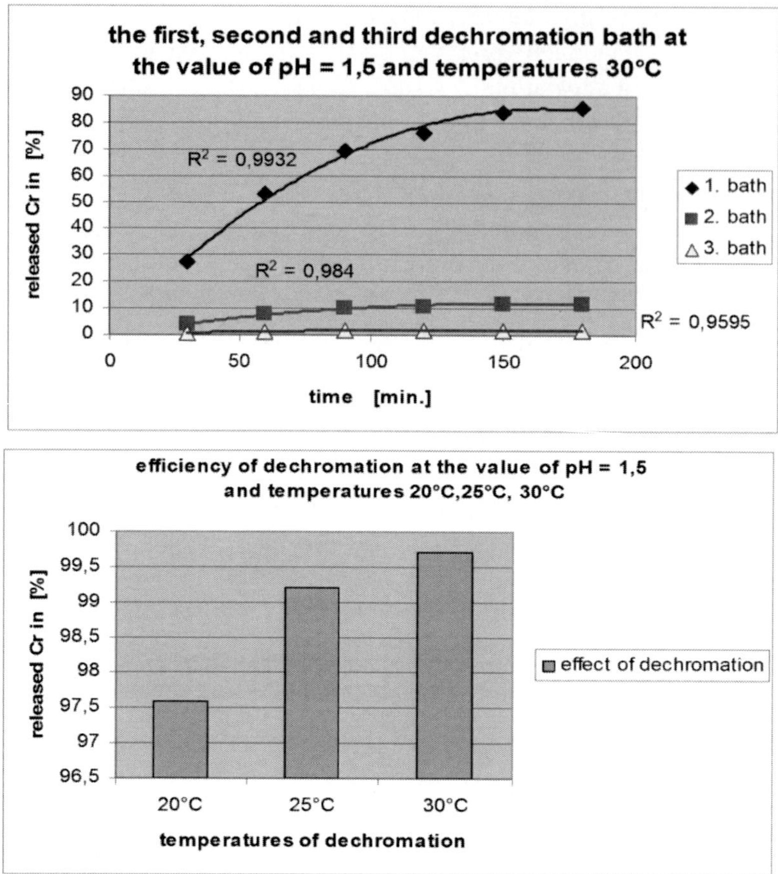

Figure 1. Experimentally determined dependencies of the released Cr in [%] in the first, second and third dechromation bath at the value of pH 1,5 and temperatures of 20, 25 and 30 °C.

Figure 2. Laboratory prepared dechromed shavings by proposed dechromation technology without oxidation of Cr^{3+} to Cr^{6+}.

The Influence of Addition of Collagen Jelly Samples No. 1, 2, 3 on Lifetime of Adhesive Mixture

Obtained results of lifetime determination of UF adhesive mixtures with collagen jelly samples No. 1, 2, 3 are presented in the table 1. The lifetime of adhesive mixtures was followed 48 hours.

From obtained results follow, that sample No. 1 has shortened lifetime > 24 h < 48 h and the lifetime of samples No. 2 and 3 is comparative with the reference sample of UF adhesive > 48 h.

The Influence of Addition of Collagen Jelly Samples No. 1, 2, 3 on Gel Time of Adhesive Mixture

Obtained results of gel time determination of UF adhesive mixtures with collagen jelly samples No. 1, 2, 3 are presented in the figure 3. The gel time was determined in the laboratory test-tube at the temperature of 100 °C.

Table 1. Lifetime of UF adhesive mixture with collagen jelly samples No. 1, 2, 3.

No. of sample	lifetime of adhesive mixture [h]
0	> 48
1	> 24 < 48
2	> 48
3	> 48

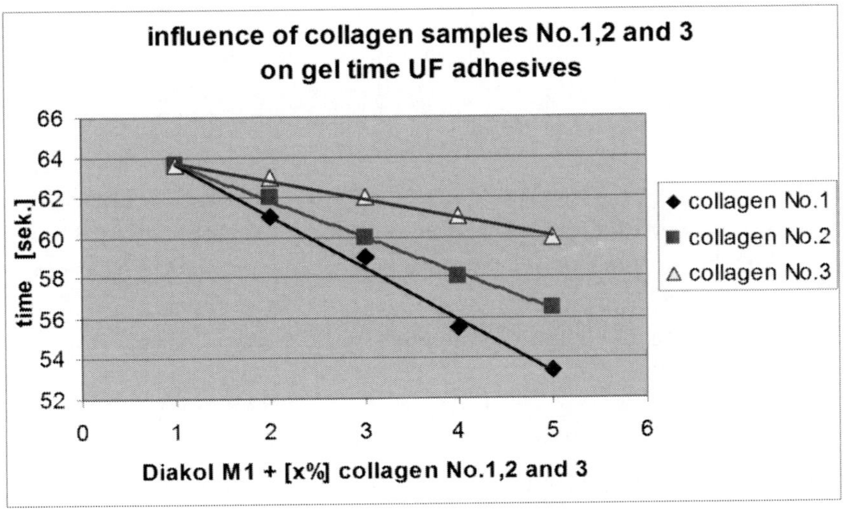

Figure 3. Experimentally determined dependency of the influence of collagen jelly samples No. 1, 2, and 3 on gel time of UF adhesive.

Determination of the gel time proved, that collagen jelly is suitable as a modifier of UF adhesive. Collagens No. 1, 2, 3 lightly accelerate the condensation reaction. Collagen jelly No. 1 obtained from first dechromation bath – least hydrolysed, the most significantly accelerates polycondensation reaction in comparison with the reference sample. Hydrolysis of collagen is increased by the impact of temperature, vitriol acid is consumed by reaction in dechromation bath, what reduces the gel time.

The Influence of Addition of Collagen Jelly Samples No. 1, 2, 3 on the Content of Free Formaldehyde in UF Resin Condensed at the Temperature of 120 °c and Time 15 min

Obtained results of the influence of amount of collagen jelly samples No. 1, 2, 3 on formaldehyde emission from UF adhesive mixtures are in the figure 4.

Determination of fd emission proved, that collagen jelly is suitable as modifier of UF adhesive for lowering of formaldehyde emission. Collagen jelly No. 1 obtained from first dechromation bath – least hydrolysed, the most significantly accelerates polycondensation reaction in comparison with the reference sample. Hydrolysis of collagen is increased by the impact of temperature, while comes to collagen hydrolysis and particular decay of amino-acids and also to lowering of amide nitrogen, what was confirmed by the decrease of iso-electric point. The decrease of reactive NH_2 groups in collagen No. 2, 3 is expressed as lowered ability to bind of formaldehyde.

The Influence of Addition of Collagen Jelly Samples No. 1, 2, 3 on the Viscosity of UF Adhesive Mixture

Obtained results of the influence of amount of collagen jelly samples No. 1, 2, 3 on the change of viscosity UF adhesive mixture. Viscosity of adhesive mixtures was measured with Höppler viscosimeter in laboratory conditions at temperature of 20 °C, results are in the figure 5.

Determination of the viscosity change of UF adhesive mixtures confirmed, that collagen jelly is suitable as modifier of UF adhesive for viscosity treatment. Collagen jelly can replace extenders as e.g., technical flour. Collagen jelly No. 1 obtained form first dechromation bath – least hydrolysed, most significantly treats viscosity in comparison with the reference sample. The decrease of mole weight of collagens No. 2, 3 is expressed as lowered ability to treat the viscosity of UF adhesives.

Strength Properties of Plywood

Strength properties of glued joints bonded with UF adhesive mixture with collagen jelly were tested on beech plywood. Plywood were prepared according to EN standardised procedure.

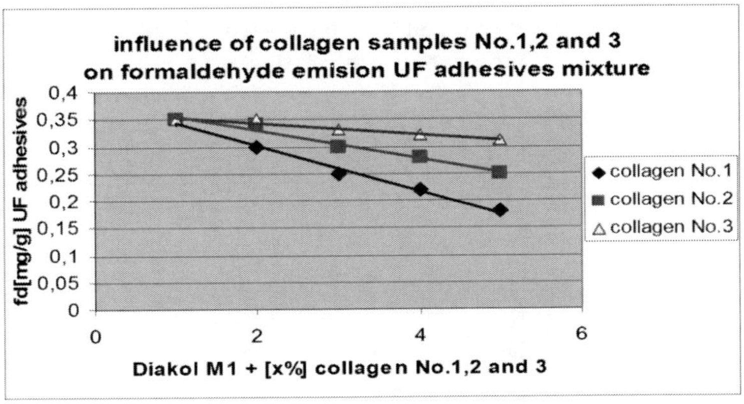

Figure 4. Experimentally determined dependency of the influence of collagen jelly samples No. 1, 2, and 3 on formaldehyde emission in UF adhesive mixture in [mg/g] condensed at the temperature of 120 °C and time 15 min.

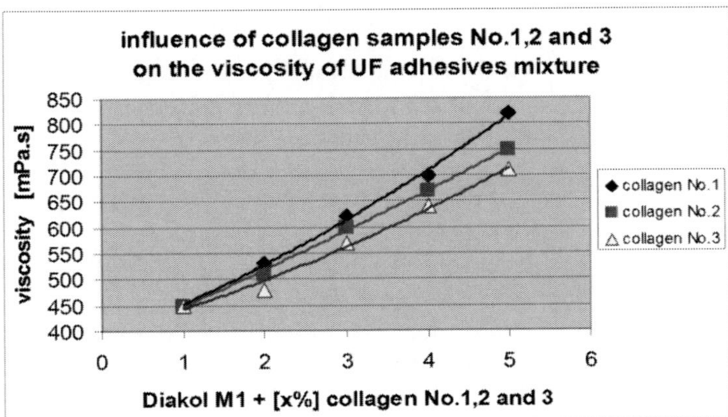

Figure 5. Experimentally determined dependency of the influence of collagen jelly samples No. 1, 2, and 3 on the viscosity of adhesive mixtures in [mPa.s].

Adhesive mixtures: Diakol M1 + collagen jelly used for viscosity treatment, marked as 1, 2, 3 added in amount of 20 weight parts on 100 weight parts of adhesive.

0 – reference sample: technical flour used for viscosity treatment in the rate of 20 weight parts on 100 weight parts of adhesive Diakol M1.

Results of shear strength of glued joints after dry climatisation and after soaking proved, that all samples fulfil required standard values.

- shear strength of glued joints partially decreased at collagen jelly samples No. 2 and 3 in comparison with the reference sample after soaking in water,
- shear strength of glued joints partially increased in comparison with the reference sample after dry climatisation at collagen jelly samples No. 1, 2, 3. Results of measurements of influence of amount of added collagen jelly samples No. 1, 2, 3 on shear strength of plywood bonded with UF adhesive mixtures are presented on the figure 6.

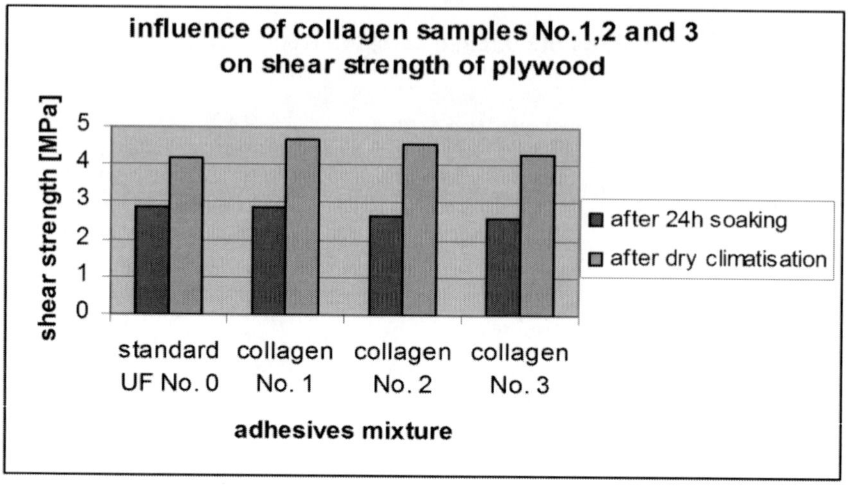

Figure 6. Results of tests of the influence of collagen jelly samples No. 1, 2, and 3 on shear strength of plywood after dry climatisation and after 24 h soaking in water 20 ± 3°C in [MPa].

Testing of shear strength of UF adhesive mixtures proved, that collagen jelly samples No. 1, 2, 3 are suitable as modifiers – extenders for UF adhesives for viscosity treatment. There is possible to replace 100% of technical flour by collagen jelly.

Conclusion

Project RESTORM, parts WP 1 and WP 6 solved the evaluation of biopolymers from chromium shavings by their application into contemporary used polycondensation adhesives.

Specific conclusions following from this work:

- obtained results bring new knowledge about possibilities of dechromation of chromium shavings without oxidation on toxic Cr^{6+} – *perspective possibility to process leather-tanning waste by proposed technology,*
- experimentally verified influence of collagen on properties of UF adhesive mixtures: lifetime, viscosity, gel time, emission of formaldehyde and shear strength of plywood – *perspective possibility to improve ecologic parameters of wood products.*

Acknowledgment

This paper was processed in the frame of the APVV projects No. APVV-351-010 as the result of author's research at significant help of APVV agency, Slovakia.

References

[1] Blažej, A; et al. *Štruktúra a vlastnosti vláknitých bielkovín*. Bratislava: VEDA, 1978.

[2] Blazej, A; et al. *Technologie kůže a kožešin*. Praha: SNTL, Bratislava: ALFA, 1984, 20-25, 208- 215.

[3] Cabeza, FL. Isolation of protein products from chrome leather waste. In: *Journal of the Society of Leather Technologists and Chemists.*, Vol. 83 (1), 1997, 14-19.

[4] Klásek, A. Způsoby odchromování usňových odpadů. In: *Kožedělné odpady a jejich ekonomické využití*. Brno: Zborník prednášok ČSVTS, 1983, 25-33.

[5] Kolomazník, K; Shánelová, K; Dvořáčková, M. Modifikované aminoplasty proteínovými hydrolyzáty pro lepení dřeva. In: *Pokroky vo výrobe a použití lepidiel v drevopriemysle*. Vinné, TU Zvolen, 1999, 91, ISBN 80-228-0790-7.

[6] Matyašovský, J; et al. Modification of polycondensation adhesives with animal proteins. Part II. In: *Annals of Warsaw Agricultural University. Forestry and Wood Technology.*, No. 55, 2004, 354-359, ISSN 028-5704.

[7] Sedliačik, M. *Nové kompozície polykondenzačných lepidiel a ich aplikácie v drevárskom priemysle*. TU Zvolen, 1992, 202, ISBN 80-228-0207-7.

[8] Sedliačik, M; Sedliačik, J. *Technológia spracovania dreva II. Lepidlá a pomocné látky*. TU Zvolen, 1998, 247, ISBN 80-228-0399-5.

[9] Sedliačik, J. Optimalizácia procesu lepenia hygienicky nezávadných preglejovaných materiálov. *Kandidátska práca*. TU Zvolen, 2000, 112 p.

[10] Sedliačik, J; Sedliačik, M. *Lepidlá a lepenie dreva*. TU Zvolen: 2003. 196, ISBN 228-1258-7.

In: News in Chemistry, Biochemistry and Biotechnology ISBN: 978-1-63117-273-1
Editors: G. E. Zaikov, G. Nyszko, L. P. Krylova et al.

Chapter 9

PREPARATION AND PROPERTIES OF ANIMAL PROTEIN HYDROLYSATES FOR OPTIMAL ADHESIVE COMPOSITIONS

Peter Jurkovič[1*], Ján Matyšovský[1], Peter Duchovič[1] and Igor Novák[2]
[1]VIPO a.s., Partizánske, Slovakia
[2]Polymer Institute, Slovak Academy of Sciences, Bratislava, Slovakia

ABSTRACT

Determination of mathematical models of the kinetics of polycondensation of proteinous hydrolysates reactions with selected cross-linking agents with the regard to the content of free formaldehyde and phenol in final products. Optimisation of adhesive compositions with the respect to their applicability in the wood processing industry.

Keywords: Animal proteins, biopolymers, formaldehyde, polycondensation adhesives, plywood, shear strength properties

INTRODUCTION

Dried collagen hydrolysates were laboratory prepared at Liptospol Liptovský Mikuláš, Slovak producer of leather and leather glue, Gelima Liptovský Mikuláš, Slovak producer of food and technical gelatine and CSIC Barcelona in Spain, leather glue prepared by oxidation method from chrome tanned shavings, with the aim to compare their influence on formaldehyde emission, physical and mechanical parameters of board materials. Hydrolysates were used for trials for preparation of adhesive mixtures with the application of biopolymers and supplementary additives.

* Corresponding author: VIPO, a.s. Partizánske, ul. gen. Svobodu 1069/4, 958 01 Partizánske, Slovakia, Email: pjurkovic@vipo.sk.

Analytic analysis of powdered collagen hydrolysate from Cr-shavings (sample from CSIC Barcelona) proved, that the presence of chromium by the method of atomic absorptive spectrophotometry was not determined at sensitivity of the method less than 0,0012 ppm of Cr. Series of application trials of described biopolymers were carried out in conditions at Technical University Zvolen with the aim to evaluate the influence of selected mixtures on ecological, physical and mechanical parameters on board materials – plywood. Preparation of collagen samples for experimental applications into other types of adhesives (polyvinylacetate - PVAC, polyurethane - PUR), possibilities of direct application of modified hydrolysate of collagen and keratin, as the input raw material for preparation of polycondensation resins.

Determination of Mathematical Modelling (Kinetics) of Poly-Condensation Reactions Control Algorithms and Reactor Dynamics

Obtaining, processing and interpretation of kinetics thermodynamic data related to polycondensations reaction urea-formaldehyde (UF), phenol-formaldehyde (PF), melamine-formaldehyde (MUF) and other similar resins modified by the addition of protein hydrolysate.

In conditions of VIPO, following works were realised for assurance of required polycondensation kinetics of UF and PF adhesives with the addition of biopolymers and their influence on physical and mechanical parameters and formaldehyde emission:

- the way of preparation of collagen and keratin hydrolysates,
- selection of analytic parameters of biopolymers evaluation, content of inorganic salts, viscosity,
- determination of optimal concentration of biopolymer in adhesive mixtures,
- the way of biopolymer modification,
- temperature and time of polycondensation, condensation time.

For application into polycondensation adhesives, there were prepared collagen hydrolysates by acid hydrolysis (HCl, H_2SO_4, HCOOH, $Al_2(SO_4)_3$, etc.), alkaline hydrolysis (NaOH, $Ca(OH)_2$, etc.), enzymatic hydrolysis (alkaline protease, tripsin), lyotropic agents (urea, $CaCl_2$, etc.).

For application into polycondensation adhesives, there is optimal technology with the addition of proteolytic enzyme, eventually lyotropic agent – urea. Collagen hydrolysate has the value of pH neutral, minimal content of inorganic salts and required concentration minimal 40 % of the dry content matter. Measurement of condensation time of adhesive mixtures confirmed significant worsening of kinetics – rate of polycondensation (condensation time in the test-tube at the temperature of 100°C was prolonged up to 100 % in comparison with the standard). For improving of polycondensation kinetics, collagen hydrolysates were modified with organic acid (HCOOH, inorganic acid HCl, H_2SO_4, $Al_2(SO_4)_3$), while the pH was gradually adjusted to the value of 1,2,3,4,5. At the value pH 4, optimal condensation times were reached and comparable with the standard (57 – 65 sec, at temperature of 100 °C), hydrolysate modified to the value pH less than 3 in UF mixtures caused shortening of the workability time, approx. than 15 min at pH 1 up to approx. 3 hours at pH less than 3.

Collagen hydrolysate with the dry content matter 40 % must be viscous liquid at the temperature of processing, and not semi-rigid gel and for standardisation of the time of hydrolysis (molecular weight) there is necessary a measurement of the viscosity.

Laboratory trials confirmed, that the addition of modified collagen hydrolysate up to 5 % related to dry content of adhesive do not grows worse physical and mechanical parameters of prepared products and at the same time significantly reduces the formaldehyde emission.

Temperature of polycondensation of UF adhesive mixtures with the addition of hydrolysate was optimised in the range of 120 – 140 °C, the temperature of 160 °C during 30 and 60 minutes caused the destruction of the hardener with following increasing of the formaldehyde content in hardened resins.

For application of biopolymers into PF adhesives, there were prepared keratin hydrolysates by oxidation and reduction technology in alkaline medium. Concentrated hydrolysates with the dry content matter of 20 – 30 % and pH min. 10,5 were evaluated at dosing up to 10 % as expressly positive – convenient physical and mechanical parameters, convenient viscosity of adhesive mixtures and sufficient storage life.

Presented possibilities of application of biopolymers describe the kinetics of polycondensation of commercially produced adhesives, (Diakol M1 – UF adhesive and Fenokol A – PF adhesive), in the dependence on the way of modification, temperature and time. The other possibility of biopolymer application at the synthesis of polycondensation adhesive is preparing at the present time.

DETERMINATION OF ADHESIVES COMPOSITIONS AND OPTIMISATION OF PROTEIN HYDROLYSATE COMPOSITIONS

Modification of recipes for preparation of adhesive compounds with respect to the results of previous analyses and trials with aim of obtaining to best possible quality of adhesive joints in wood processing applications. Preparation of adhesive compounds and necessary mechanical and chemical testing.

For the completion of results with the sample of hydrolysate from CSIC Barcelona there was realised the series of comparative trials, with the aim to consider ecological, physical and mechanical parameters of adhesive mixtures of the three types of collagen biopolymers:

- acid hydrolysis – VIPO Partizánske,
- enzymatic hydrolysis – UTB Zlín,
- oxidation method – CSIC Barcelona,

with the evaluation of the influence of collagen hydrolysate prepared by oxidation method from Cr-shavings on formaldehyde emission.

Comparative measurements of powdered samples of collagen hydrolysates from CSIC Barcelona – oxidation method, sample from Liptospol Liptovský Mikuláš (technology VIPO – acid hydrolysis) and Gelima – Weishardt Liptovský Mikuláš (standard – producer of food gelatines) were realised, while it can be stated that:

- dry content matter of samples is almost the same and do not shows larger deviations,
- values of pH of water solutions are comparable and we can consider them as very slightly acid or neutral,
- temperature of gelation was in the range of 12 – 15 °C,
- viscosity of 2 % solutions at 20°C after 24 hours was:
    1. sample from CSIC Barcelona – 5,50 mPas,
    2. sample from Gelima – 5,07 mPas,
    3. sample from Liptospol – 4,76 mPas ,

and its comparison with standards of similar parameters from Gelima and Liptospol:

- viscosity at concentration of 6,67 % and temperature of 60 °C – 29,7 mPas,
- the strength of gel – 97 Bloom,

confirmed, that hydrolysate from CSIC Barcelona has comparable parameters with collagen applied to UF and PF adhesives in previous trials.

Cr^{6+} was not determined in samples qualitatively with diphenylcarbazide, either presence of overall Cr by the method of atomic absorptive spectrophotometry on equipment "Shimadzu AA 6601 F". Following analysis were realised at UTB Zlín.

In laboratory conditions, there were consequently prepared liquid modified forms enzymatically and with lyotropic agent from collagen powder hydrolysates. Neutral or slightly acid with the value of pH 5 – 6,5 was adjusted with inorganic acid to the value of pH = 4 and consequently adhesive mixtures were condensed at temperature of 140 °C during 30 and 45 min. After polycondensation and conditioning, samples were ground and formaldehyde emission determined colorimetrically from water extract prepared by absorption and also extraction method. Results of emissions confirmed the decrease of formaldehyde at absorption determination (decrease about approx. 30 %).

Reached results are presented in the table 1.

We have prepared the reference trials with the application of 3 hydrolysates, which will be applied in eco-adhesives. The CSIC Barcelona glue powder was applied as:

- 40 % solution (substitution of 5 % adhesive),
- second case as fine powder (substitution of 5 % adhesive) – fine powder was impossible to homogenise in adhesive and the decrease of formaldehyde was minimal, from this reason laboratory trials continued only with 40% water gels.

Table 1. Properties of prepared collagen hydrolysates

Sample	VIPO Partizánske	UTB Zlín	CSIC Barcelona
dry content matter %	89,3	91,9	87,3
viscosity 20°C of 2% solution (after 1 hour) [mPa.s]	15,57	12,68	11,08
extinction 405/495nm (transparency of solution)	0,072/0,015	1,03/0,734	1,241/0,913
pH 2% solution	6,2	6,9	6,75
Cr^{6+}	negative	negative	negative

Table 2. Formaldehyde content in hardened adhesive mixtures

Sample	pH of collagen hydrolysate	content of fd /mg/litre/
Standard UF adhesive + hardener	–	0,131
+ Hydrolysate VIPO	6	0,054
+ Hydrolysate VIPO	4	0,047
+ Hydrolysate CSIC	6	0,057
+ Hydrolysate CSIC	4	0,047

Results of the formaldehyde content in hardened adhesive mixture are presented in the table 2.

ACKNOWLEDGMENT

This publication was prepared as part of the project „Application of Knowledge-based Methods in Designing Manufacturing Systems and Materials“ co-funded by the Ministry of Education, Science, Research and Sport of the Slovak Republic within the granted stimuli for research and development from the State Budget of the Slovak Republic pursuant to Stimuli for Research and Development Act No. 185/2009 Coll. and the amendment of Income Tax Act No. 595/2003 Coll. in the wording of subsequent regulations in the wording of Act. No. 40/2011 Coll.

REFERENCES

[1] Blažej, A. et al. Technologie kůže a kožešin. SNTL Praha, 1984.

[2] Matyašovský, J; Kopný, J; Meluš, P; Sedliačik, J; Sedliačik, M. Modifikácia polykondenzačných lepidiel bielkovinami. In: Pokroky vo výrobe a použití lepidiel v drevopriemysle. TU Zvolen, 2001, 37–42.

[3] Matyašovský, J; Kopný, J; Jurkovič, P; Sedliačik, J. *Modification of polycondensation adhesives with animal proteins*. In: Annals of Warsaw Agricultural University. Forestry and Wood Technology. No 53. SGGW Warszawa, 2003, 228 – 231.

[4] Matyašovský, J; Kopný, J; Jurkovič, P; Sedliačik, J; Kasala, J. *Modification of polycondensation adhesives with animal proteins*. Part II. In: Annals of Warsaw Agricultural University. Forestry and Wood Technology. No 55., SGGW Warszawa, 2004, 354–359.

[5] Restorm – Wp1, 24 month Progress Meeting Report. Ecoadhesives. Priebežná hodnotiaca správa projektu 5. rámcového programu EÚ. Freiberg, 12 p.

[6] Sedliačik, j; Lepidlá a ich aplikácia. Vedecké štúdie. TU Zvolen, 2002, 51 p.

[7] Sedliačik, J. Procesy lepenia dreva, plastov a kovov. TU Zvolen, 2005, 220 p. ISBN 80-228-1500-4.

[8] Sedliačik, M; Sedliačik, J. Utilisation of leather proteins for glue compositions. In: *Wood science and engineering in the third millenium*. Transilvania university of Brasov, 2004, s. 159–160. ISBN 973-635-385-0.

[9] Sedliačik, M; Sedliačik, J; Matyašovský, J; Kopný, J. Bielkoviny ako nadstavovadlo pre močovinové a fenolické lepidlá. Drevo 56, No. 8, 2001, 164–165.

In: News in Chemistry, Biochemistry and Biotechnology ISBN: 978-1-63117-273-1
Editors: G. E. Zaikov, G. Nyszko, L. P. Krylova et al.

Chapter 10

A Review: Preparation, Characterization and Applications of Magnesium Stearate, Cobalt Stearate and Copper Stearate

Mehmet Gönen[1], Theresa O. Egbuchunam[2], Devrim Balköse[3]*, Fikret İnal[3] and Semra Ülkü[3]

[1]Department of Chemical Engineering, Suleyman Demirel Universitesi, Isparta, Turkey
[2]Department of Chemistry, Federal University of Petroleum Resources,Effurune, Nigeria
[3]Department of Chemical Engineering, İzmir Institute of Technology Gulbahce, Urla, İzmir, Turkey

Abstract

Metal soaps, such as zinc, calcium, copper, magnesium are insoluble or sparingly soluble in water. Because of this property, they are commercially important compounds and find applications in industry, such as driers in paints or inks, components of greases, stabilizers for plastics, in fungicides, catalysts, waterproofing agents, fuel additives, components of creams and additive in drug formulation and etc. Magnesium stearate is in widespread use as gelling, sanding and anti-sticking agents, stabilizer, lubricant, emulsifier and plasticizer for polymers, in the paint, food, rubber, paper and pharmaceutical industries. Copper stearate is used mainly for rot-proofing textiles, ropes, etc. It is also used in paints since they are soluble in oils, white spirits, etc. Quartz crystals coated with $CuSt_2$ was used in the detection of volatile organic compounds. Cobalt stearate has applications in producing Co nests, mesoporous silica, as adhesion promoter.

Keywords: Metal soaps, magnesium stearate, cobalt stearate, copper stearate, PVC thermal stability

---

* Corresponding author: Email: devrimbalkose@iyte.edu.tr.

INTRODUCTION

Metal soaps are compounds of long-chain fatty acids with metals having different valences. Depending on the nature of cation and alkyl chain length, the physical properties of metal carboxylates may vary considerably. For instance, the general surface active materials, sodium and potassium carboxylates, are soluble in water, metal soaps, such as zinc, calcium, copper, magnesium are insoluble or sparingly soluble in water. Because of this property, they are commercially important compounds and find applications in industry, such as driers in paints or inks, components of greases, stabilizers for plastics, in fungicides, catalysts, waterproofing agents, fuel additives, components of creams and additive in drug formulation and etc.[1]. Metal soaps are produced in different forms such as fine powders, flakes, and granules. They are usually produced using precipitation or fusion techniques. Although precipitation method produces very light, fine powders with a high surface area, fusion technique produces flakes or pellets. Another issue relating to the product purity is that in precipitation process products with a high purity can be obtained at the expense of washing and filtering cost[2,3]. In addition to above mentioned applications, a number of other uses of polyvalent metal soaps have been suggested. Current interest in low dimensional compounds has led to a number of investigations on the potential application of metal soaps in this area, particularly as Langmuir-Blodgett (LB) multilayers[3].

The synthesis and characterization of metal stearates have commended considerable attention recently owing to their wide range of potential applications. Despite their wide application in industry, the fundamental characteristics of heavy metal soaps and their roles in various industrial preparations need to be investigated systematically as characterization and structural elucidation of the soaps at room temperature are of considerable importance in elucidating the structure of greases, flatting agents, coatings, and other products made from these soaps. In all these fields, understanding of the phase state of the soaps, and the changes which they may undergo as a result of processing steps or of the action of solvents, may lead to greatly improved products or processes.

MAGNESIUM STEARATE

Magnesium stearate ($MgSt_2$), ($Mg(C_{17}H_{35}COO)_2$) is a fine white odorless bulky powder with a very high covering capacity [4]. Magnesium stearate is in widespread use as gelling, sanding and anti-sticking agents, stabilizer, lubricant, emulsifier and plasticizer for polymers, in the paint, food, rubber, paper and pharmaceutical industries [5]. Magnesium soaps are also used as batting agents to reduce the gloss of paints and varnishes and also to thicken paints. Its production throughout the world is essentially based either on the reaction of stearic acid with a magnesium compound such as carbonate, oxide or on the reaction of magnesium chloride with sodium or ammonium stearate in aqueous solution leading to the precipitation of the dihydrate, $C_{36}H_{70}MgO_4 \cdot 2H_2O$. In the field of drug manufacturing, where it is mainly used as a solid lubricant, its lubricating capacity and overall activity in the various pharmaceutical forms in which it is incorporated may vary. Lubricants are essential to the production of all tablet formulation. As with other classes of pharmaceutical excipients, lubricating agents aid in the manufacture of tablets and ensure that the finished products are of appropriate quality. $MgSt_2$, with its low friction coefficient and large "covering potential", is an ideal lubricant widely used in tablet manufacturing [6]. Aerosol performance of micronized drug powders was increased when they were coated with magnesium stearate. The agglomerate strength of the powders was decreased by the coating process [7].

The variability in the physical characteristics of $MgSt_2$ creates problems in its applications. Commercial $MgSt_2$ is a mixture of magnesium salts of different fatty acids, mainly stearic and palmitic, and of others in lower proportions. The magnesium weight fraction in the dried substance is 4% at the least and 5% at the most. The fatty acid fraction contains at least 40% of stearic acid, and 90% of stearic and palmitic acids altogether [5]. $MgSt_2$ exists in different hydration levels, such as anhydrous form, monohydrate, dihydrate and trihydrate. The endotherm observed at 120 °C in DSC curve of anhydrous magnesium stearate corresponds to destruction of the lamellar (LAM) mesophase, and melting of the ordered arrangements of the alkyl chains. Up to 190°C, an ordered hexagonal phase with molten alkyl tails exists. At higher temperatures a disordered phase is present [8]. The peak seen in DSC curve around 100°C is due to removal of hydrate water in DSC curve of magnesium stearate monohydrate. Melting endotherm was observed at higher temperatures. Magnesium stearate monohydrate, dihydrate and trihydrate adsorbed moisture at 96% relative humidity. When the different samples were outgassed at 105° C under vacuum, peaks related to water removal disappeared in DSC curves and melting peaks were observed at 120°C and 130 °C for magnesium stearate mononohydrate and dihydrate, respectively. For trihydrate two stage melting endotherm starting at 115° C was observed. X-ray diffraction peaks at two theta values of 3, 5 and 9° were present [9]. Formation of stable semisolid lipogels prepared from magnesium stearate and water in liquid paraffin depends on the type of $MgSt_2$ used and preparation technique. $MgSt_2$ was essentially in crystalline state in semisolid lipogels producing α-crystalline lamellar phases [10].

COBALT SREARATE

$CoSt_2$ is synthesized by double decomposition of cobalt acetate with sodium stearate according to the procedure reported in the literature and the thermal characterization and other physico-chemical properties of cobalt stearate have been reported [11,12]. The prepared cobalt stearate was characterized in terms of its solubility and thermal behavior amongst others. The solubility of $CoSt_2$ was determined in polar/non-polar and protic/non-protic solvents and the results revealed that cobalt stearate is water insoluble but soluble in all the organic solvents like THF, DMF, xylene and toluene. The FTIR spectra of $CoSt_2$ exhibited absorbance at 1560 cm^{-1} due to asymmetric vibration stretching of the carboxylic group coordinated to the metal ion. The TG curve showed a single step decomposition with the initial temperature of degradation at 291.3°C. The cobalt content in the stearate was found to be 6.24% and from the results of the elemental analysis and the molecular formula of $CoSt_2$ was $Co(OOCC_{17}H_{35})_3.2H_2O$ [12]. $CoSt_2$ exists in three different crystalline phases (Cr1, Cr2 and Cr3), one mesophase (M) and isotropic liquid phase (I). The transition temperatures between the phases are 308.1, 380.9 and 404.4 K for Cr2 to Cr1, Cr1 to M and M to I phases, respectively [13].

$CoSt_2$ had asymmetric and symmetric vibrations of carboxylate groups at 1589 and 1440 cm^{-1} indicating it existed as bridging (polymeric) complexes [14].

Cobalt stearate ($CoSt_2$) is used as pro-oxidants for polyethylene. The last few decades have seen a tremendous increase in the use of polyethylene, particularly in the agriculture and packaging sectors. This has resulted in its increased production and associated plastic litter problem as polyethylene in its pure form is extremely resistant to environmental degradation. An excellent way to render polyethylene degradable is to blend it with pro-oxidant additives, which can effectively enhance the degradability of these materials [11]. Common pro-oxidants include transition metal salts with higher fatty acids, cobalt stearate being a typical example. The pro-oxidant activity of cobalt has been attributed primarily to (a) its ability to generate free radicals on polyethylene and (b) decompose the resulting hydro peroxides. The incorporation of these additives is expected to decrease the lifetime of polyethylene in general.

Cobalt stearate has applications in producing Co nests [15], mesoporous silica [16], as adhesion promoter[14]. Cobalt stearate assembled to micelles acted as soft template for the formation of primary nanorods during solvothermal processing of cobalt acetate and stearic acid. The nanorods then assembled to hollow cobalt spheres with a dense shell. These Co spheres transformed to Co nests constructed by netlike frameworks. Co nests are effective catalyst in hydrogenation of glycerol [15]. Tuning of porous structure of silica containing cobalt was possible using pink $CoSt_2.2H_2O$ as co-template in the synthesis [16]. $CoSt_2$ was used as adhesion promoter in curing of rubber [14].

COPPER STEARATE

Copper stearate ($CuSt_2$) is prepared by the inter-action of the corresponding soap with copper sulfate solution. It is used mainly for rot-proofing textiles, ropes, etc. It is also used in paints since they are soluble in oils, white spirits, etc. Quartz crystals coated with $CuSt_2$ was used in the detection of volatile organic compounds [17]. The vibration frequency of the crystal changed as the organic compounds were adsorbed on $CuSt_2$. A super hydrophobic copper surface with 153° contact angle was obtained by coating with $CuSt_2$ by applying DC voltage to copper electrodes immersed in stearic acid solution [18].

C-O antisymmetric stretching vibration of $CuSt_2$ was at 1588 cm^{-1}.[19] C-O antisymmetric and symmetric vibrations were at 1583 and 1417 cm^{-1} for midchain monomethyl branched C17 copper soap with distinct hexagonal columnar mesophase [20].

POLYVINYL CHLORIDE THERMAL STABILIZER

Metal soaps are the most used heat stabilizers for polyvinylchloride (PVC). The carboxylate group of the metal salt substitutes the tertiary or allylic chlorine atoms and stops the initiation of dehydrochlorination. Magnesium stearate was used in PVC thermal stabilization [21]. Copper-containing layered double hydroxide effected the thermal and smoke behavior of poly(vinyl chloride) [22]. Metal dicarboxylates were effective in retarding the dehydrochlorination reaction of PVC [23.]

Conclusion

The preparation, characterization and applications of metal soaps of stearic acid prepared from second group element magnesium and transition metal elements cobalt and copper were reviewed in the present study. They were mostly prepared by double decomposition reactions. They were crystalline solids with lamellar structure. While magnesium was used mainly for its lubricating property, the catalytic and surface properties of copper stearate and cobalt stearate allowed them to be used as templates for mesoporous compounds and additives to polymers as either prooxidants or fire retardents.

References

[1] Elvers, B; Hawkins, S; Schulz, G. *Ullmann's Encyclopedia of Industrial Chemistry*; Wiley: Weinheim Germany, 1990.

[2] Gonen, M; Balkose, D; Inal, F; Ulku, S. Zinc Stearate Production by Precipitation and fusion Processes. *Ind. Eng. Chem. Res.,* 2005, *44*(6), 1627-1633.

[3] Gonen, M; Ozturk, S; Balkose, D; Ulku, S. Preparation and Characterization of Calcium Stearate Powders and Films Prepared by Precipitation and Langmuir-Blodgett Techniques. *Ind. Eng. Chem. Res.,* 2010, *49,* **1732-1736**

[4] Lower, ES. Magnesium stearate: A review of its uses in paints, plastics, adhesives and related industries, *Pigment & Resin Techn*., 1993, 10(12), 7 – 25.

[5] Bracconi, P; Andres, C; Ndiaye, A. Structural properties of magnesium stearate pseudopolymorphs: effect of temperature. *Int. J. Pharm*., 2005, 262(1-2), 109-124.

[6] Patel, S; Kaushal, AM; Bansal, AK. Lubrication Potential of Magnesium Stearate studied on Instrumented rotary tablet press. *AAPS PharmSciTech*., 2007, 8(4), E1-E8.

[7] Zhou, QT; Qu, L; Larson, I; Stewart, PJ; Morton, DAV. Improving aerosolization of drug powders by reducing powder intrinsic cohesion via a mechanical dry coating approach. *Int. J. Pharm*., 2010, 394(1-2), 50-59.

[8] Wakabayashi, K; Register, RA. Phase Behavior of Magnesium Stearate Blended with Polyethylene Ionomers. *Ind. Eng. Chem. Res*., 2010, 49(23), 11906-11913.

[9] Koivista, M; Jalonen, H; Lehto, VP. Effect of Temperature and humidity on vegtable grade Magnesium Stearate. *Powder Techn*., 2004, 147, 79-85.

[10] Sheikh, KA; Kang, YB; Rouse, JJ; Eccleston, GM. Influence of hydration state and homologue composition of magnesium stearate on the physical chemical properties of liquid paraffin lipogels. *Int. J. Pharm*., 2011, 411(1-2), 121-127.

[11] Roy, PK; Surekha, P; Rajagopal, C; Choudhary, V. Thermal degradation studies of LDPE containing cobalt stearate as pro-oxidant. *Polymer Letters*, 2007, 1(4), 208–216.

[12] Roy, PK; Surekha, P; Rajagopal, C; Choudhary, V. Effect of cobalt carboxylates on the photo-oxidative degradation of low-density polyethylene, *Poly. Degrad. Stab*., 2006, 91, 1980-1988.

[13] Van Hecke, GR; Nakamoto, T; Clements, TG; Sora, M. Adiabatic calorimetry of the metallomesogen purple cobalt stearate $Co(O_2CC_{17}H_{35})_2$. *Liq. Cryst*., 2003, 30(7), 831-837.

[14] Jona, E; Ondrusova, D; Pajtasova, M; Simon, P; Michaler, J. A study of curative interactions in the presence of cobalt(II) stearate. *J. App. Polym. Sci.*, 2001, 81(12), 2936-2943.

[15] Liu, QY; Guo, XH; Li, Y; Shen, W. Synthesis of Hollow Co Structures with Netlike Framework. *Langmuir*, 2009, 25(11), 6425-6430.

[16] Grosshans-Vieles, S; Tihay-Schweyer, F; Rabu, P; Paillaud, JL; Braunstein, P; Lebeau, B; Estournes, C; Guille, JL; Rueff, JM. Direct synthesis of mesoporous silica containing cobalt: A new strategy using a cobalt soap as a co-template. *Microporous Mesoporous Mat.*, 2007, 106(1-3), 17-27.

[17] Filippov, AP; Strizhak, PE; Il'in, VG. Quartz crystal microbalance modified with Cu(II) stearate and octadecylamine co-ordination chemical compounds for detection of volatile organic compounds. *Sens. Actuators B.*, 2007, 126(2), 375-381.

[18] Huang, Y; Sarkar, DK; Chen, X. G.A one-step process to engineer superhydrophobic copper surfaces. *Mater. Lett.*, 2010, 64(24), 2722-2724.

[19] Satake, I; Matura, R. Studies with copper soaps, Colloid-Zeitshrift, 1960, 176, 31-38

[20] Corkery, WR. A variation on Luzzati's soap phases. Room temperature thermotropic liquid crystals, Phys. Chem. *Chem. Phys.*, 2004, 6, 1534-1546

[21] Simon, P; Oremusava, P; Valko, L; Kovarik, P. Influence of metal stearates on thermal stability of poly(vinyl chloride)II. Magnesium stearate. *Chem. Pap.*, 1991, 45(1), 127-134.

[22] Zhu, H; Wang, W; LiU, T. Effects of Copper-Containing Layered Double Hydroxide on Thermal and Smoke Behavior of Poly(vinyl chloride). *J. App. Polym. Sci.*, 2011, 122(1), 273-281.

[23] Liu, Y-B; Liu, W-Q; Hou, M-H. Metal dicarboxylates as thermal stabilizers for PVC, Poly. *Degrad. St.*, 2007, 97, 1565-1571.

In: News in Chemistry, Biochemistry and Biotechnology ISBN: 978-1-63117-273-1
Editors: G. E. Zaikov, G. Nyszko, L. P. Krylova et al.

Chapter 11

WATER SORPTION OF POLYVINYL CHLORIDE–LUFFA CYLINDRICA COMPOSITES

Hasan Demir[1*] and Devrim Balköse[2]
[1]Osmangazi Korkut Ata Universitesi Kimya Mühendisliği Bölümü, Osmangazi, Turkey
[2]Department of chemical engineering İzmir Institute of Technology, İzmir, Turkey

ABSTRACT

Natural Luffa Cylindrica fibers were modified with 0.1M sodium hydroxide (NaOH) for removing lignin and hemicellulose. Natural and modified Luffa fibers were characterized by using IR spectroscopy. Composites were produced with PVC plastisol and natural Luffa fiber. Natural Luffa fiber is a highly hydrophilic substance. This feature increased the water sorption capacity of the composites. Flexible PVC-luffa cylindrica composites had higher liquid water sorption capacity (0.3-0.6%) compared to that of flexible PVC (0.1%). There was no volume change of composites due to liquid water sorption.

Keywords: Luffa fibers, flexible PVC, water vapour adsorption, liquid water sorption

INTRODUCTION

Thermoplastics reinforced with special wood fillers are enjoying rapid growth due to their many advantages. Light weight, reasonable strength and stiffness are some of these advantages. The composite is presenting flexible, economical and ecological properties. Wood is polymeric composite consisting primarily of cellulose, hemicellulose and lignin. Lignin behaves a barrier and surrounds cellulose to hinder attack from enzymes and acids[1,2]. Hemicellulose and lignin cause problems when wood is used as a filler[3].

Luffa sponge products are readily available in the cosmetic and bath section of department stores, discount stores, pharmacies and specialty shops. Many environmentally

* Corresponding author: Email: demirrhasan.hd@gmail.com.

conscious consumers appreciate that luffa products are biodegradable, natural and renewable resources. The tough fibers can also be processed into industrial products such as filters, insulation, and packing materials, [4]. Luffa fibers consist of 51.2% cellulose, 13.7% lignin, 11.2 % hemicellulose, 1.8% ash, 6% moisture at room temperature[5]. Siquear et al. reported that luffa cylindrica contained 60.0-63.0% cellulose, 19.4-22% hemicellulose and 10.6-11.2 % lignin[6] Microcrystalline cellulose and cellulose nanocrystals were obtained from luffa fibers[6].

Luffa fibers were used as a filler in polypropylene and as nucleating agents in PVC foams [7,8]. Composites having 0.3 volume fraction of luffa fibers in polyester matrix absorbed 15% liquid water. The water diffusion coefficient in composites was found as $9.7 \times 10^{-6} mm^2/s$ [9]. Microcrystalline cellulose PVC composites with 40 phr isononylphtalate were biodegradable since soil microorganisms could consume cellulose as a source of nutrient. The micropores formed by cellulose degradation allow water in the composites. The weight loss increased with time and reached to 10% after 8 weeks for 30 phr microcrystalline cellulose content[10].

Since luffacylindrica fibers had a network structure, it is expected that when composites are prepared from them two continious phases, polymer and the interconnected cellulose phase will be obtained. The hydrohilic continious network phase of the composites can transport water or water vapour at a controlled rate from high water content medium to low water content medium. Thus, this type of materials are controlled water release agents.

In this study, water sorption properties of Luffa fibers and its composite with PVC plastisol were aimed to be investigated. Samples were characterized by using infrared spectroscopy, optical micrography, SEM, differential scanning analysis. Water and water vapour sorption at 25°C were investigated.

Materials and the Methods

LuffaFibers

Luffa cylindrica were obtained from local specialty shop. The Luffa fibers washed with water to remove the adhering dirt. They were dried in an oven at 70°C for 2h. After drying, they were cut with Waring Blendor for reducing dimensions to 2-3 mm. Some fibers were modified with 0.1 M sodium hydroxide (NaOH) solution at boiling temperature for 10 min. Sodium hydroxide was obtained from Sigma Co. Modified fibers were washed with distilled water until all sodium hydroxide was removed. After washing, they were dried in an oven at 70°C for 2 h. Natural and modified luffa fibers were characterized by using KBr disc technique with Shimadzu IR-470 spectrophotometer. Differential scanning calorimetric curves of the samples in equilibrium with 75% relative humidity air at 25°C was obtained by Seteram DSC92 calorimeter. The samples were heated in 25 to 250°C range at10°C/min heating rate.

Composite Preparation

Composites made from luffa fiber as filler and a PVC plastisol as polymer matrix which contains 100 parts poly (vinyl chloride), 60 parts Dioctyl phthalate (DOP), 5 parts Epoxidized soybean oil, and 5 parts zinc stearate. Composites were prepared in aluminum caps with 4 cm diameter. Luffa Fibers were cut into shape of the aluminum caps and pressed on PVC plastisol inside the caps. Two composites, composite I and II were prepared by using fiber network from inside and outside of the luffa gourd. Inner fibres were thicker from outer fibres. Composites were put into an oven at 150°C for gelation of plastisols into a plastic mass. Plastic discs having 3.8 (composite I) and 4.0% w/w (composite II) luffa fibers were obtained by this method. A control plastigel without any Luffa fibers was prepared in the same manner.

Microscopy

Expansion of fiber diameter on wetting was also observed with time by optical microscopy. The micrographs of the fibers were taken using Orthomat Polarizing microscope in transmittance mode after wetting with a drop of water. Morpholgy of natural and modified luffa was observed by using scanning electron microscope with Philipps XL-305 FEG. Interface, between luffa fiber and PVC plastisol matrix, were also observed.

Water Absorption

Natural and modified fibers and composites were immersed into static distilled water bath for observing absorption of water. The samples were wiped with tissue paper to remove surface water before weighing. Water uptake of samples (x %) at time t was calculated from

$$x\% = \frac{w_t - w_{t0}}{w_{t0}} x100 \qquad (1)$$

where

W_t : Weight of sample at time t
W_{t0}: Weight of sample at t = 0

Water Vapour Adsorption of Fibers

Water vapour adsorption isotherms of fibers at 25°C were obtained by using Omnisorp 100CX after outgassing the fibers at 110°C under 0.01 Pa pressure.

RESULTS AND DISCUSSION

Morphology of Natural and Modified Fibers

Natural luffa fibers were composed of cellulose, lignin and hemicellulose. Water absorption into Luffa fibers became harder with lignin and hemicellulose structure. The lignin and hemicellulose could be removed with chemical processes. Natural luffa fiber is processed with sodium hydroxide for dissolving lignin and hemicellulose. As seen in Figure 1 lignin layer was removed from surface of the fiber by NaOH treatment.

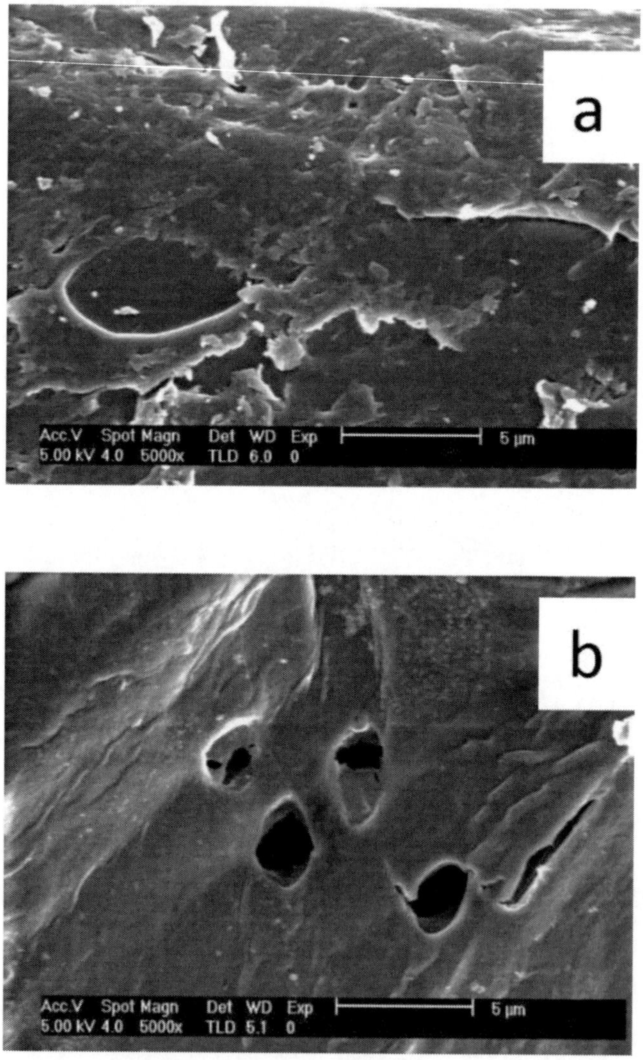

Figure 1. SEM micrographs of (a) natural (b) modified fibers.

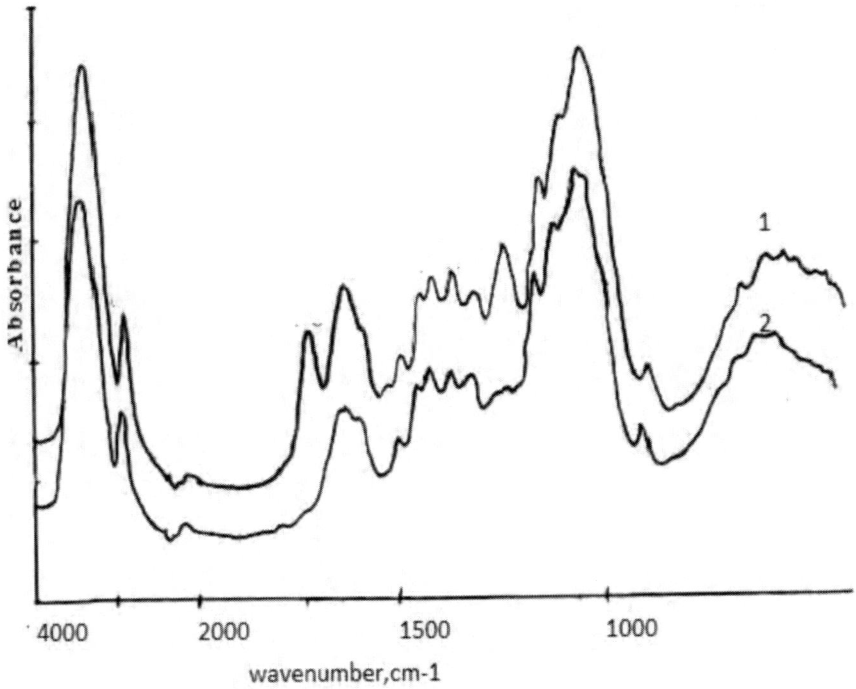

Figure 2. FTIR spectra of natural and modified luffafiber.

IR Spectra of LuffaFibers

Infrared spectra of modified and natural fibers are shown in Figure 2. The bands at 1070, 1115 and1165 cm^{-1} represent cellulose backbone vibrations of the polymer chain. Broad region of O-H vibration bond around 3450 – 3300 cm^{-1} is also characteristic peak for cellulose solids. The peak at 1740 - 1730cm^{-1} indicate the vibration of C=O stretching of carboxyl groups [1]. IR spectrum of delignified fibers does not have the band at 1740 – 1730 cm^{-1} due to removal of lignin and had lower intensity band at 1640cm^{-1}.

Differential Scanning Calorimetry

DSC curves for fibers equilibriated at 25°C at 75% relative humidity and heated from 25 to 250°C with 10°C/min heating rate are shown in Figure 3. Using the graphs, heat of vaporization of water, which was absorbs by fiber, was 2456.6J/g and 2421J/g for natural and modified luffa fibers respectively. For free water that is 1714J/g at 25°C. Obviously heat of vaporization of adsorbed water is more than that for free water. During the heating, mass losses of samples are 15.4 and 14.5% for natural and modified luffa fibers respectively. Heat of desorption of water from fibers was higher than heat of evaporation of free water.

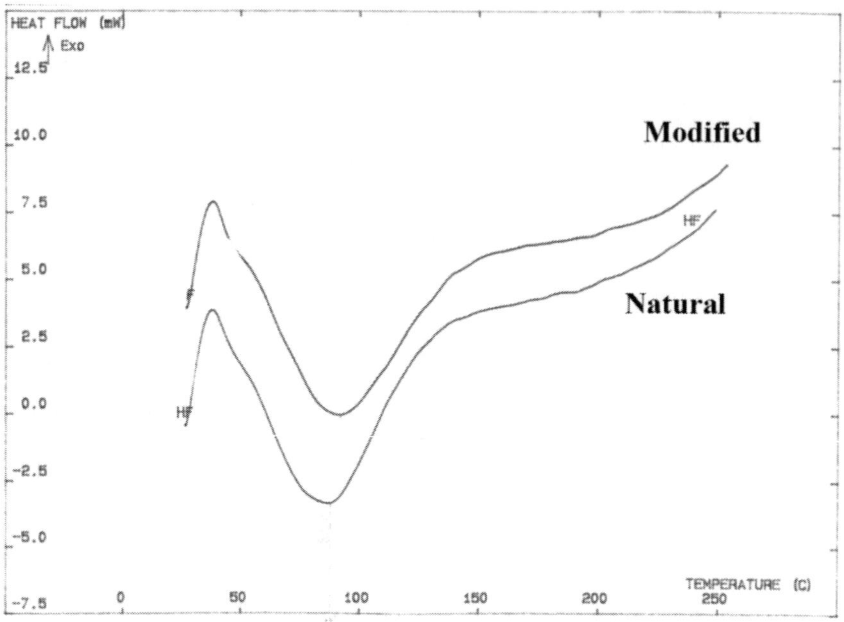

Figure 3. DSC curves of natural and modified luffa fiber.

Liquid Water Sorption of Fiber

Rate of water uptake versus time graph of fibers are shown in Figure 4. Modified luffa fiber absorbs water much more than natural luffa fiber. Removal of lignin made the fibers had more affinity to liquid water. Fibers dimensions increased with time due to water absorption. Figure 4 shows expansion of fiber diameter with respect to time. Diameter of modified luffa fiber expands slower than natural luffa fiber at the start of the process. But after 3.5 minutes, modified luffa fiber diameter expands more than natural fiber.

After that expansion of fiber diameter reaches equilibrium. Modified luffa fiber absorbs water much more than natural luffa fiber since it is more hydrophilic. While natural luffa fiber absorbs 213% water, modified luffa fiber takes up 281% water. At equilibrium 26.9% and 58.8% swelling occurred for natural and modified fibers. Liquid water files the pore spaces of the fibers and cause relaxation of the structure.

Water Vapour Adsorption of Fibers

The water vapour adsorption of the fibers show a different behavior than liquid water adsorption. While the natural fibers adsorp 6.9% water vapour modified fibers adsorp less 4.9 % at 95% relative humidity at 25°C as seen in Figure 5. The shape of the isotherm indicated cluster formation of water molecules in emty spaces of the fibers. Modified fibers adsorbed less water vapour than raw fibers.

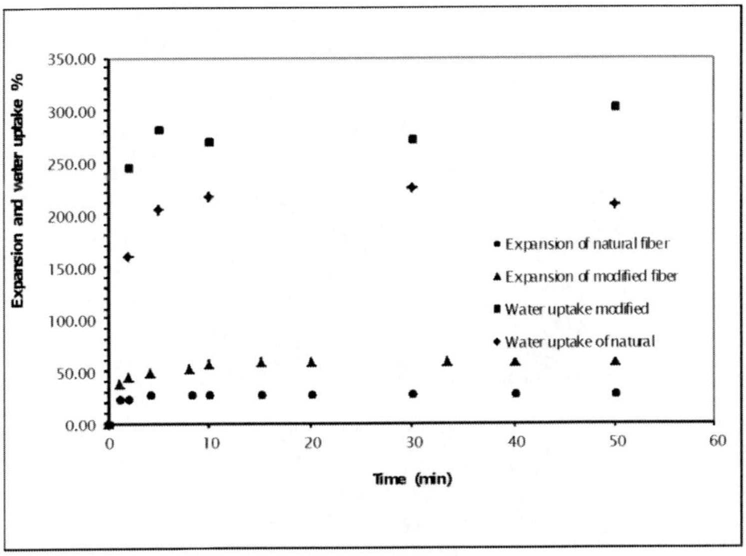

Figure 4. Expansion in water and water uptake of natural and modified fibers.

Fiber Plastigel Interphase

There were a small space between the fibre and the matrix of the composites as seen in SEM micrograhs in Figure 6. The surface of the fibres should be made more compatable with the matrix by silanation or malleation for enhancement of interphase.

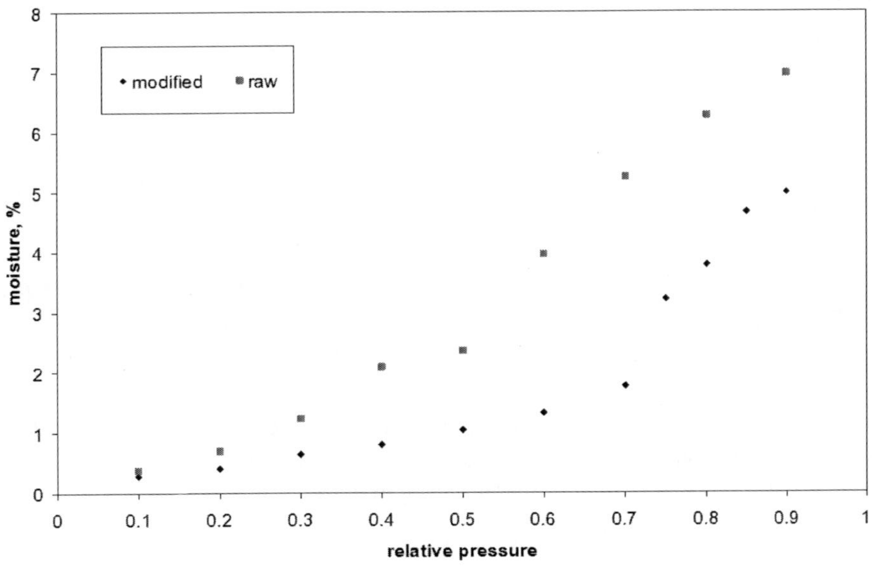

Figure 5. Water vapour adsorption isotherm of natural and modified fibers at 25° C.

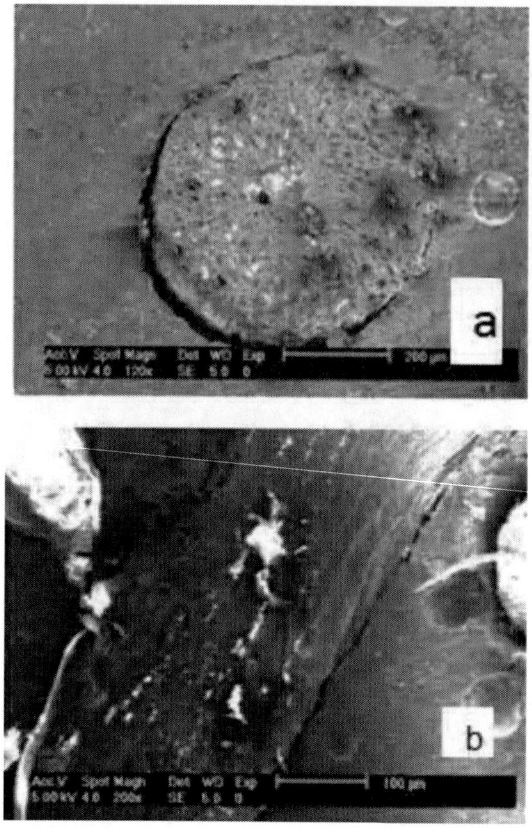

Figure 6. SEM micrographs of composites crossections (a) crossection of fiber (b) surface of fiber in composites.

Liquid Water Absorption of Composites

Water uptake ratios were calculated using Equation 1 for composites and plastigel. Figure 7 shows the water uptake percentages of pure plastisol and composite I and II. In the figure, plastigel absorbs water rapidly and reaches equilibrium. Plastisol water uptake curve shows deviation due to time. After 10 min, plastigel weight was decreased, since some dioctyl phthalate (DOP) was dissolved in water. Composite I and II's water uptake ratios were higher than that of pure plastisol. Consequently, sorption property of luffa fiber affects structure of composites. Composite I shows higher water uptake ratios than composite II. It could be depended fiber structure into the composites. However, composite I and II indicate similar water uptake path. Flexible PVC-luffa cylindrica composites had higher liquid water sorption capacity (0.3-0.6%) compared to that of flexible PVC (0.1%). There was no volume change of composites due to liquid water sorption.

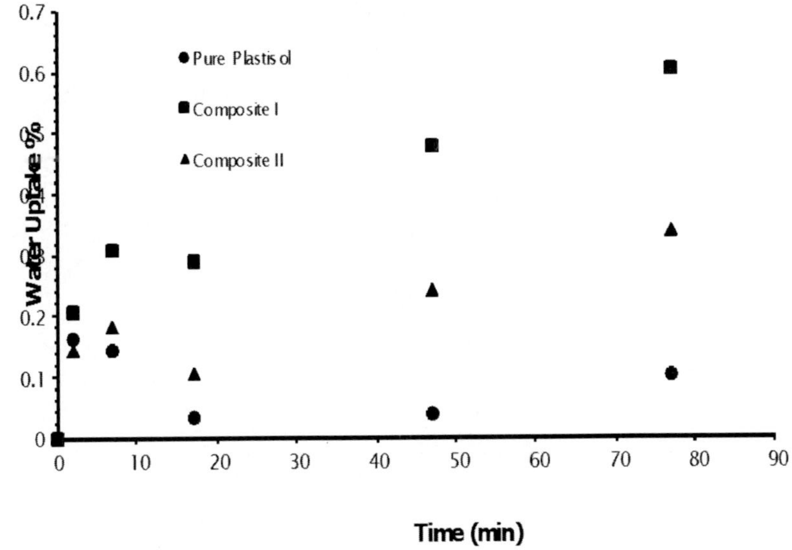

Figure 7. Liquid water uptake of pure plastisol and composites versus time.

Liquid Water Diffusivity in Fibers and Composites

By assuming Fickian diffusion in fibers and composites the diffusivity of the liquid water was determined from the initial experimental rate data and Equation 2.

$$M_t / M_e = 4/L(Dt/\pi)^{1/2} \quad (2)$$

Where M_t and M_e are weight increase at time t and at equilibrium, L is the half thickness of slab or the radius of the fiber.

It was found as $1.5x10^{-10}$ m^2/s, $6.4x10^{-9}$ m^2/s, $2.9x10^{-10}$ m^2/s, 3.4 $x10^{-10}$ m^2/s $1.075x10^{-10}$ m^2/s for the natural fiber, modified fiber, plastigel, composite I and composite II respectively.

Conclusion

Infrared spectra of fibers showed that lignin was removed with modification process. Succesful modification are known to distrupt lignin barrier to increase the reactive sites of cellulose and increase pore volume as well as available surface area. DSC curves predicted that natural and modified Luffa fibers had a high water content. H eat of desorption of water, 2456.6 and 2421 J/g for natural and modified Luffa fibers respectively was higher than heat of evaporation of free water,1714 J/g at 25°C.

The results showed that the rate of water absorption of water was higher in the luffa PVC composites than PVC plastigel. Flexible PVC-luffa cylindrica composites also had higher liquid water sorption capacity (0.3-0.6%) compared to that of flexible PVC (0.1%). While luffa fibers swell in water to a high extent, there was no volume change of composites due to liquid water sorption. Further studies are being made with modified fibers.

References

[1] Cheng, W. Pretreatment and enzymatic hydrolysis of lignocellulosic materials, *MS thesis*, West Virginia Uni., 2001.

[2] Annadurai, G; Juang, RS; Lee, DJ. Use of cellulose-based wastes for adsorption of dyes from aqueous solutions. *J. Haz. Ma.*, 2002, B92, 263-274.

[3] Avella, M; Bozzi, C; Erba, R; Focher, B; Morzetti, A; Martuscelli, E. Steam-exploded wheat straw fibers as reinforcing material for polypropylene-based composites. Characterization and properties. *Angew. Macromol. Chem.*, 1995, 233(4075), 149-166,

[4] Davis, JM; DeCourley, CD. Luffa sponge gourds: A potential crop for small farms. 560-561. In: *J. Janick and J.E. Simon*, Eds., New Crops. Wiley, New York, 1993.

[5] Baltazar, A; Jimenez, A; Bismarc, A. Wetting Behavior, moisture uptake and electrokinetic parameters of lignocellulosic fibers. *Cellulose*, 2007, 14, 115-127.

[6] Siquera, G; Bras, J; Dufresne, A. Luffa Cylindrica as a lignocellulosic source of fiber, microfibrillated cellulose, and cellulose nanocrystals. *Bioresources*, 2010, 5, 727-740.

[7] Demir, H; Atikler, U; Tihminlioglu, F; Balköse, D. The effect of fiber surface treatments on the mechanical and water sorption properties of PP-Luffa composites. *Journal of Composite Part A,* 2006, 37, 447-456.

[8] Demir, H; Sipahioglu, M; Balköse, D. Ülkü S. Effect of additives on flexible PVC foam formation, *Journal of Materials Processing Technology,* 2008, 195, 144-153.

[9] Boynard, CA; D'Almedia, JRM. Water absorption by sponge guord(luffa cylindrica)-polyester composite materials. *Journal of materials Science Letters*, 1999, 18, 1789-1791.

[10] Chuayhijit, S; Su-uthai, S; Charachinda, S. Poly(vinyl chloride) film filled with microcrystalline cellulose prepared from cotton fabric waste: properties and biodegradability study, *Waste manegement research*, 2010, 28, 109-117.

In: News in Chemistry, Biochemistry and Biotechnology ISBN: 978-1-63117-273-1
Editors: G. E. Zaikov, G. Nyszko, L. P. Krylova et al.

Chapter 12

CONTROL OF THE PARTICLE SIZE AND PURITY OF NANO ZINC OXIDE

Filiz Ozmıhçı Omurlu* and Devrim Balköse
İzmir Institute of Technology Chemical Engineering Department
Gülbahçe köyü Urla, İzmir Turkey

ABSTRACT

Effects of template, mechanical mixing and/or ultrasound mixing on the size of the ZnO crystals obtained by precipitation at 30 °C from aqueous zinc chloride and potassium hydroxide solutions were investigated by 2^k factorial design. Precipitation method is employed to synthesize nano zinc oxide particles. Monodisperse nano ZnO having 29 nm particle size was produced by adding triethyl amine and applying simultaneously mechanical and ultrasound mixing. The surface area and the density of the powder were 21 m^2/g and 4.8 g/cm^3. It contains 5.2% impurities present as CO_3^{-2} and bound OH^- groups. Volumetric resistivity was found as 1.3 x 10^7 ohm cm. Absorption spectrum of the powder showed absorption peak at 353 nm. The room temperature fluorescence spectrum of the powder revealed a strong and sharp UV emission band at 391 nm due to free exciton or bound exciton of ZnO and a weak and broad violet emission band at 405 nm due to zinc vacancies.

Keywords: Nano zinc oxide, triethylamine, precipitation, electrical resistivity, luminescence

INTRODUCTION

Nano crystalline materials have found an increasing research area on the material science, chemical and electronic engineering during the past years. ZnO is composed of tetrahedrally coordinated O^{2-} and Zn^{2+} ions, stacked along the c-axis. It is a semiconducting material with a band gap of about 3.2 eV and a large exciton binding energy of 60 meV [1, 2]. It is an

* Corresponding author: Filiz Ozmihci Omurlu İzmir Institute of Technology Chemical Engineering Department 35430 Gülbahçe köyü Urla /İzmir Turkey Email: filizozmihci@iyte.edu.tr.

important material due to its unique properties of near-UV emission value and has applications as electrical conductive and optically transparent additive in a polymer matrix [3-6].

There are many different methods for the preparation of nano ZnO powder and precipitation method is a good choice in the industrial point of view because of the low growth temperature and its potential for scale-up [7-9]. This synthesis method has the advantage of preparing highly crystallized particles with narrow size distribution and high purity without further treatment at higher temperature. Size and morphology can be controlled by controlling reaction temperature, reaction time and additives [10-15].

Precipitation method was used to prepare ZnO nano sheets by sono chemical method using $ZnCl_2$ and NaOH as a precursor under constant stirring and at pH 13 [16]. Different shapes of ZnO powders were prepared with sonochemical synthesis, which were in nanorod, trigonal, and dentritic shapes. X-ray diffraction patterns of the synthesized powders were in good agreement with the hexagonal wurtzite structure of ZnO[16].

Particles with different morphologies and sizes were obtained by adjusting the templates [17]. Nano ZnO particles was synthesized by Wei and Chang at room temperature and at 50°C under ultrasonic condition by hydrothermal method by using cetyl trimethyl ammonium bromide and triethanolamine surfactants [16]. While the bulk ZnO obtained by using only ultrasonic water bath treatment at 50 °C had 454 nm particle size, the size was reduced approximately to 28-60 nm when surfactants were used [18]. Flower like ZnO microstructures were obtained from aqueous zinc nitrate, sodium hydroxide and triethylamine at 180°C. Triethylamine played a dual role both as the complexing agent and the alkaline reagent [19]. ZnO was also obtained from aqueous zinc acetate and triethylamine solutions [20]. Lai et al. also investigated the hydrothermal synthesis of ZnO powder with the assistance of ultrasonic treatment. At ambient conditions, the aqueous solution of precursors that contains zinc acetate and sodium hydroxide was very clear. However after ultrasonic treatment the clear solutions become cloudy and a white precipitate was observed. Due to acoustic cavitation, H_2O decomposed into H^- and OH^- radicals. The radicals react with Zn^{+2} ions to form ZnO and water molecules. The ultrasonic energy also converts $Zn(OH)^{-2}$ to $Zn(OH)_2$ and $Zn(OH)_2$ to ZnO [21].

Zinc oxide can also be obtained by hydrothermal transformation of zinc hydroxide chloride. For instance, Zhang and Yanagisawa studied metal hydroxide salts (MHS) with layered structures. The common synthesis methods of the metal hydroxide salts include coprecipitation method and the obtained products usually have the lamellar morphologies such as films, sheets, and plates. In their paper, zinc hydroxide chloride (ZHC) sheets were synthesized the by a simple hydrothermal method. After thermal treatments, the ZHC sheets were transformed to sheet like dense ZnO [22].

ZnO can exhibit unique optical, photocatalytic, piezoelectric, and pyroelectric properties, produces an efficient blue-green luminescence, and displays excitonic ultraviolet (UV) laser action. ZnO has a relatively high absorption band starting at 380 nm [23] and extending into the far-UV. In addition to its excellent UV absorption characteristics, ZnO has several other advantages as a UV and visible light emitting additive material, it does not migrate, it is not degraded by absorbed light and in many cases it may improve mechanical, optical and electrical properties of polymers which they are added.

Material synthesis involves the control of the particle size and morphology since the electrical and optical properties of materials depend both on the size and the shape of the particles. Therefore, morphologically controllable synthesis of ZnO having nano or microstructures is crucially important to answer the demand for exploring the potentials of ZnO [24].

The present investigation was focused on the preparation of mono dispersed nano zinc oxide powder by a hydrothermal precipitation method at low temperature. The effects of mechanical mixing, sonication and using a template on the particle size were investigated by the aid of statistically designed experiments. The particle size distribution, morphology and crystal structure of the powders were determined. Pure nano zinc oxide powder having the smallest particle size was characterized in more in more detail by the measurement of the volumetric resistivity, the density and the optical properties.

MATERIALS AND METHODS

Materials

Analytical grade chemicals, zinc chloride ($ZnCl_2$) (98%; Aldrich), potassium hydroxide(KOH) (Pancreac) and triethylamine (TEA)($(CH_3CH_2)_3N$), Merck), were used for the preparation of zinc oxide powders throughout the experimental study. Millipore ultrapure fresh water (18 ohm cm) was used in all steps of the synthesis. TEA was used as a template.

Experimental Design

The temperature, the concentration of the precursors and the template type were held constant in preparation of the nano ZnO powder. Addition of the template (0.02 $moldm^{-3}$concentration), the sonication (for 30 minutes period) and the mechanical mixing (at 500 rpm) were the main factors. The particle size of the powders was chosen as the response. The experimental factors and the categorical levels are as given in Table 1. The experiments were performed by considering a 2^3 full factorial design consisting of 8 experiments as shown in Table 2. The analysis of variance full factorial design was carried out using Design of Expert 8.0.1.0.

Table 1. Experimental factors and levels investigated for optimum particle size of ZnO powder

Factors name	Factors Symbol	Low	High
Template	a	0	1
Mechanical Mixing	b	0	1
Sonication	c	0	1

Table 2. Full factorial experimental design

Template	Sonication	Mechanical Mixing	Experimental Mean Particle Size (nm)	Predicted Mean particle size(nm)
1	1	1	29	78
0	0	0	650	433
0	0	1	1312	392
1	0	1	77	54
1	1	0	738	225
0	1	1	148	695
0	1	0	137	603
1	0	0	373	304

Typical ZnO Synthesis Method

100 cm^3 solution having 0.2 mol dm^{-3} KOH and 0.02 mol dm^{-3} template TEA was added instantly to 100 cm^3 0.1 mol dm^{-3} $ZnCl_2$ solutions. Control experiments without template TEA addition were also made. Ultrasonic treatment was applied by immersing the beaker containing the reactants in an ultrasonic bath (Elma; Transsonic 660/H) at 30 °C for 30 minutes. Mechanical mixing at 500 rpm was made using IKA RW 20 mechanical mixer. The solid and liquid phases were separated by centrifuging using Hettich, Rotofix 32. The solid phase was then washed for three times with water and dried at 50°C for 15 h.

Characterization of ZnO Powders

The phase identification and the crystal size of ZnO powders were determined by X-Ray diffractometer (Philips X'Pert diffractometer, Cu-K_α radiation). The powder morphology was determined by SEM with Philips XL-30S FEG. The particle size distribution of the powders dispersed in water was determined by Zeta Sizer (Malvern Instruments 3000 HSA).

Detailed Characterization of Nano Zinc Oxide Powder Obtained by the Template Addition, the Mechanical and the Ultrasonic Mixing

Helium pycnometer (Quantachrome Co. Ultrapycnometer 1000) was used to determine the density of the powder. The N_2 adsorption/desorption analysis were performed to determine the surface area of the powder (ASAP Micromeritics 2000). The impurities in the monodisperse nano ZnO powder was determined by FTIR spectroscopy using Shimadzu FTIR-8201 by KBr disc method. ZnO pellets having 2.5 cm diameter and 2 mm were prepared from the nano ZnO powder by pressing under 10 MPa pressure. Silver contacts were formed by thermal evaporation of silver on both surfaces of the ZnO pellet for the resistivity measurement. The volumetric resistivity of the pellet was determined by changing potential between -50 V and +50 V and recording I-V data with Keithley 2420. Absorption spectrum of a dilute suspension of ZnO powder was obtained by using the UV-Vis spectrometer Perkin

Elmer Lambda 45. The Fluorescence spectrum was obtained by using the fluorescence spectrometer Varian Cary Eclipse by using a ZnO pellet. The emission data were recorded in the 390 and 600 nm range after exciting the sample at 380 nm for 15 s.

RESULTS AND DISCUSSION

Factorial Design for Particle Size of ZnO

Table 2 gives the list of kinetic results for full factorial experimental design. The effects of the sonication time, the mechanical mixing and the template addition on the particle size of the powders were found according to fitted regression model with 0.5 confidence interval to experimental particle size data. Equation 1 is the fitted model for the particle size, as a function of the presence of template, a, the sonication, b and the mechanical mixing, c.

$$d= 433-128.8a-170.0b-41.5xc+249.3ab-209.8ac-133.0bc+29.8abc \qquad (1)$$

A factor was designated by "1" or "0" if it was present or not in the system respectively. Therefore, the values either 1 or 0 should be used for the variables a, b, c when predicting the particle size using Equation 6.

The model results for particle size indicate that, template addition, sonication and mechanical mixing have a negative effect on the particle size. All the main effects and interaction effects on particle size of the powder are significant. However, sonication has the largest negative effect on the response and the template affects the particle size more than the mechanical mixing. The fitted model predicts that the interaction between the template and the sonication is the most important interaction parameter for increasing the particle size. Both "sonication and mechanical mixing" and "template addition and mechanical mixing" make a synergistic effect for minimizing the particle size. To obtain minimum particle size the template addition, the sonication and the mechanical mixing should be applied simultaneously during precipitation of the powder as seen in Table 2. The predicted particle size of each sample from Equation 6 is as reported in Table 2. However the model predicts the smallest particle size for the template addition and the mechanical mixing case.

Purity of ZnO Particles

Factorial design method has shown that the simultaneous template addition, sonication and mechanical mixing will result in the minimum sized nanoparticles. However the purity of ZnO is another fulfillment that should be met. Thus the effects of these three variables on the purity of the product ZnO were also investigated.

Effect of Template

In the present study, to investigate the effect of template on particulate properties triethylamine (TEA) was added to the reaction medium. Figure 1 gives the XRD patterns of the powders prepared with and without the template for the case without any mixing. The powder synthesized in the absence of template was a complex compound as depicted in Figure 1a. The XRD pattern belongs to a metal hydroxide salt with layered structure [22]. The product was confirmed to be $Zn_5(OH)_8Cl_2H_2O$ (ZHC) (JCPDS Card No: 07-0155) and no other impurity phases were found. In the X-ray diffraction diagram of the TEA added powder in Figure 1b there are diffraction peaks at 2θ values of 31.6 °, 34.26 °, 36.1 °, 47.35 °, 56.4 °, 62.66 °, 66.2 °, 67.76 °, 68.86 °, 72.2 ° and 76.78 °. In the XRD pattern of ZnO powder reported in JCPDS Card No: 79-0207 there are peaks at 2θ values of 31.7 °, 34.4 °, 36.3 °, 47.5 °, 56.6 °, 62.3 °, 66.5 °, 67.9 °, and 69.1 °. Thus the powder synthesized with TEA addition was pure ZnO. The ratio of the intensity of the peak of 002 planes at 34.3° to the intensity of the peak of the 101planes at 36.1° is 0.44 for the bulk wurtzite [25]. The sample prepared with the template had hexagonal ZnO crystals preferentially oriented in 002 direction since this ratio is 0.65. SEM images and particle size distributions of the template added and template free powders are given in Figures 2 and 3 respectively. As shown in Figure 2a hexagonal shaped sheets were obtained when there was no template in the medium. The sheets are 1.5-2 μm in size and their thickness is around 10 nm. Template free sample had 61, 25, 14 mass % of Zn, O and Cl respectively as determined by EDX analysis. Using the EDX and XRD data it was concluded that zinc hydroxy chloride (ZHC) sheets were synthesized when there was no template. The mean particle size of ZHC sheets was around 650 nm as seen in Figure 3a.

As depicted in Figure 2b the template (TEA) added powder was agglomerated to flower like particles similar to the ones made with triethylamine at 180 °C by previous workers[19]. The flower like particles formation at 30°C in the present study indicated that the thermal treatments at high temperatures are not necessary for this purpose. The average particle size of the powder was found around 373 nm as shown in Figure 3b and as reported in Table 2.

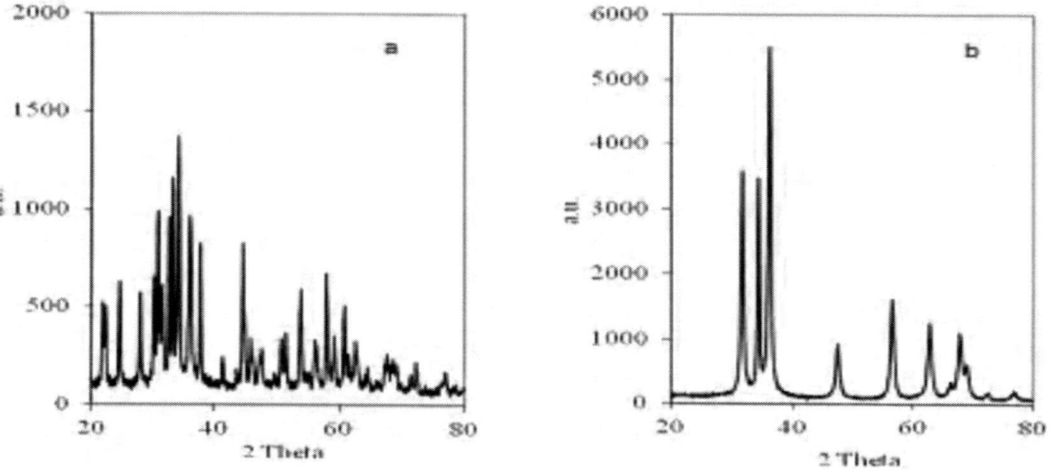

Figure 1. XRD pattern of powders prepared without any mixing and a) without TEA b) with TEA.

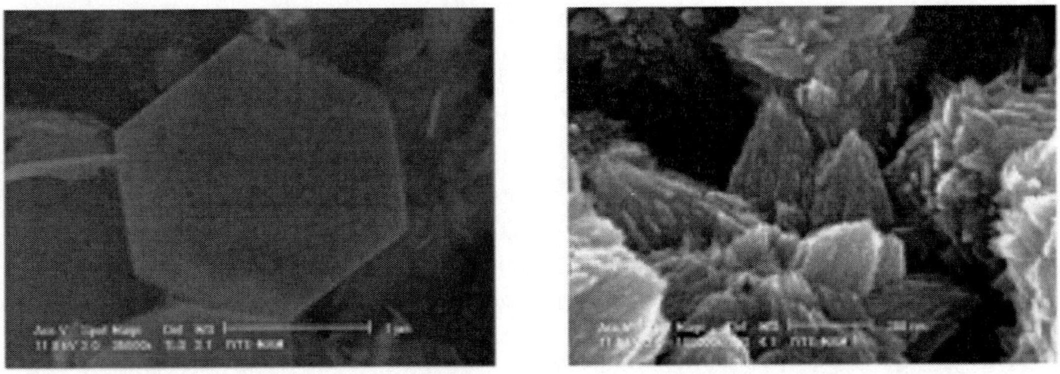

Figure 2. SEM image of powder precipitated without any mixing and a) without TEA b) with TEA.

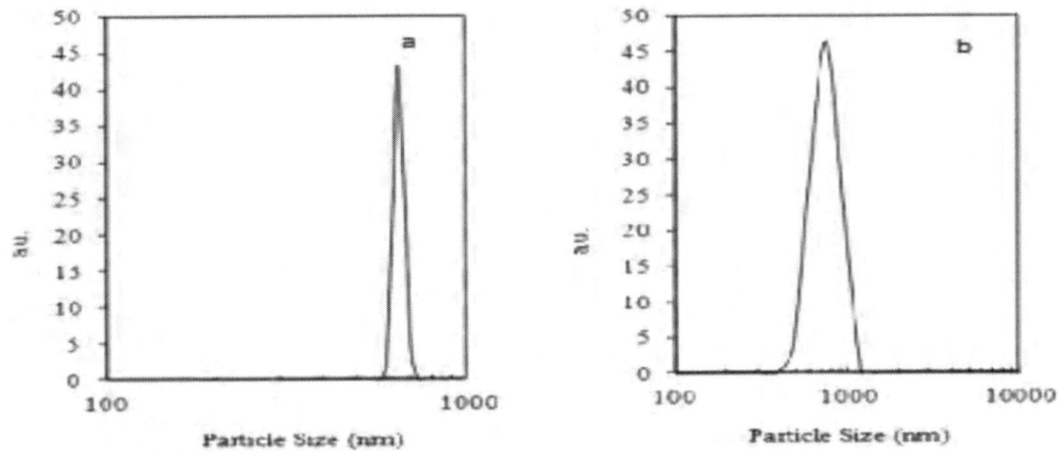

Figure 3. Particle size distribution of powder without any mixing and a) without TEA b) with TEA.

Template addition had a direct influence on the type of the product obtained. In mixing $ZnCl_2$ and KOH solutions consecutive reactions shown in Equations 2-5 occur.

$$ZnCl_2(aq) + KOH(aq) \rightarrow \mathrm{ZnOHCl} \downarrow + KCl \quad (2)$$

$$ZnOHCl \downarrow \mathrm{(s)} + \mathrm{KOH(aq)} \rightarrow \mathrm{ZnO} \downarrow + \mathrm{H_2}O + KCl \quad (3)$$

$$\mathrm{2ZnCl_2}(aq) + 4KOH(aq) \rightarrow Zn(OH)_4^{-2} Zn^{+2} + 4KCl \quad (4)$$

$$Zn(OH)_4^{-2} Zn^{+2} \rightarrow 2ZnO \downarrow + 2H_2O \quad (5)$$

Zinc hydroxychloride (ZnOHCl) sheets formed if the template was not used as shown in Equation 2. Large sheets of ZnOHCl were obtained due to faster growing of crystals than nucleation. TEA molecules associate in one dimensional chains [19] and act as templates for small sized ZnO particles. TEA also reduces the surface tension of water [26]. Template TEA creates nucleation centers by complexation with zinc ions and large numbers of $Zn(OH)_4^{-2}Zn^{+2}$ nuclei form and during slow crystal growth they are transformed to ZnO by reactions shown in equations 4-5.

Two experiments were made to understand the effect of template addition for the case of applying both the sonication and the mechanical mixing. The first powder was prepared without the template using sonication and mechanical mixing and the other one was prepared with the template using both sonication and mechanical mixing. XRD patterns, SEM images and the particle size distributions of the powders are as given in Figure 4, 5 and 6 respectively.

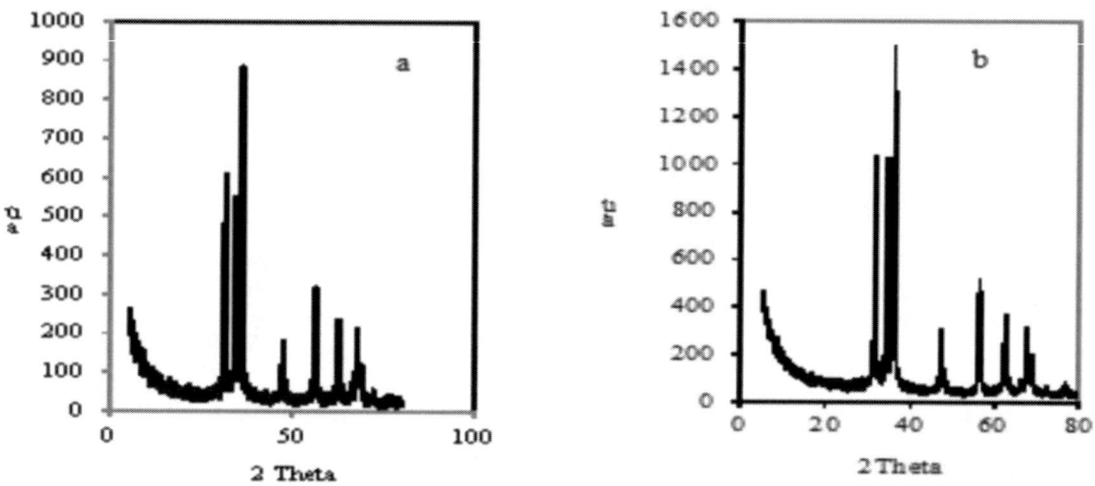

Figure 4. XRD pattern of the powder prepared with mechanical mixing and sonication a) without TEA b) with TEA.

Figure 5. SEM image of the powder prepared with mechanical mixing and sonication a) without TEA b) with TEA.

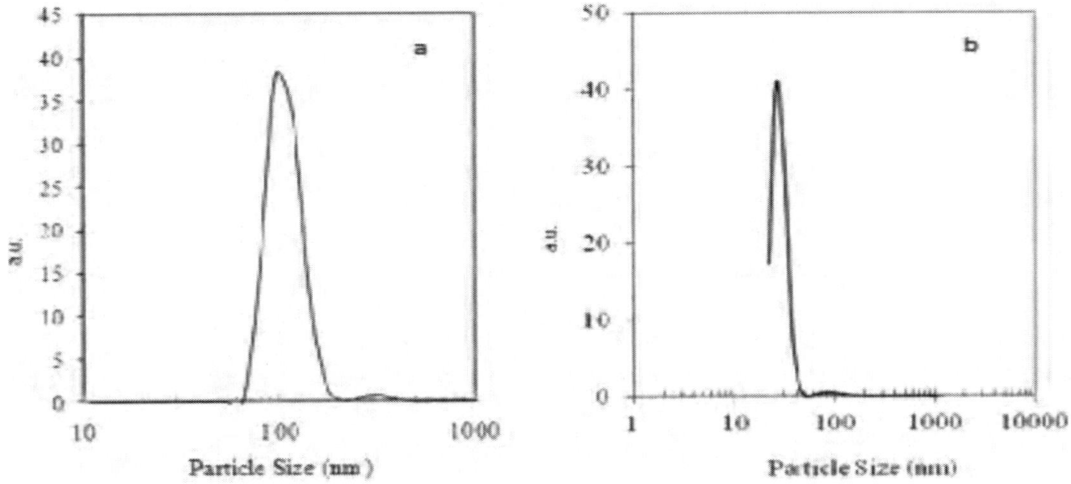

Figure 6. Particle size distribution of the powder prepared with mechanical mixing and sonication a) without TEA b) with TEA.

XRD patterns of the powders in Figure 5 were found to be similar to each other and to that of ZnO. The ratio of the intensity of the peak of 002 planes at 34.3^{o} to the intensity of the peak of the 101planes at 36.1^{o} is 0.61 and 0.7 for template free and template added samples. This indicated hexagonal crystals of template added ZnO were more oriented in 002 direction than the ZnO without template. However the SEM image of TEA free sample shown in Figure 5a is polydisperse in particle size. Very small and very large particles were present in sheet like form. However only a small fraction of larger particles was present when there was no template as seen in Figure 6a. The average particle size was 148 nm when there was no template. On the other hand primary particles were agglomerated to form particles having the shape of a droplet are observed in the SEM image of the TEA added powders in Figure 5b. Particle size distribution of the TEA added powder seen in Figure 6b confirms the monodispersity and the average particle size was found as 29 nm.

The application of sonication and mechanical mixing simultaneously reduced the particle size compared to unmixed case and the difference between particle sizes of the powders obtained with and without template addition was also reduced. Without template zinc hydroxychloride sheets were obtained when there was no mixing. When the reactants were mixed, due to faster growth of crystals than their nucleation, sheet like precipitates were formed. On the other hand TEA created nucleation centers and large numbers of nuclei formed and slower crystal growth occurred reducing the particle size.

Mixing Effect

Sonication and mechanical mixing were used to understand the mixing influence on particle size and morphology. The reaction temperature (30°C) and concentrations of the precursors and pH (10.5) were held constant and no template was used. Figure 7, 8, and 9 show the XRD patterns, SEM image and particle size distribution of precipitates obtained by applying only sonication and mechanical mixing respectively.

The XRD patterns of the precipitates had sharp peaks at 2θ values 31.6 °, 34.26 °, 36.1 °, 47.35 °, 56.4 °, 62.66 °, 66.2 °, 67.76 °, 68.86 °, 72.2 ° and 76.78 °. The peaks observed were identical with the characteristic XRD pattern of ZnO powders (JCPDS Card No: 79-0207). The ratios of the intensity of the peak of 002 planes at 34.3° to the intensity of the peak of the 101planes at 36.1° are 0.68 and 0.5 for sonified and mechanically mixed samples. This indicated hexagonal crystals of ZnO obtained by sonication and mechanical mixing were oriented in 002 direction. Sheet like and polydisperse crystals are seen in the SEM image of the sonified precipitate in Figure 8a. Mechanically mixed precipitate's SEM image in Figure 8b shows aggregated sphere like crystals. The particle size distribution of sonified and mechanically mixed samples seen in Figure 9a and 9b indicated that mechanically mixed powder had larger crystal size. Sonified powder's size distribution in Figure 9a is bidisperse. A small fraction of the particles were larger in size. Monodisperse size distribution was obtained for only mechanically mixed particles as seen in Figure 9b.The mean particle sizes for only sonified and only mechanically mixed samples were found as 137 nm and 1312 nm respectively. The above results showed that applying only sonication and only mechanical mixing was not enough to have a monodisperse nano sized ZnO powder.

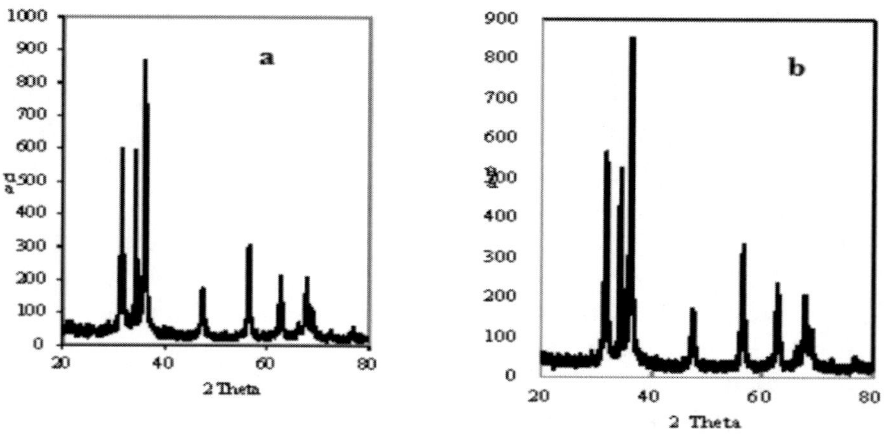

Figure 7. XRD pattern of precipitates obtained without template and with a) only sonication b) only mechanical mixing.

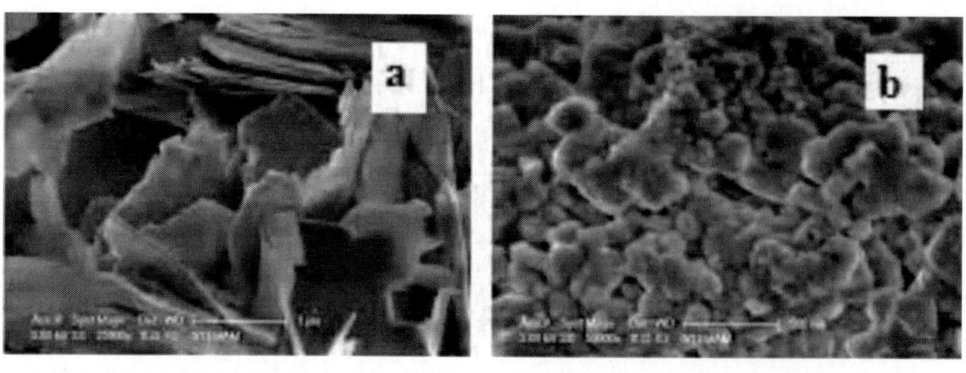

Figure 8. SEM image of precipitates obtained without template and with a) only sonication b) only mechanical mixing.

Experiments were also done to analyze the effect of mixing on template added precipitates. XRD pattern, SEM image and particle size distribution of the template (TEA) added precipitates are given in Figure 10, 11 and 12 respectively for sonication and mechanical mixing applied samples.

The template added precipitates XRD patterns give the pattern of typical ZnO as seen in Figure 10. In the SEM image of sonified powder in Figure 11 small and big shapeless particles and flake like structures are seen. Monodisperse particles smaller than100 nm are seen in the SEM image of the mechanically mixed powder in Figure 11b. The particle size distribution of sonified powders seen in Figure 12a was bidiperse and the mean particle size was determined to be 738 nm. This value is at the same order with the size (454 nm) of ZnO particles synthesized at 50 °C under ultrasonic conditions Wei and Chang [16]. However, the particle size distribution range of the mechanically mixed powders found between 30-300 nm and the mean particle size of the mechanically mixed powder was 77 nm.

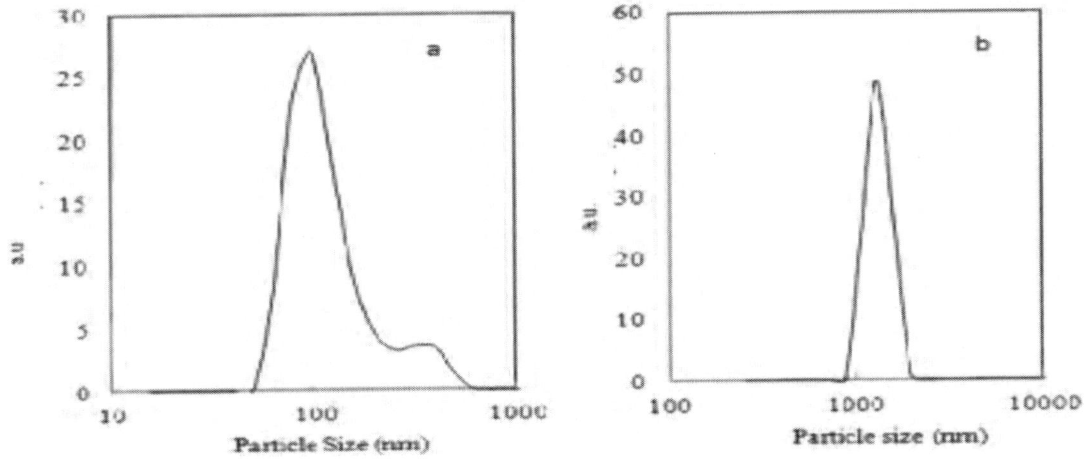

Figure 9. Particle size distribution of the precipitates obtained without template and with a) only sonication b) only mechanical mixing.

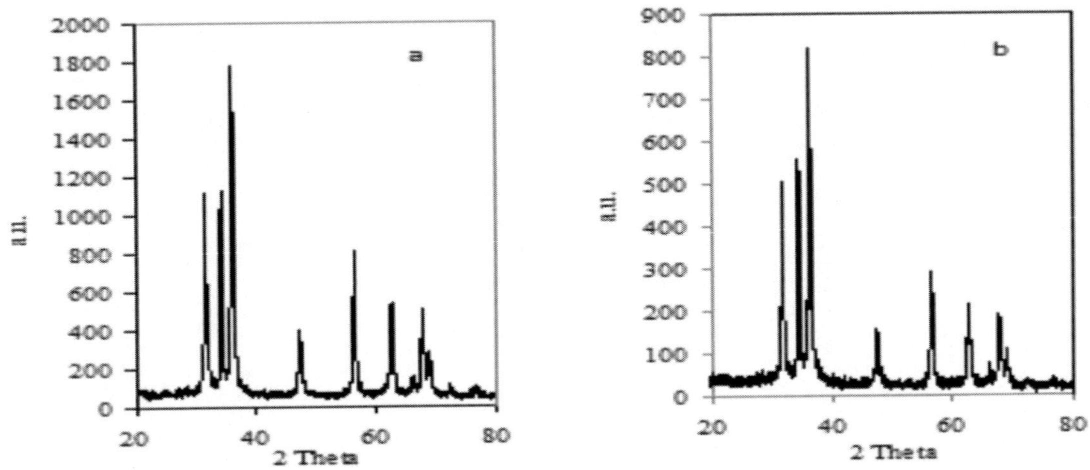

Figure 10. XRD pattern of template added precipitates prepared with a) only sonication b) only mechanical mixing.

Figure 11. SEM image of template added precipitates prepared with a) only sonication b) only mechanical mixing.

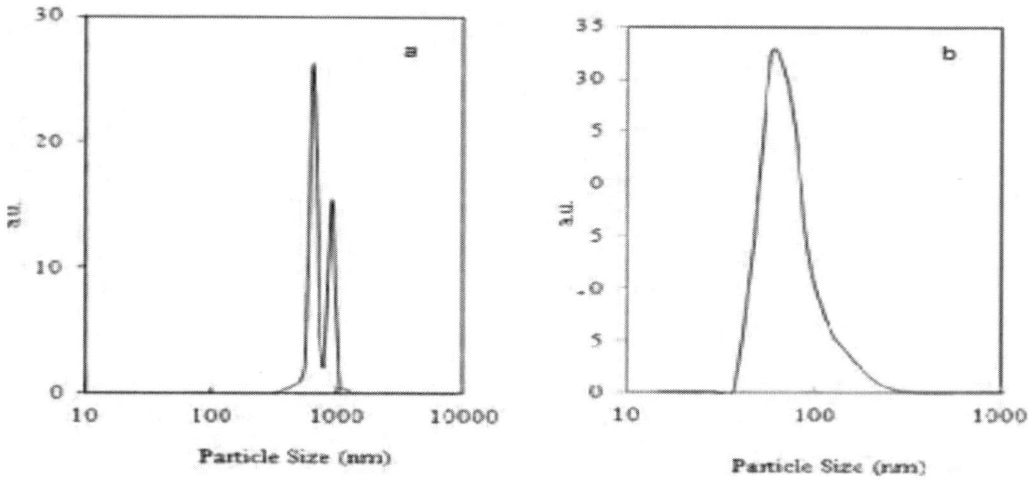

Figure 12. Particle size distribution of template added precipitates prepared with a) only sonication b) only mechanical mixing.

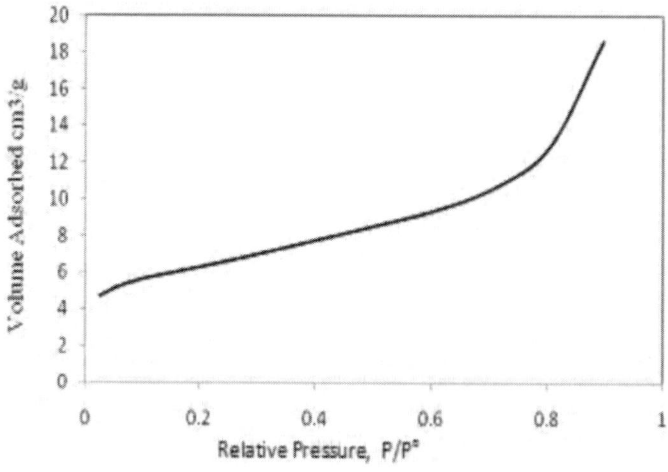

Figure 13. N_2 adsorption isotherm of ZnO at 77K.

Characterization of Nano ZnO Powder Synthesized by Template Addition, Mechanical Mixing and Sonication

The N_2 adsorption isotherm of nano ZnO powder is given in Figure 13. BET surface area of the nano ZnO powder was determined to be 21 m^2/g using the data in Figure 13. If spherical particles are assumed and N_2 gas is adsorbed on the external surface of the particles, this corresponds to a particle size of 28 nm, confirming the average particle size, 29 nm determined by Zeta Sizer. The density of the powder was 4.8 g/cm^3 as determined by helium pycnometry. This value is lower than the density of pure zinc oxide, 5.1 g/cm^3. There were impurities in ZnO powder causing the density to be lower. Energy dispersive X-ray analysis (EDX) showed that the surface composition of the powders was 81% Zn, 14 % O and 5 % C in mass. On the other hand pure ZnO should have 80% Zn and 20% O. Since there is no C-H stretching vibration peak at 2985 cm^{-1} in the FTIR spectrum of the powder shown in Figure 14, there was no TEA in the samples. Thus the presence of C in the powder could be due to the adsorbed CO_2 from atmosphere.

Peaks were present at 3400 cm^{-1} and 1660 cm^{-1}corresponding to hydrogen bonded OH stretching and bending vibration of H_2O respectively in the FTIR spectrum of the ZnO sample in Figure 14. The peaks at 908 cm^{-1}, 707 cm^{-1} belonged to OH group which may due to presence of $Zn(OH)_2$. The broad peak in the range 1517 and 1390 cm^{-1} could be attributed to ν_3 stretching mode of carbonate ions. There were also peaks at 835 cm^{-1} (ν_2 mode of carbonate), at 737 (sh) and 710 cm^{-1} (ν_4 mode of carbonate) in the spectrum. The source of carbonate ions in nano zinc oxide could be the adsorbed CO_2 from atmosphere during preparation and drying of the particles. The basic pH of the precipitation medium caused absorption of CO_2 from air. The large surface area of the nano particles also allowed the adsorption of CO_2 from air during drying of the particles[23]. CO_2 adsorption on ZnO was also reported by other workers [27, 28]. Zn-O stretching vibrations at 473cm^{-1} and 532 cm^{-1} had the highest absorption value and indicated that the powder was mainly ZnO.

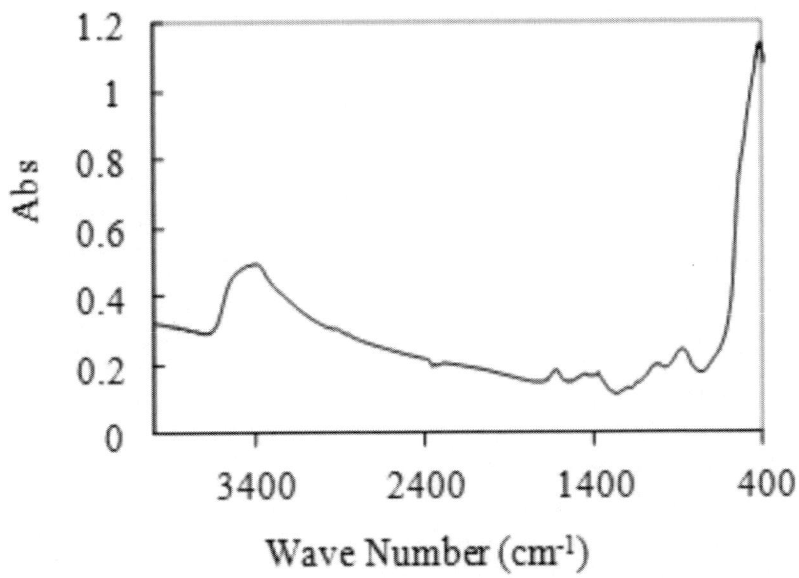

Figure 14. FTIR spectrum of nano ZnO powder.

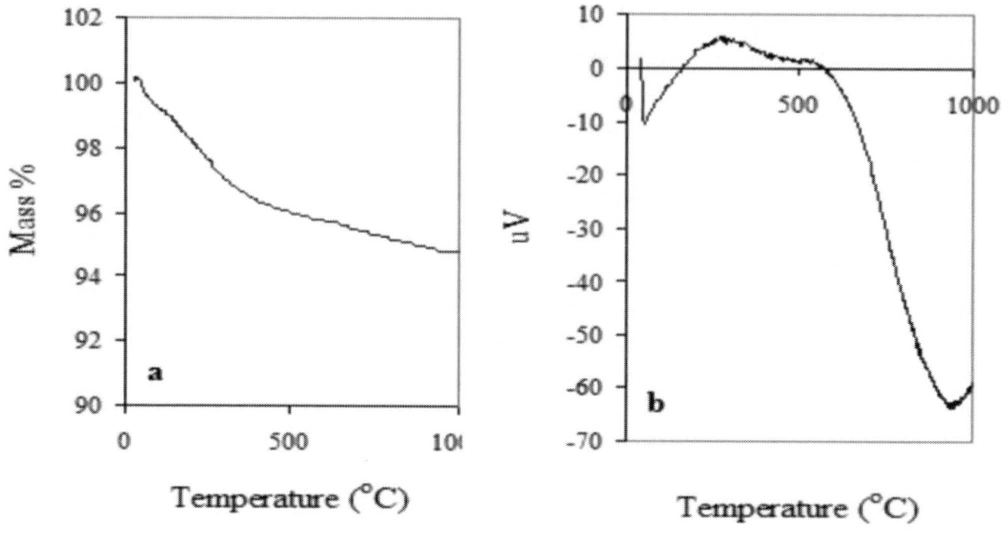

Figure 15. Thermal characterization of nano ZnO powder a) TGA b) DTA.

The TG and DTA curves of the nano ZnO powder dried at 50 °C are shown in Figure 15. The mass loss of the nano ZnO powder was 5.2% at 1000° C as seen in TG curve of the nano powder. This could be due to elimination of water and CO_2 from the sample on heating.

The DTA curve has three endothermic peak maxima at 59 ° C, ~430 ° C, and 940 ° C, which are related to the release of adsorbed water, dehydration of $Zn(OH)_2$ and decomposition of the other impurities such as carbonates in the powder and sintering of ZnO particles respectively. Presence of $Zn(OH)_2$ was also detected by the FTIR spectroscopy and DTA. Thus the mass loss in TGA was due to drying of ZnO and the dehydration of $Zn(OH)_2$ and evolution of adsorbed CO_2. The endotherm at 940°C could be due to the sintering of ZnO particles to each other. The melting point of ZnO is 2200°C, but the surface of the nano particles melts at a much lower temperature due to the high surface to volume ratio and sintering occurs at much lower temperatures.

Resistivity of the powders

Figure 16 shows the current versus sweeping voltage (I-V) for nano ZnO powder. The curve was linear with a very high "0.9984" correlation coefficient. The resistivity value was calculated according to Ohm's law using the inverse of the slope of the I-V line. The volumetric resistivity was found as 1.3x 10^7 ohm cm. The resistivity of zinc oxide thin films was reported as 2.8 x 10^{-4} ohm cm [29] and for films prepared by spray pyrolysis and 1.4 to 2 x 10^{-4} ohm cm independent of the preparation method [30]. The resistivity of the nano ZnO pellet was much higher than those of the thin films. However the prepared zinc oxide was a semi conductive material that can be used in moderately conductive applications.

Light absorption by the powder

Zinc oxide has ability to absorb UV light. The peak maximum value is 353 nm in absorption spectrum of nano ZnO powder as seen in Figure 17a. UV-A light was strongly absorbed by the ZnO powder as observed by previous investigators [23].

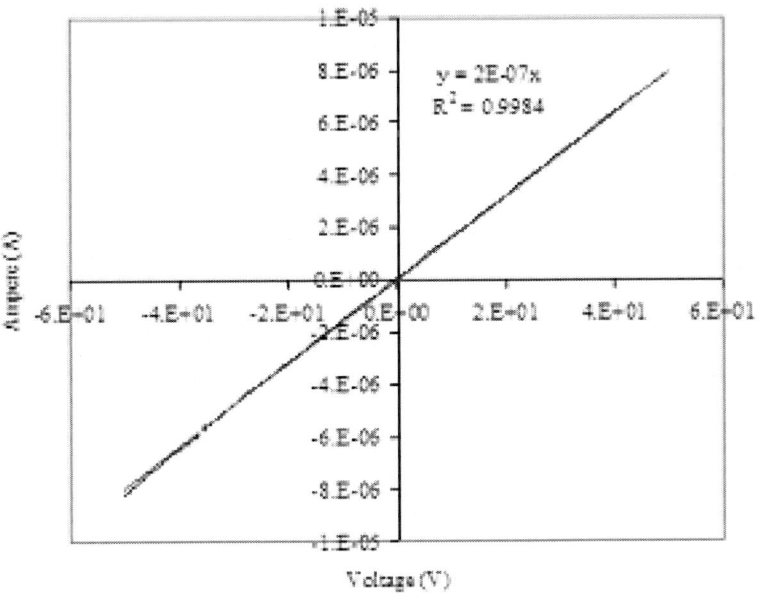

Figure 16. Sweeping voltage versus current values for nano ZnO powder.

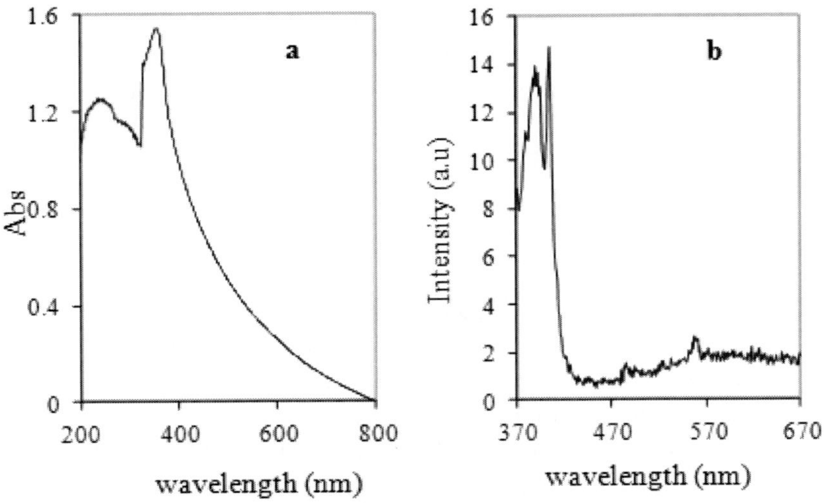

Figure 17 a. Absorption and b. fluorescence spectra of nano ZnO powders.

Light emission by the powder

Two peaks are observed in the fluorescence spectrum of the dry pressed pellet of ZnO powder as seen in Figure 17b. The peak at 391 nm corresponds to free exciton or bound exciton of ZnO in the UV region. A violet luminescence which is attributed to the zinc vacancies is observed at 405 nm. This wavelength is higher than that of ZnO powders which has strong UV luminescence at 398 nm obtained by combustion technique [23]. However the green and yellow luminescence which was mentioned for complex defects [28] was not observed for this powder. There was also no blue band emission attributed to singly ionized oxygen vacancies [31].

CONCLUSION

Experimental design was used to find out the most important variables affecting the size of the particles in ZnO preparation. It was found that minimum sized particles were obtained by TEA addition, sonication and mechanical mixing. Template addition creates nucleation centers and a large number of nuclei forms and crystal growth stops at nano size level due to depletion of the ions in solution. Thus nano particles of ZnO were obtained. Mixing influenced the homogenous dispersion of the chemicals and nano ZnO crystals with a very narrow size distribution oriented in 002 direction were obtained.

The nano powder was synthesized using TEA under mechanical stirring and ultrasonic treatment simultaneously at 30 ^{o}C. The crystals of the powder had 29 nm size. The XRD pattern gave the characteristic peaks of ZnO. However there were some peaks related with $Zn(OH)_2$ and CO_3^{-2} in its FTIR spectrum. It was 95% ZnO.

Moderately conductive nano ZnO powder was obtained having 1.3 x10^7 ohm cm electrical resistivity. Absorption spectrum of the powder showed absorption peak at UV-A region. The room temperature fluorescence spectrum of the powder revealed a strong and sharp UV emission band at 391 nm and a weak and broad violet emission band at 405 nm showing to free exciton or bound exciton of ZnO in the UV region and zinc vacancy, respectively.

The ZnO powder obtained by TEA addition, sonication and mechanical mixing can be used as a polymer additive to produce statically dissipating composites with luminescence properties.

REFERENCES

[1] Wang, ZL. *Journal of Physics*: Condensed Matter, 2004, 16, R829–R858.

[2] Satyanarayana, VNT; Kuchibhatla, AS; Karakoti, D; Seal, S. One dimensional nanostructured materials. *Progress in Materials Science*, 2007, 52, 613- 699.

[3] Bangal, M; Ashtapure, S; Marathe, S; Ethiraj, A; Hebalkar, N; Gosavi, SW; Urban, J; Kulkarni, SK. Semiconductor Nanoparticles. *Hyperfine Interactions*, 2005, 160, 81–94.

[4] Tang, E; Tian, B; Zheng, E; Fu, C; Cheng, G. Preparation of Zinc Oxide Nanoparticle via Uniform Precipitation Method and Its Surface Modification by Methacryloxypropyltrimethoxysilane. *Chem. Eng. Comm.*, 2008, 195, 479–491.

[5] Vergés, M; Martínez-Gallego, M; Lozano-Vila1, A; Díaz-Álvarez, JA. Microscopy Techniques Applied to the Study of Zinc Oxide Microcrystalline Powders Synthesis. In: *Current Issues on Multidisciplinary Microscopy Research and Education,* Méndez-Vilas A., Labajos-Broncano L., Eds., FORMATEX Microscopy Book Series (N° 2), 2005.

[6] Triboulet, R; Sanjosé, VM; Zaera, RT; Tomas, MCM; Hassani, S. Zinc Oxide – A Material for Micro- and Optoelectronic Applications. In : *The Scope of Zinc Oxide Bulk Growth*. Terukov E., Ed., VCH: Springer: Netherlands, 2005.

[7] Karpinski, PH; Wey, JS. Precipitation processes. In: *Handbook of Industrial Crystallization* (2nd ed.), Myerson A.S., Ed., Butterworth-Heinemann, Wolburn, 2002.

[8] Zhang, W; Yanagisawa, K. Hydrothermal Synthesis of Zinc Hydroxide Chloride Sheets and Their Conversion to ZnO. *Chemical Materials*, 2007, 2 19, 2329-2334,

[9] Yoshimura, M; Byrappa, K. Hydrothermal processing of materials: past, present and future. *Journal of Materials Science*, 2008, 43, 2085-2103.

[10] Sue, K; Kimura, K; Murata, K; Arai, K. Effect Of Cations and Anions on Properties of Zinc Oxide Particles Synthesized in Supercritical Water. *J. of Supercritical Fluids*, 2004, 30, 325–331.

[11] Loh, KP; Chua, SJ. Zinc Oxide Nanorod Arrays: *Properties and Hydrothermal Synthesis*, Springer, Berlin, 2007 .

[12] Ohara, S; Mousavanda, T; Umetsua, M; Takamia, S; Adschiria, T; Kurokib, Y; Takatab, M. Hydrothermal Synthesis of Fine Zinc Oxide Particles Under Supercritical Conditions. *Solid State Ionics*, 2004, 172, 261–264.

[13] Deng, HM; Ding, J; Shi, Y; Liu, XY; Wang, J. Ultrafine Zinc Oxide Powders Prepared By Precipitation/Mechanical Milling *Journal Of Materials Science*, 2001, 36, 3273– 3276.

[14] Xiao, Q; Huang, S; Zhang, J; Xiao, C; Tan, X. Sonochemical synthesis of ZnO nanosheet *Journal of Alloys and Compounds*, 2008, 59, L18–L22.

[15] Yu, J; Yu, X. Hydrothermal Synthesis and Photocatalytic Activity of Zinc Oxide Hollow Spheres. *Environmental Science and Technology*, 2008, 42, 4902–4907.

[16] Wei, Y; Chang, P. Characteristics of Nano Zinc Oxide Synthesized Under Ultrasonic Condition. *Journal of Physics and Chemistry of Solids*, 2008, 69, 688–692.

[17] Chen, D; Jiao, X; Cheng, G. Hydrothermal synthesis of zinc oxide powders with different morphologies. *Solid State Communications*, 2000, 113, 363-366.

[18] Ammala, A; Hill, AJ; Meakin, P; Pas, SJ; Turney, TW. Degradation Studies of Polyolefins Incorporating Transparent Nanoparticulate Zinc Oxide UV Stabilizers *Journal of Nanoparticle Research*, 2002, 4, 167–174.

[19] Yi, R; Zhang, N; Zhou, HF; Shi, RR; Qiu, Z; Liu, XH. Selective synthesis and characterization of flower-like ZnO microstructures via a facile hydrothermal route. *Materials Science and Engineering B-Advanced Functional Solid-State Materials*, 2008, 153, 25-30.

[20] Colon, G; Hidalgo, MC; Navio, JA; Melian, EP; Diaz, OG; Rodriguez, JMD. Photoluminescence and photocatalysis of the flower-like nano-ZnO photocatalysts prepared by a facile hydrothermal method with or without ultrasonic assistance. *Applied Catalysis B-Environmental*, 2008, 83, 30-38.

[21] Lai, YL; Meng, M; Yu, YF; Wang, XT; Ding, T. Photoluminescence and photocatalysis of the flower-like nano-ZnO photocatalysts prepared by a facile hydrothermal method with or without ultrasonic assistance. *Applied Catalysis B-Environmental*, 2011, 105, 335-45.

[22] Xie, J; Li, P; Li, Y; Wang, Y; Wie, Y. Solvent-induced Growth of ZnO Particles at Low Temperature. *Materials Letters*, 2008, 62, 2814–2816.

[23] Tarwall, NL; Jadhav, PR; Vanalkar, SA; Kalagi, SS; Pawar, RC; Skaikh, JS; Mali, SS; Dalavi, DS; Shinde, PS; Patil, PS. Photolimunescence of zinc oxide nanopowder synthesized by a combustion method. *Powder Technology*, 2011, 208, 183-188.

[24] Dijken, AV; Meulenkamp, EA;Vanmaekelbergh, D; Meijerink, A. The luminescence of nanocrystalline ZnO particles: the mechanism of the ultraviolet and visible emission *Journal of Luminescence*, 2000, 87, 89, 454-456.

[25] Jana, A; Bandyopadhyay, NR; Devi, PS. Formation and assembly of blue emitting water lily type ZnO flowers. *Solid State Sciences*, 2011, 13, 1633-1637.

[26] Campbell, AN. Activity coefficients, densities, dipole moments, and surface tensions of the system triethylamine-methylethylketone-water. *Can. J. Chem.*, 1981, 51, 127-131.

[27] Becheri, A; Durr, M; Nostra, PL; Baglioni, P. Synthesis and characterization of zinc oxide nanoparticles: application to textiles as UV-absorbers. *J. Nanopart. Res.*, 2008, 10, 679-689.

[28] Oo, WMH; McCluskey, MD; Lalonde, AD; Norton, MG. Infrared Spectroscopy of ZnO nanoparticles containing CO_2 impurities. *Applied Physics Letters*, 2005, 86, 07311-1-07311-3.

[29] Sali, S; Boumaour, M; Tala-Ighil, R. Preparation and characteristic of low resistive zinc oxide thin films using chemical spray technique for solar cells application. The effect of thickness and temperature substrate. Revue des Energies Renouvelables CICME'08 Sousse, 2008 , 201 - 207

[30] Ellmer, K. Resistivity of polycrystalline zinc oxide films: current status and physical limit. J. Phys. D: *Appl. Phys.*, 2001, 34, 3097–3108.

[31] Tian, Y; Ma, H; Shen, L; Wang, Z; Li, YQS. Novel and simple synthesis of ZnO nanospheres through decomposing zinc borate nanoplatelets . *Materials Letters*, 2009, 63, 1071-1073.

In: News in Chemistry, Biochemistry and Biotechnology ISBN: 978-1-63117-273-1
Editors: G. E. Zaikov, G. Nyszko, L. P. Krylova et al.

Chapter 13

A Novel Supramolecular Hyaluronan/ Polyborate Systems for Tumour Treatment by Boron Neutron Capture Therapies

S. A. Uspenskii[1], P. L. Ivanov[1,2], A. N. Zelenetskii[2], M. A. Selyanin[1] and V. N. Khabarov[1]

[1] Martinex International Research Centre, Russia, Moscow, Russia
[2]Enikolopov Institute of Synthetic Polymeric Materials of Russian Academy of Sciences, Moscow, Russia

Abstract

We present a novel strategy for synthesizing drugs for boron neutron capture therapy (BNCT) based on the formation of a supramolecular organic/inorganic polymer-polymer structure on the basis of hyaluronic acid (HA). IR spectroscopy analysis of the products has demonstrated that as a result of solid-state reactions HA and polyborates obtained from borax via a multi-step neutralization process create a mesh of polychelate complexes. In these structures HA plays a role of ligand boron oxide. Such complexes are very stable in the form of aqueous solution and when injected *in vivo* and the content of boron isotope in cells meets the requirements for boron neutron capture therapy.

Keywords: Hyaluronic acid, polyborates, boron neutron capture therapy (BNCT), neoplastic cells, solid-state reaction blending (SSRB)

Targeted delivery of compounds for boron neutron capture therapy and their retention in the tumor are one of the most topical problems in the field of oncology. Hundreds of such compounds have been synthesized, but the majority of them do not produce a desired therapeutic effect, so the search for a better solution and attempts to synthesize a better compound continue. All of these “third generation” drugs are based on their high selectivity for accumulation in tumor cells. The selectivity of accumulation of ^{10}B isotope is determined by the efficacy of its delivery to target cells and the degree of its retention inside the cells [1].

The biggest obstacle towards developing these drugs is a complicated multi-step synthesis process required to produce the biomolecule-spacer-boron component system. The role of the boron component in third generation drugs is played by sodium mercaptoundecahydro-closo-dodecaborate (BSH) which has a very low yield of ^{10}B isotope. While synthesizing drugs the content of ^{10}B isotope compared to initial polyhedral boron is reduced at least by a factor of 10 [2].

The concept suggested here is the synthesis of complex supramolecular organic/inorganic polymer-polymer structure. The role of the biological carrier is played by hyaluronic acid, which is able to form a complex with polyborates as a result of solid-state reaction blending with borax. The potential of this method lies in biocompatibility of both components and a high coefficient of ^{10}B during its conversion into borax.

In addition to the above HA is a biopolymer that is easily absorbed by cellular membranes which indicates a high potential for its use as a carrier of pharmaceutical compounds, including boron for neutron-capture therapy. HA is naturally present in almost every tissue of a vertebrate organism where it plays many roles in the regulation of cellular activity: speeds up or slows down the cellular division and migration, participates in restructuring of chromatin and gene switching, and is involved in the adaptation of cells to physical and chemical influences, fertilization process, embryogenesis, angiogenesis, inflammation, regeneration, and tumor growth [3].

In this work we demonstrate the possibility of synthesis of a novel supramolecular hyaluronan/polyborate system. This system eliminates the usage of spacers which makes the production of the drug significantly easier and cheaper. In our particular case almost all of ^{10}B is converted into borax, which is highly important considering the costs and difficulties associated with obtaining the isotope in the first place [4].

Analysis of IR Spectrums of Reaction Products between HA and Borax

HA and borax (sodium tetraborate) in 1:1 and 4:1 mol ratio (samples 1 and 2, respectively) were ground in an agate mortar and placed on a Bridgman anvil [5]. Deformations were done under 1 GPa pressure and rotation of 500°. IR fourier spectra of the final product were recorded without additional processing.

Let us analyze the most important changes in the IR spectra. While analyzing the structures of the compounds we combine two methods: Fourier transform and Raman spectroscopy, with each of them complementing the other. Looking at the 1750-1200 cm^{-1} area (Figure 1) we note the appearance of a new absorption peak at 1481 cm^{-1} in sample 1. It does not exist in either of the initial compounds and therefore it is a product of the reaction. In borate chemistry such as an absorption pattern is indicative of the change in the amount of B-O- bonds in the borate network upon the addition of electron-withdrawing; and electron-donating reagents. That is, the interaction of borate with HA because of carboxyl and carbinol groups acting as proton donors with O atoms of B-O-B- bonds and borate anions B-O(-): O-H···O, O-H···(-)O, and simultaneously because of free electron pairs of heteroatoms it can be an electron donor -O(:)···B, -N(:)···B. This peak notably decreases and shifts batochromically (1461 cm^{-1}, shoulder) when increasing the content of HA in the sample, such as in sample 2. This is the area of the spectrum in which absorbs B-O bond of boric acid esters R-O-B.

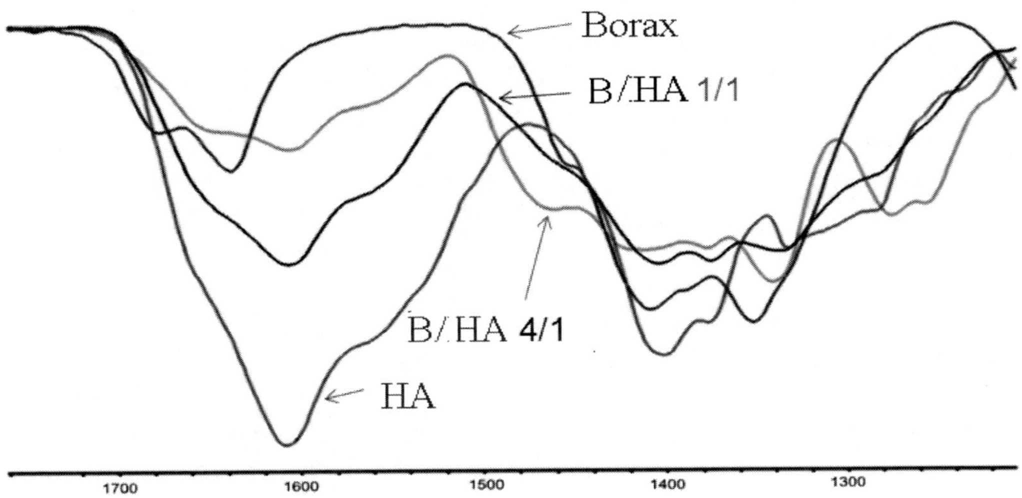

Figure 1. Carbinol and amide bands (spectrum normalized according to CH).

ν(CO)+δ(OH) bands, which are responsible for the superposition of the frequencies of stretching vibrations of the CO bond and deformational fluctuations of the OH bond occur at 1399 cm^{-1} in the HA spectrum, and in samples number 1 and 2 at 1407 and 1426 cm^{-1}. That is, aliphatic hydroxyl binds strongly to the product and is clearly associated with borates. This absorption can also be attributed to the vibrations of the C-O-B esters of boric acid. Typical peak appears in the borate sample number 1 at 1333 cm^{-1} (in borax, the peak is at 1362 cm^{-1}). This absorption is due to the B-O (-) connected to the long-chain borate links. Moreover, according to the literature the peaks at 1340 cm^{-1} correspond to the asymmetric stretching vibrations of B-O va borate different cycles. As can be seen in the product all the absorption bands vary greatly due to the interaction with the polysaccharide and the neutralization of acid by borax.

At 1260 and 1280 cm^{-1} in the spectrum of sample number 1 two peaks appear. These bands are of ν(CO)+δ(OH) type, but the bonds in the product are of a completely different character than in HA (shoulder at 1280 cm^{-1}). Borax does not show this kind of absorption pattern and neither do boric anhydride nor boric acid. However, this absorption pattern is characteristic of tri-, tetra-and pentaborate groups that are formed by partial neutralization of borax in solid state by HA.

Field of vibrations of the C-O-containing bonds is the area of stretching vibrations of acetal bonds - rings and inter-ring bridges, as well as the C-O (H) – bonds of carbinols. It can be likely assumed that in this region C-O(B) bond absorbs as well, which is a connection of alkyl-boron esters. 1200-800 cm^{-1} region in the IR method is the area of borate stretching vibraions (BO4 and borate cycles).

The spectrum in Figure 2 clearly shows the redistribution of the intensity of the absorption bands in the products compared to the spectrum of HA. Normalization for CH should raise the intensity of the bands in the IR spectra of the products to the levels that were observed in the initial acid. We can see great changes in the spectra of sample 1 and 2, and not only in intensity, but also in the position of the bands. In general, changes in qualitative. Although the spectrum of HA is visible, it appears significantly altered. The most intense bands in this area are the ones produced by pyranose cycle groups, including its carbinol groups with a maximum at 1035 cm^{-1} and a shoulder at 1082 cm^{-1}. They turn into a similar group of bands at 995 cm^{-1} with a shoulder at 1033 cm^{-1}. The product with a high HA content has a spectrum closer to that of HA. Perhaps it is not only a shift of vibrations of the HA groups, but rather a complex composition of the absorption bands and the range of products of interaction of HA with redistribution highs. In place of a shoulder at 1082 cm^{-1} in the IR spectra of sample 1 a maximum peak at 1077 cm^{-1} occurs, instead of the maximum at 1149 cm^{-1} a peak appears in the product at 1129 cm^{-1}. But this is an area of medium-intensity pendulum vibrations of r(NH3 +) ammounium. Ammonium can be present in partially deacetylated HA and in a compound with boron it can be a complex such as H^2N (+)-B (-). Therefore it seems impossible to determine how many of these groups were present in the initial acid and in the product. Borax produces the absorption maxima of borates, which in this region of the spectra closely coincide with the absorption maxima of the products. The ratio of the intensities of these peaks, however, is quite different. Finally, there is a peak at 944-945 cm^{-1} common to all. It is intense in the borax spectrum and less intense in the spectrum of HA. Their nature is also different. In borates it is produced by diborate cycle fluctuations, and in HA by ν (CC) vibrations of pyranose cycle. These groups cannot interact and their absorption patterns overlap.

There are two conclusions:

1. We are see (we can observe) a strong interaction of all ether and hydroxyl (alcohol) groups of the polysaccharide with O-atoms B-O-B bonds and the borate anion B-O (-) simultaneously through free pairs of electrons and hydrogen atoms O(:)···B, O-H···O, O-H···(-)O. In a cyclic version it means the formation of chelate complexes with nearly quantitative yield with a ratio of 1:1. The spectra of sample 2 (the one with a higher HA content) shows a greater superposition of polysaccharide bands, but the interaction can also be seen on them.

2. We can not exclude the formation of the C-O-B due to the reaction: B-O-B + HO-C- → B-OH + B-O-C.

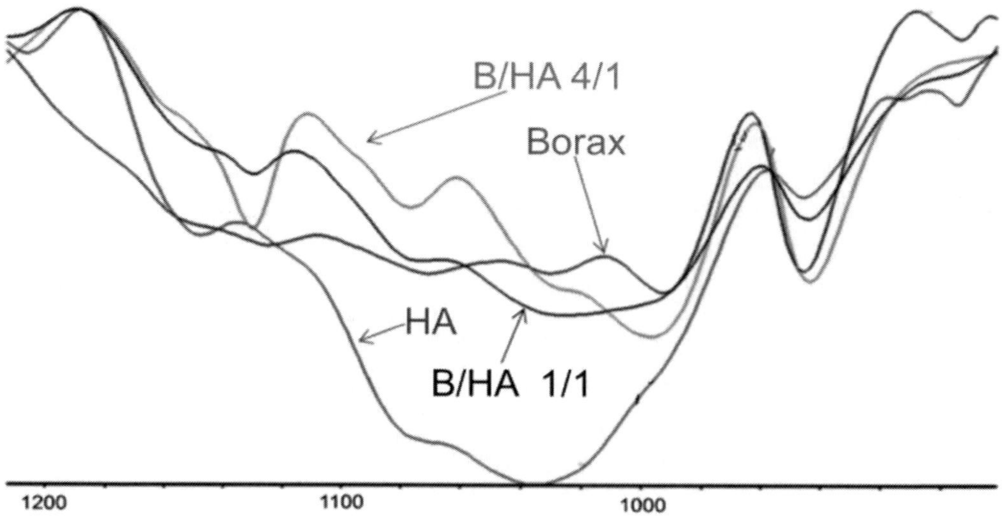

Figure 2. Field of vibrations of the C-O-containing bonds (spectrum normalized by CH).

Rheological behavior of hyaluronic acid complexes with borax was studied using a rotational viscosimeter Rheotest 2.0 in the slot "cylinder-cylinder" (R/r = 1.02) at the temperatures between 25 and 60°C (±0.2°C). Dynamic experiments have been conducted in the linear viscoelastic region, where the rate of shift j = 1.5 - 656, c^{-1} Dynamic viscosity η (Pa.s) was defined as a function of the rate of shift j (per s-1).

The dissolution of HA occurs through separation of the molecular chains and their diffusion into the solvent (water). This process is dependent on the flexibility of the molecular chain. Parts of a flexible chain can move, its units can be swapped with the molecules of the solvent, and its diffusion does not require a large expenditure of energy to overcome the intermolecular interactions. Introducing borax into solutions of HA leads to the increased rigidity of hyaluronic acid chains because of the formation of maximum possible number of hyaluronan/polyborate complexes at a 1:1 ratio, which cannot swap parts of a chain with the solvent molecules, but rather only move as a whole. This severely impedes the diffusion process due to their high molecular mass and as a result, we see almost a twofold reduction of viscosity in boron-modified HA solutions compared to unmodified hyaluronic acid (figure 3).

The above conclusion is verified through the results of rheological measurements taken on a sample with a lower concentration of HA (a diluted solution). The tests were done on an Ubbelohde capillary viscometer with a diameter of 0.54 mm at 25 ±0,2°C, varying pH and ionic strength, and had shown that the addition of borax to aqueous solution of HA leads to the reduction in the polymer's viscosity independent of pH (figure 4).

The results of viscosity measurements taken with a capillary viscometer show that the addition of borax into an aqueous solution of HA leads to the reduction of the polymer's viscosity. Lower viscosity points to the reduction of hydrodynamic volume of HA clumps, which may be a result of HA/borax complexes forming and/or screening of electrostatic interactions by the ions that are produced during the dissociation of borax.

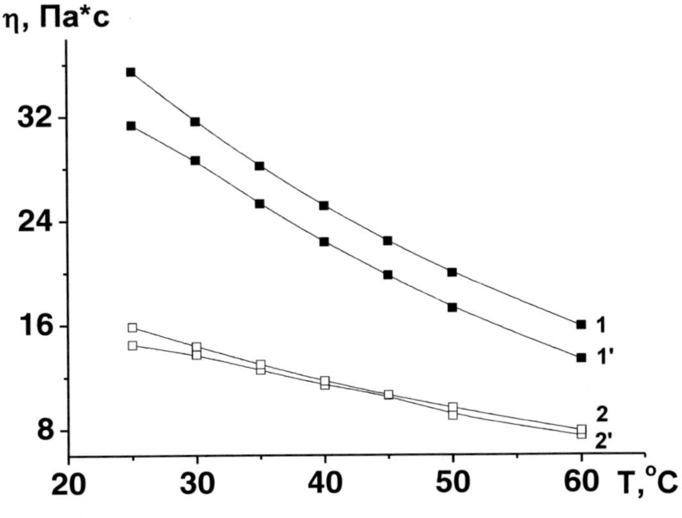

Figure 3. Flow curves of 2% HA solution in water (1.1' j=0.0 c^{-1}) and in borax solution (2,2' j=1.5 c^{-1}) at 25 – 60°C, pH = 6.50 and 9.80 and fixed shear rate.

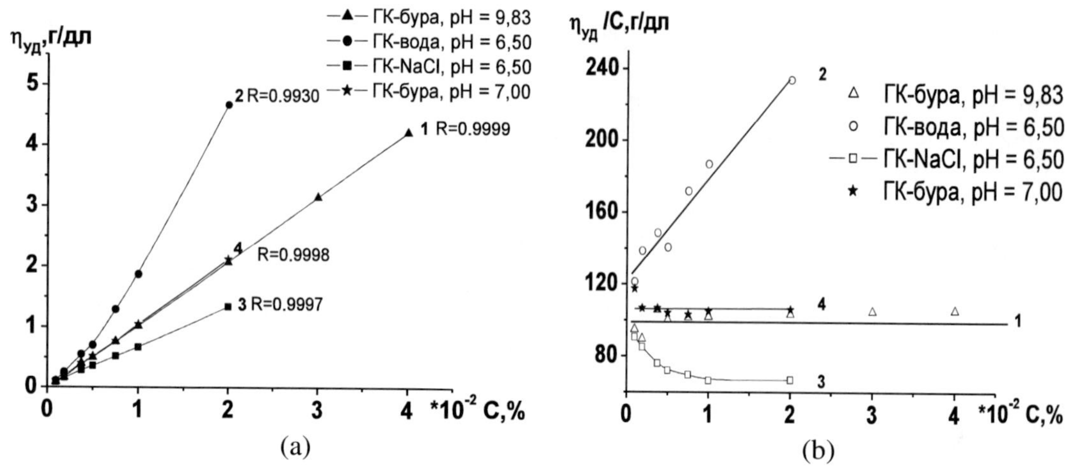

Figure 4. Concentrated dependence of specific viscosity (a) and given viscosity (b) of high molecular weight HA in water; borax solution, pH = 9,8; 0,1M NaCl; borax solution, pH = 7.0.

Investigating the effects of HA/borax compounds in vivo: Martinex International Research Centre in collaboration with Medical Radiological Research Center of Russian (Obninsk) have investigated the dispersion of boron in animal tissues after injecting a tumor with HA/borax compounds obtained using solid-state modification techniques.

For this experiment F1 male mice with body weight between 20 and 22 grams were selected. Approximately 10^6 cells of B16 mouse melanoma in suspension (0.2 ml) were injected into the right hind leg of each mouse. Twelve days following the introduction of melanoma cells the mice were injected with 0.1 ml of boron-containing compound "Borhyal" intratumorally (volume of the tumors was between 0.8-1.2 cm^3). The animals were split into 7 groups with 7 mice in each group and tissue samples were obtained from mice decapitated under narcosis 0.25, 0.5, 1, 3, 6, 9, 12 and 24 hours after the injection of Borhyal. The samples were taken from the tumor, blood, skin, muscle, liver, kidney, spleen, and lungs of the mice and then tested for presence of boron.

The highest boron content in the tumor (55 μg/g) was observed 15 minutes after the injection. One hour after the injection the boron content decreased in half and 3 hours after injection decreased further to less than 1/5 of the initial measurement, which indicates that this compound must be injected less than 1 hour prior to NCT. The most favourable time for NCT is between 15 and 30 minutes since the time of injection, when the concentration of boron in the tumor is above 30 μg/g and is higher than in surrounding tissues. When compared to blood and muscle tissue (Figure 5) the gradient of boron content gets as high as 5 and 2.5, respectively. High boron concentration found in the skin could be linked to partial leakage of the compound from the injection point onto the surface of the skin. No significant boron content was observed in liver, kidney, spleen, and lungs indicating a quick "exit" from the organism of test specimens [6].

Considering results outlined above, we deem logical to further discuss the use of this compound for use in neutron capture therapy.

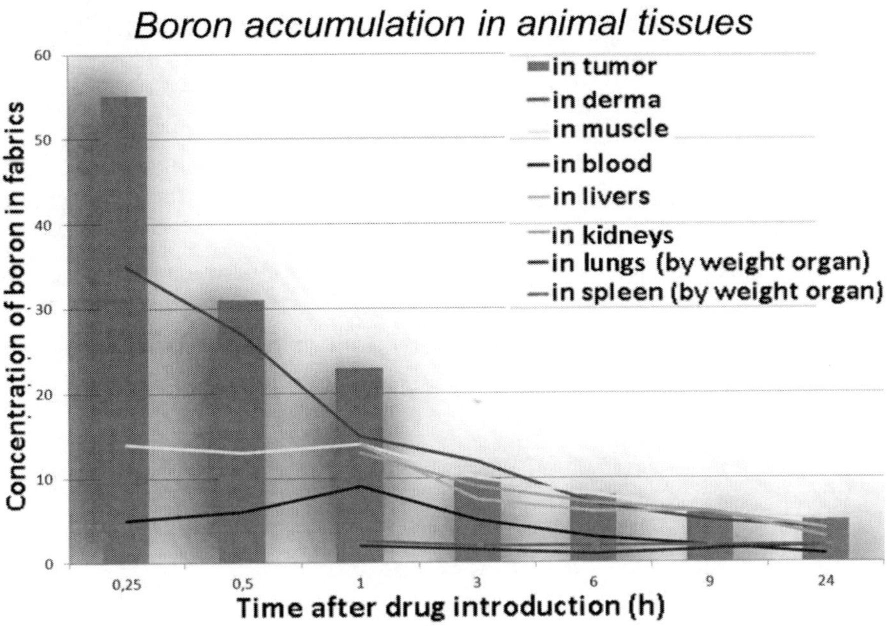

Figure 5. Comparison of the concentration of boron in blood and tumor (colums) at different times after introtumoral injection.

CONCLUSION

- Hyaluronan/polyborate complexes are obtained through a solid-state synthesis process that is relatively simple, economical, and has a high potential for development of polysaccharide-based biomaterials.
- The stability of obtained polycomplexes is comparable to covalently bonded compounds because they contain polychelate fragments spread out along the entire chain of the macrocomplex.
- The formation of borate chelate complexes is verified through an IR analysis in a solid state and rheological behavior of the aqueous solution.
- Investigating the behavior of both kinds of HA/^{10}B compounds (obtained using the solid-state method as well as those mixed in aqueous solution) in vivo have shown high bias towards accumulation in tumor cells.
- Due to low toxicity of both components and high content of boron in resulting compounds, hyaluronan/polyborate complexes and the suggested synthesis method appear to be a valid approach to BNCT.

REFERENCES

Ivanov, P.L., Korjakin, C.H., Habarov, H.V., et al. Synthesis and study of new compounds for neutron capture therapy based on hyaluronic acid and boron-10. *Pharmaceutical Chemistry Journal*, in press.

Khabarovsk, B.H., Boykov, I.L., Villager, M.A., Hyaluronan: preparation, properties, applications in biology and medicine. *Practice of medicine*, M.: 2012. - 224s.

Nemodruk, A.A., Karalova, Z.O.C. Analytical chemistry of boron. *AA Nemodruk*. M. 1964. - 282c.

Orlov A.B., *Synthesis of conjugates of polyhedral boron compounds with lactose as a new potential agents for boron neutron capture therapy of cancer*. Ph.D. diss: 02.00.03. M., 2005.-94c

Sivaev I.B., Bregadze, WI., Boron neutron capture therapy of cancer. Chemical aspect. *Russian chemical journal*. 2004. XLVIII, No. 4, p.109-125

Volkov, V.I., Zelenetsky, A.N., Ivanov, submarines, etc. The process for producing a boron-containing hyaluronic acid. *Patent of the Russian Federation*, No. 2445978 g.2012.

In: News in Chemistry, Biochemistry and Biotechnology ISBN: 978-1-63117-273-1
Editors: G. E. Zaikov, G. Nyszko, L. P. Krylova et al.

Chapter 14

THE ANALYSIS OF THE COMMON FACTORS OF INACTIVATION AND STABILIZATION OF GLUTATHIONE PEROXIDASE I WITH THE USE OF POLYACRYLIC ACID AS A WAY OF RECEIVING PREPARATIONS FOR CURING THE DISEASES OF THE CENTRAL NERVOUS SYSTEM

I. S. Panina, L. Y. Filatova*, A. V. Kabanov and N. L. Klyachko
M.V. Lomonosov Moscow State University, Department of Chemistry,
Division of Chemical Enzymology, Leninskiye Gory, Moscow, Russia

ABSTRACT

An investigation of thermal inactivation kinetics of glutathione peroxidase I – the enzyme, playing the key role in the system of the anti-oxidant defence of the organism has been made. Oligometric glutathione peroxidase has been shown to be inactivated according to a monomolecular mechanism at 37°C. An effective means of stabilizing the enzyme by the polyacrylic acid has been offered.

Keywords: Reactive oxygen species, glutathione peroxidase, polyelectrolyte, kinetics of inactivation, stabilization

LIST OF ABBREVIATIONS

ROS	reactive oxygen species
GP	glutathione peroxidase
GR	glutathione reductase

* Corresponding author: telephone number 84959393476, fax number 84959395417, E-mail: luboff.filatova@gmail.com.

G-SH	reduced glutathione
NADPH	nicotinamide adenine dinucleotide phosphate
EDTA	ethylenediaminetetraacetic acid
PAA	polyacrylic acid

INTRODUCTION

Oxidative processes, proceeding in the organism with the participance of the reactive oxygen species (ROS) have been attracting more and more scientists recently. ROS are highly reactive chemical species, such as free radicals containing oxygen (O_2^{-}•, HO_2•, HO•, NO•, ROO•) and molecules, able to easily produce free radicals (H_2O_2, ROOH, ROOR). ROS damage many biomolecules due to the non-specific oxidation and initiation of chain reactions.

Excessive activation of free-radical oxidation reactions occurs in case of various diseases (atherosclerosis, Alzheimer's and Parkinson's diseases, diabetes, cataract, oncological diseases, premature ageing) [1]. Enzyme systems and low molecular compounds participate in the ROS defence.

Low molecular anti-oxidants are oxidized by the reactive oxygen species, yet they do not prevent the formation of ROS, but only fight with the negative consequences [2].

The enzymatic system of the anti-oxidant defence of the organism is more effective. Superoxide dismutase, catalase and glutathione peroxidase are the most important anti-oxidant enzymes necessary for the normal life of the mammals' organisms [3, 4, 5, 6, 7]. Glutathione peroxidase catalyzes the hydrogen peroxide and organic peroxides' decomposition process with the simultaneous glutathione oxidation, which gives priority to this enzyme in the anti-oxidant defence of the organism [8, 9].

The development of the methods of receiving highly effective and stable anti-oxidant preparations on the basis of glutathione peroxidase for the curing and prevention of the central nervous system diseases and other dangerous organism affections is becoming very important. The aim of the present work is to develop the methods of receiving stable preparations on the basis of glutathione peroxidase I.

MATERIALS AND METHODS

Materials

The preparation of glutathione peroxidase from bovine erythrocytes (lyophilized powder, activity 713 U/mg of the protein), glutathione reductase (GR) from the baker's yeasts (suspension in 3.6 M $(NH_4)_2SO_4$, pH 7.0, containing 0.1 mM of the dithiothreitol), reduced glutathione (GSH) and nicotinamide adenine dinucleotide phosphate (NADPH), all that of «Sigma» company, hydrogen peroxide of «Chemapol» company, ethylenediaminetetraacetic acid (EDTA), potassium phosphate dibasic, potassium hydroxide of «Sigma-Aldrich» company, hydrochloric acid of «Reachem» company, polyacrylic acid (PAA) with the molecular weight of 5.1 kDa of «Aldrich» company.

Methods

The manufacturing of glutathione peroxidase complexes with the polyacrylic acid

The equal volumes of the polyacrylic acid and the enzyme solutions in the potassium phosphate buffer (0.02 M) with a pH 7.0 were mixed in such a way that the molar ratios of the PAA/enzyme were 1:1, 10:1, 100:1.

The received solutions were left for different periods of time (1 hour, 12 hours or twenty-four hours) at 4°C for the preparation of complexes.

Table 1. The quantitative parameters of glutathione peroxidase inactivation in the presence of the polymer at 37°C and pH 7.0

PAA/enzyme molar ratio	K_{in}, hr^{-1}
0	0.027±0.002
1	0.022±0.002
10	0.022±0.001
100	0.021±0.001

The analysis of the influence of the polyacrylic acid on the activity of glutathione peroxidase

The activity of glutathione peroxidase was determined spectrophotometrically from the decrease of absorption of nicotinamide adenine dinucleotide phosphate at 340 nm under the method [10].

1 ml of the potassium phosphate buffer (0.5 M KH_2PO_4 with pH 7.0, containing 0.5 mM EDTA), 10 μl 0.0084 M NADPH, 10 μl 0.15 M GSH, 3 μl of the glutathione reductase (2.2 mg/ml), 1.5 μl 0.136 M H_2O_2 were placed into the quartz cell. The reaction was initiated by the addition of 5-8 μl of glutathione peroxidase solution (0.3 or 1 mg/ml) or its mixture with the polyelectrolyte and the change of optical dencity was registered at 340 nm and 37°C.

The analysis of the stability of glutathione peroxidase and its complexes with the polymer at 37°C

The stability of the enzyme or its complexes with the polyacrylic acid at 37°C was analyzed spectrophotometrically by means of selecting aliquots during certain periods of time with the further measurement of the activity under standard conditions.

Native electrophoresis

The native electrophoresis of glutathione peroxidase was provided according to the methods of Bio-Rad [11].

The analysis of glutathione peroxidase and its complexes with the polymer by means of dynamic light scattering

The change of the size of glutathione peroxidase particles during incubation at 37°C was controlled with the use of «Zetasizer Nano». The enzyme solutions (0.3 and 1 mg/ml) in the potassium phosphate buffer (0.02 M KH_2PO_4, pH 7.0) were put through the filters «Millipore» with a diameter of the pores 0.22 μM with a further thermostating at 37°C and the dimension of the sizes of the particles through certain periods of time. In a similar way the sizes of glutathione peroxidase particles in the complexes with the polymer were measured.

Results and Discussion

The inactivation curves of glutathione peroxidase in the semi-logarithmic coordinates are presented in Figure 1. From the figure one can conclude that the enzyme inactivation is subordinate to a monomolecular mechanism at glutathione peroxidase concentrations of 0.3-1 mg/ml. The value of the first order inactivation constants (K_{in}) is equal to 0.027±0.002 hr^{-1}. Neverthehless, the kinetic curves do not provide sufficient information about the enzyme inactivation process. To check the dissociation-association processes of glutathione peroxidase molecules the enzyme solutions with various inactivation levels were analyzed with the use of the native electrophoresis and the dynamic light scattering methods.

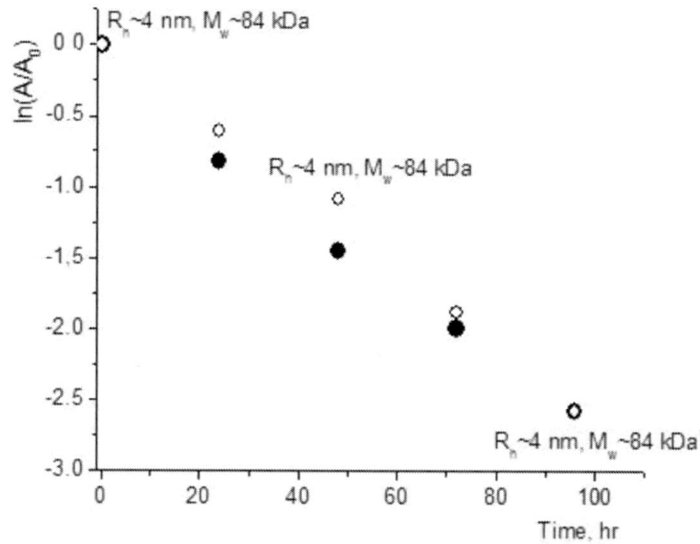

Figure 1. The thermal inactivation of glutathione peroxidase under the concentrations of the enzyme 0.3 (white signs) and 1mg/ml (black signs). The conditions of the experiment: 37°C, 0.02 M the potassium phosphate buffer, pH 7.0.

Native electrophoresis data showed that the molecular weight of the enzyme does not change during inactivation. The absence of the protein aggregation processes in the solution has also been confirmed by means of dynamic light scattering. The value of the effective hydrodynamic radius of the non-inactivated glutathione peroxidase under the concentrations of 0.3 and 1 mg/ml is about 4 nm. No considerable change of the particles' sizes has been observed during the incubation of the enzyme solutions under the concentrations of 0.3 and 1 mg/ml at 37°C. Thus, glutathione peroxidase is inactivated according to a monomolecular mechanism at 37°C.

It is known that the addition of the polyelectrolytes is a good way of suppressing conformational changes of protein globules. To stabilize glutathione peroxidase the polyacrylic acid with the molecular weight of 5.1 kDa being highly movable informatically and non-toxic has been chosen. On the surface of glutathione peroxidase molecules there are positively charged areas, with which the molecules of the negatively charged PAA may cooperate.

The value of the inactivation constant which was compared to that for the individual enzyme has been accepted as a quantitative measure of glutathione peroxidase inactivation process in the presence of the polyelectrolyte. It has been stated that in the presence of 1, 10 and 100- fold excess of polymer the enzyme inactivation takes place under the first order.

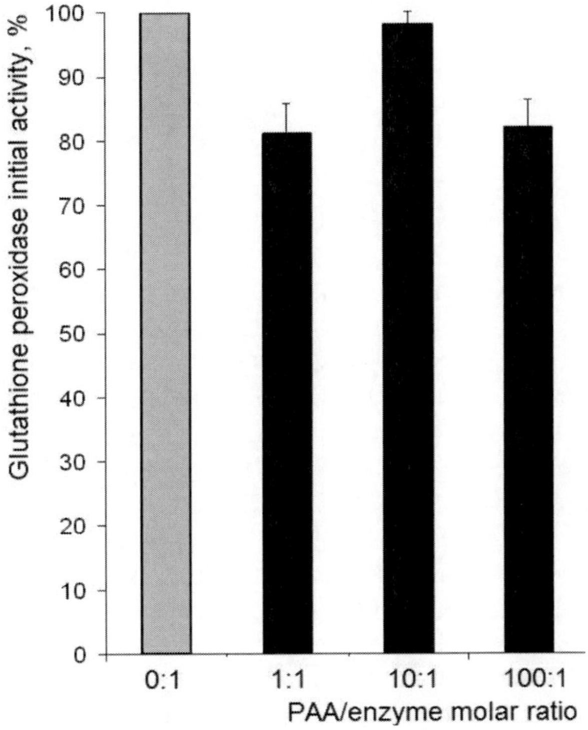

Figure 2. The influence of the polyacrylic acid on glutathione peroxidase activity at pH 7.0 and 37°C.

The values of inactivation constants of glutathione peroxidase and the enzyme in the presence of the polyacrylic acid are presented in table 1. From table 1 one can see that the polyacrylic acid with a molecular weight of 5.1 kDa stabilizes glutathione peroxidase: the value of the enzyme inactivation constant in the presence of the PAA decreases approximately to 20% and does not depend on the polymer/enzyme molar ratio within the range from 1:1 to 100:1. It has been shown by means of dynamic light scattering that the value of the effective hydrodynamic radius for glutathione peroxidase molecule is equal to 4 nm, for the polyelectrolyte R_h is equal to 2-4 nm. Under the conditions when the stabilization effect is observed (during the interaction of GP with the polyacrylic acid) particle sizes increase up to the value of the effective hydrodynamic radius of about 50 nm, which testifies the possible achievement of the stabilization effect due to the complex formation.

It should be noted, that the inclusion of glutathione peroxidase into the complexes with the polyacrylic acid is accompanied by the preservation of the enzyme's activity (Figure 2).

CONCLUSION

1. A supposition has been made that glutathione peroxidase inactivation takes place according to a monomolecular mechanism at 37°C.
2. The stabilization effect of the polyacrylic acid with molecular weight of 5.1 kDa (37°C) without a loss of the enzyme activity has been observed.
3. It has been proved that the stabilization effect of the polymer is preconditioned by the formation of the enzyme-polyelectrolyte complexes.

The work has been fulfilled within the framework of the project of Russian Ministry of Education 11.G34.31.0004.

REFERENCES

[1] Sies, H; Helmut, M. *Exper. Phys.*, 82 (2), 291 (1997).
[2] SculacheV, VP. *Soros Educational Journal*, 3, 4 (1996).
[3] Mills, G. *Arch. of Biochem. Biophys.*, 86, 1 (1960).
[4] Nagababu, E; Chrest, F; Rifkind, J. *Biochim. Biophys. Acta*, 1620 (3), 211 (2003).
[5] Rotruck, J; Pope, A. *Science*, 179, 588 (1973).
[6] Flohe, R. *Free Rad. Biol. Med.*, 27 (9 – 10), 951 (1999).
[7] Arthur, J. *CMLS Cell. Mol. Life Sci.*, 57, 1825 (2000).
[8] Okovity, SV. *Farmind-pract*, Moscow, 2003. 85 p. (in Russian).
[9] Muller, F; Lustgarten, M; Jang, Y. *Free Radic. Biol. Med.*, 43 (4), 477 (2007).
[10] Koller, L; South, P; Exon, J; Whitbeck, G. *Can. J. Comp. Med.*, 48 (4), 431 (1984).
[11] http://www.bio-rad.com/webroot/web/pdf/lsr/literature/10007296.PDF

In: News in Chemistry, Biochemistry and Biotechnology ISBN: 978-1-63117-273-1
Editors: G. E. Zaikov, G. Nyszko, L. P. Krylova et al.

Chapter 15

COMPARISON OF TWO BIOREMEDIATION TECHNOLOGIES FOR OIL POLLUTED SOILS (RUSSIA)

V. P. Murygina*[*], S. N. Gaidamaka and S. Ya. Trofimov
Moscow State University, Chemistry Faculty, Department of Chemical Enzymology, Moscow, Russia

ABSTRACT

This paper deals with two bioremediation technologies of bogs, accidentally polluted with oil, which are applied in the Northern part of Russia in the Komi Republic and the Western Siberia. One of the technologies is a typical ex-situ and the other is in-situ one without gathering of oil out of the surface of the bog and milling of the moss. So, different results were obtained after bioremediation of bogs with an oil-oxidizing preparation Rhoder there.

Keywords: Bogs, moss, soil, oil pollution, bioremediation, degradation, oil-oxidizing preparation.

1. INTRODUCTION

The basic oil production places in Russia are situated in the Northern parts of the Komi Republic and in the Western Siberia. Vast scale territories polluted by oil often are located in difficultly passable bogs. Penetration depth of oil in such bogs doesn't exceed of 0.3-0.6m and oil is usually propped up with a permafrost layer. Areas contaminated with oil are huge: from 1-2 to 10 hectares or more. Climate of the regions is similar with severe prolonged winter and cool short summer. In these two regions different remediation technologies are applied for restoration of bogs polluted with oil. This is due to many factors: economic, geologic, availability of contaminated sites for facilities, volume of oil spills, sizes of areas contaminated with oil and so on.

[*] Corresponding author: phone: +7(495) 939-5083, fax: +7(495) 939-5417, e-mail: vp_murygina@mail.ru.

In the Komi Republic before the bioremediation a lot of oil is carefully gathered from the surface of polluted bogs. The surfaces of the bogs are often washed from the residual oil by water with pumps, sometimes with addition of surfactants [1,2]. Gathered oily water and previous collected oil are transferred to refinery plant. In winter or in early spring until bogs thaw out, the top layer of 5-7cm of highly contaminated soil is usually cut off and excavated, washed off on special devices from oil and all hydrocarbons (HC) are transferred to refinery plant. However, it is impossible to wash off the soil completely till the most probably concentration of oil. And the soil with the residual oil in it is returned to the same original location. The bioremediation with or without oil-oxidizing microorganisms and fertilizers and following phytoremediation are usually performed on this soil.

In the Western Siberia in-situ bioremediation of the soil is preferred because the excavation of a top layer of the soil polluted by crude oil from huge areas of impassable bogs is technically difficult and economically inefficient. Besides, the sheet of water settles down under a moss layer of thickness about 3-10 m on the most part of impassable bogs. With the best case such bogs are once milled at the very beginning of summer, until the permafrost layer completely thawed. At once a large amount of fertilizers, lime, seeds of oats and any oil-oxidizing preparation is brought into the moss [3]. At worst for example behind the Polar Circle of the Western Siberia the polluted bogs are left without any treatments.

In the present study the efficiency of two remediation technologies for bogs polluted with oil which were applied in Russia are compared. This study presents an approach for development of a new bioremediation technology for impassable bogs polluted with oil in the Western Siberia without application of consecutive stages of a classical remediation of technical, agrochemical and biological ones.

2. Materials

2.1. The Oil-Oxidizing Preparation Rhoder

The Rhoder consists of two bacterial strains belonged to the genus Rhodococcus, (R. ruber Ac-1513 D and R. erythropolis Ac-1514 D), isolated from soils polluted with crude oil. The strains are non-pathogenic and non-mutagenic to humans, animals, plants and bacteria. The Rhoder is approved for widespread using in Nature and it has been successfully used for bioremediation of oil refinery sludge, soils, wetlands and water surfaces polluted with oil [4-11] and the Rhoder is used in these described field-scale tests.

2.2. Background

2.2.1. The Komi Republic, Usinsk Town

In 2008 soil polluted with a residual oil was taken from a special device of washing off the oil sludge and returned to the same place from which previously was excavated. This place was an area with the size of 136*82 m2 near Usinsk town. Bioremediation of soil was started in early July, 2008. The weather at the beginning of July, 2008, was dry and hot for two weeks, the temperature varied from 17°C to 33°C and then it went down. Rains were rare that was atypical for this region at that time of a year. In August the temperature went down till 15-11°C and rains began.

The subsoil of the site for the washed off soil was presented by genetic types of a lake and marine alluvial precipitations. The alluvial sediments situated into surfaces were presented also with brown peat clay sand and light-gray-brown sequence of semi-loam. Marine deposits were presented by loam of gray-blue clay with inclusions of gravel and pebble of the large size.

2.2.1.1. Bioremediation

Bioremediation of the soil was carried out with addition of 100 kg of dry fertilizer into the soil and treated it with a disk harrow. Then a working solution of the Rhoder with the most probably number (MPN) of active hydrocarbon oxidizing (HCO) cells of $2.5*10^6$ cells/ml with addition of 0.1% fertilizer was sprinkled with a water cart on the soil surface and then the soil was milled again. Three such treatments with the interval of 2-3 weeks were performed and the Rhoder in a total quantity of 30kg of a dry powder with HCO bacteria cells of $1.0*10^{10}$ in 1g of powder was used for the bioremediation of this area. The soil was milled after each introduction of the Rhoder.

2.2.1.2. Sampling for Analysis

Soil samples for analysis were selected from five points of the site before the bioremediation and before each next treatment with the Rhoder and a half month after the completion of treatments. Samples taken from one point of the site were mixed carefully and homogenized samples (about 0.5 kg) were passed to analyze. Every time 5 samples were taken for analysis.

2.2.2. The Western Siberia, Muravlenko Town

In 2011, an impassable bog with a size about 0.8 hectares polluted with spring accidental oil spill and halved by high knolls was offered for the bioremediation with using the Rhoder only. This bog located near the Muravlenko town. Typical marsh plants (moss, cloudberry, wild rosemary) existed on the knolls, which were practically not affected by the oil spill. Large spots of the oil were situated on swampy impassable depressions. Vegetation (moss, sedge) on these depressions perished almost completely. A layer of the oil with a thickness about 1 cm and more was presented on the water surfaces on these depressions. The penetration of the oil into the moss was about 40-45 cm. The oil contamination of the bog was unequal. The bog had a slight bias towards a sand bank which had been made to prevent spreading of the oil pollution and, in fact, turned into the road. Two previously digged pits to collect oil with the pump were presented on the bog. However, oil was gathered poorly, and these pits still had much of oil. The thickness of the oil on the water surfaces of the pits were more than 1 cm. The oil under an air temperature below 10oC became viscous on the surface of the water of the depressions and pits. The oil from the surface of the polluted bog was not collected additionally. The soil was not mixed by a disk harrow or other devices. An attempt to perform the bioaugmentation with the Rhoder was undertaken without additional gathering of the oil and without application of milling because of technical complication of doing classical ex-situ remediation on the impassable bog polluted with oil. It was needed to minimize expenses on the bioremediation.

The weather during the bioremediation of this bog was not favorable, the air temperature did not exceed 10 - 14°C, and it was raining from time to time.

2.2.2.1. Bioremediation

The oil polluted bog was treated three times with intervals for 3 weeks by the working solution of the Rhoder with the MPN of hydrocarbon oxidizing cells of $1.0*10^8$ per 1 ml by the sprinkling from the fire-engine vehicle, previously washed with water. The Rhoder was used in total quantity of 120kg as a liquid concentrate with HCO bacteria cells of $1.0*10^{11}$ cells/mL.

2.2.2.2. Control of the Soil Toxicity

After the third application of the Rhoder the seeds of oats and perennial grasses were sown on the cleaning area to determine a toxicity of the soil and perform the phytoremediation. Half of seeds were previously treated with a solution of the hamates "Extra" (Russia) to identify an impact of the hamates on a stability of herbs germination on the bog after bioaugmentation with the Rhoder. The oil contaminated bog was divided in two parts with using landmarks. One part of the site was sowed with seeds previously treated with the solution of hamates (right), second (left) one was sowed of the seeds without any treatment. On the right and left halves of the bog two plots (size of $1.5*5.0$ m^2) were done and covered with a non-woven material to assess the impact of it on the bog restoration. This non-woven material usually is recommended in the farming and gardening sectors to protect sprouts of plants from adverse environmental conditions.

2.2.2.3. Sampling for Analysis

Soil samples were collected before and after finishing of the bioremediation from 12 points of the bog contaminated with oil from the depths of 0-10cm and 10-25cm, and 25-40cm (by using GPS) for microbiological, chemical and agrochemical analysis. Each sample had weight about 150 g.

3. METHODS

3.1. Chemical and Agrochemical Analyses

Several samples (8 samples) of moss (No. 1,2,3,12,18,22,24,26) from the bog in the Western Siberia, Muravlenko town, were excessively polluted with crude oil. The oil from these samples were at first extracted by chloroform (150 mL) in chemical flasks (each flask with a capacity of 400 ml), which were shaken for 30 minutes at room temperature. Received solutions of the oil were transferred to the other flasks through waterless sodium sulfate (Na_2SO_4) to remove remains of water. The chloroform was evaporated at 75°C. Each sample of the moss excessively polluted with oil was then extracted three times as described above. The chloroform extracts in flasks were heating at 105°C till a constant weight. Samples of the moss after oil extraction were dried at 75°C, weighed and the oil was calculated per 1kg of a dry moss. Chemical analyses of HC in all other samples and in the dried samples of the moss after previous chloroform extraction at room temperature were carried out by a gravimetric method with using of Soxhlet's apparatus and column chromatography with Silica gel. GC and HPLC methods were used too [12].

Saturated HCs from each sample after column chromatography with Silica gel (1 μL of hexane fraction) were analyzed on GC Cristallux 4000m by company Meta Chrom (Russia), column was OV-101, 50m*0.22mm*0.50mkm, detector was FID, gas-carrier was nitrogen. Detector temperature was 300ºC, initial column temperature was 80ºC, velocity of heating was 12º per minute till the temperature 270ºC. Time of analysis was 40 min. Mixture of Undecane, Dodecane, Tetradecane, Hexadecane and Squalane were used as external standards in concentrations of 5μg/μL for each substance.

HPLC analyses were carried out on Knauer HPLC with the ultra-violet detector, on the reversed-phase column of Diasfer 110-C18 for HPLC, length of the column was 250 mm, the diameter was 4 mm, the grains were 5 microns. Samples for analyses on HPLC were prepared after drying of hexane fractions and following extraction each sample with 1mL of acetonitrile during 20 min under shaking and then analyzed. Phenantrene, Pyrene and Benzo(e)pyrene were used as external standards in concentrations of 10μg/mL for each substance in acetonitrile [13].

Humidity and pH in samples of the moss were determined with standard agrochemical methods and the soluble nitrogen and phosphorus were determined calorimetrically [14].

3.2. Microbiological Analyses

The MPN of microorganisms in the treated soils and in the Rhoder were estimated by method of ten-fold dilutions with using the meat-peptone agar for heterotrophic bacteria (HT) and selective agar for detection of actinomicetes, pseudomonas, nitrifying, ammonifying, oligotrophic microorganisms and fungi [15].

Modified liquid Raymond media with crude oil was used to determine of hydrocarbon oxidizing (HCO) bacteria [16] (g/L): Na_2CO_3 - 0.1; $CaCl_2$ *6 H_2O - 0.01; $MnSO_4$ *7 H_2O - 0.02; $FeSO_4$ - 0.01; Na_2HPO_4 *12H_2O - 1.0; KH_2PO_4 - 1.0; $MgSO_4$ *7 H_2O - 0.2; NH_4Cl - 2.0; NaCl - 5.0; pH = 7.0. Raymond media was prepared on the distilled water with consistently bringing the components. Then 4.5 ml of media was placed in each test tube. 50mg of crude oil were added to each test tube and all test tubes were sterilized under 121ºC during 30min.

Two drops of 0.05% solution of Twin 80 were added in the first test tubes with the soil samples (about 1g) and carefully stirred up to wash away most fully cells of HCO bacteria from soil particles to prepare ten-fold dilutions. After preparing ten-fold dilutions of the investigated soil samples (or the Rhoder) 0.5 ml of dilutions was passed into test tubes with Raymond media. Test tubes were incubated at 28°C for two weeks and results were considered on a dispersion of the oil or disappearance of the oil film, or by a turbidity or oil inflation that testified an ability of microorganisms from samples of soil to utilize oil. Total number of HCO microorganisms in samples was determined by the last number of the test tube in which dispersion of oil or inflation or turbidity in the liquid media or disappearance of oil was observed.

Microorganisms were identified by microscopic observation: cells morphology, motility, Gram-coloring, capsule- and spore-forming, acid-fast, conidial stage. Morphology examination involved the shape and color of colonies, mycelia and pseudo mycelia, growth on the selective agar media and a biochemical characterization.

4. Results and Discussion

4.1. Bioremediation of Washed Off Soil Polluted with Residual Oil in the Komi Republic

4.1.1. Microbiological Monitoring

Preliminary microbiological analysis of samples of the soil polluted with residual oil after washing off (from 14.05.2008) had showed that the MPN of different groups of microorganisms varied from 10^3 to 10^4 CFU/g of soil (Table 1). Composition species of soil microorganisms was presented of genus Bacillus, Pseudomonas, Rhodococcus, Nitrosomonas, Nitrosococcus, Nitrobacter, Azotobacter, Aspergillus, Penicillium, and Fusarium. In soil samples, which were selected before directly starting of bioremediation (03.07.2008), the MPN of the same groups of microorganisms was higher by 2-4 orders, caused by a positive effect of a warm weather. However, the MPN of HCO microorganisms ($7.2*10^5$ CFU/g of soil) was still insufficient for effective degradation of HC in soil polluted with residual oil and an introduction of the oil-oxidizing preparation Rhoder was justified.

Introduction of Rhodococcus from the Rhoder into the soil during the bioremediation (three times) had increased the MPN of HCO bacteria responsible for degradation of oil (Table 1), and the process of oil decontamination of the soil was activated. In addition, the introduction of the Rhoder did not adversely affect on the indigenous microorganisms that was evident in the obtained results (Table 1). The peak of MPN of all analyzed groups of microorganisms was observed in the middle of July till early August 2008, and the MPN of microorganisms was remained relatively high for a long time.

Table 1. MPN of different types microorganisms which presented in the washed soil with residual oil before and during bioremediation with the Rhoder

Sample	MPN, CFU/g of soil						
	HT	Ammo-nifying	Nitrifying	Oligo-nitro-philic	Pseudo-monas	Mold	HCO
14.05.2008	$5.2*10^4$	$1.4*10^4$	$5.2*10^4$	$1.7*10^4$	$1.2*10^4$	280	$1.0*10^3$
03.07.2008	$1.5*10^8$	$1.7*10^8$	$2.9*10^8$	$2.2*10^7$	$1.3*10^6$	$2.5*10^6$	$2.7*10^5$
16.07.2008	$6.2*10^8$	$1.9*10^8$	$8.9*10^8$	$2.2*10^7$	$1.0*10^7$	$4.1*10^6$	$8.1*10^7$
04.08.2008	$6.0*10^7$	$3.9*10^7$	$5.0*10^7$	$1.3*10^6$	$1.1*10^8$	$4.1*10^6$	$2.1*10^6$
29.09.2008	$5.7*10^7$	$2.9*10^7$	$5.0*10^7$	$6.9*10^6$	$1.2*10^7$	$1.4*10^6$	$7.8*10^6$

Note: the averages are the total number of microorganisms identified in samples taken from five points of the plot.

4.1.2. Agrochemical Monitoring

Preliminary agrochemical analysis of the washed off soil (May 2008) showed a relatively high content of nitrogen (as ammonia) and phosphorus which were available to plants and microorganisms (Table 2). Additional introduction of the fertilizer (100 kg/ha) before the bioremediation of the soil had optimized concentration of the fertilizer for microorganisms and plants [14]. Soil moisture was sufficient for the life of microorganisms responsible for the biodegradation of HC and restoration of the biological activity of the soil. The pH of the soil was close to neutral and was supported by the introduction of the lime before the bioremediation and before the third treatment with the Rhoder and subsequent phytoremediation on the bog (Table 2).

4.1.3. HC Degradation during the Bioremediation

The initial concentration of the oil in the washed off soil before bioremediation was 126.1g/kg of dry matter (DM). Group composition of HC (average value) in this soil before application the Rhoder and during the bioremediation and after the phytoremediation of the soil is shown on Figure 1 (A). Three times application of the Rhoder decreased the concentration of the pollutant in the soil an average by 43.6%. The concentration of the saturated HC was decreased by 61.5%, the aromatic HC by 30.4%, resins and asphaltenes only by 3.4% (Figure 1A).

GC analysis of the soil from the bog before and after bioremediation confirmed the obtained results. According to results of GC analysis the degradation of oil products in soil was by 74-84 % (Figure 1B-C).

Analysis of aromatic HC on HPLC in the washed off soil before and after the bioremediation showed a significant reduction of these compounds, though a small peak similar to benzo(e)pyrene was found in samples of the soil after the bioremediation with the Rhoder (Figure 1D). An appearance of such peak looked like the benzo(e)pyrene associated with the formation of intermediate compounds by microorganisms, which would be eventually degraded by cells of the Rhoder. Previously in laboratory experiments sometimes the appearance and then disappearance of a like benzo(e)pyrene substance in soils was observed during the bioremediation with the Rhoder.

Table 2. Monitoring of agrochemical parameters of the soil before and in the process of bioremediation with the Rhoder

Date of sampling	pH	Humidity, %	Nitrogen $N\text{-}NH_4^+$ mg/kg of soil	Phosphorus PO_4^{3-} mg/kg of soil	Salinity of soil, %	Contamination, g/kg
14.05.08	6.73±0.24	65.80±7.13	12.92±5.59	56.39±10.11	-	126.0
03.07.08	6.62±0.33	76.70±9.00	13.13±3.97	56.39±10.11	10.9	101.0
04.08.08	6.48±0.06	73.47±10.24	31.02±2.23	89.45±27.74	-	73.0
29.09.08	6.84±0.21	66.46±4.55	16.40±2.42	75.58±23.22	6.6	69.8

Note: not determined.

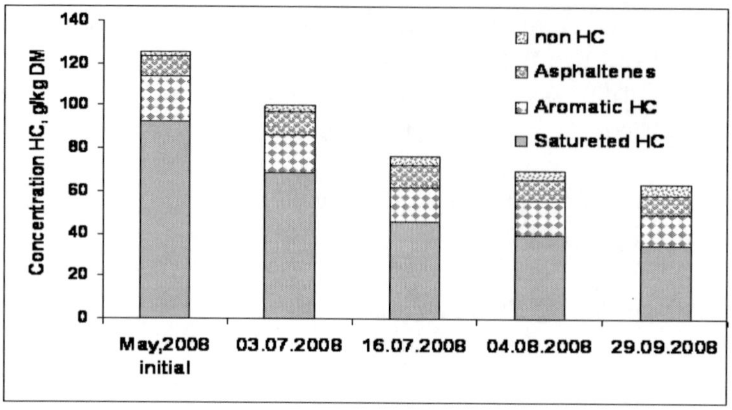
Concentration HC, g/kg DM
140
120
100
80
60
40
20
0
non HC
Asphaltenes
Aromatic HC
Satureted HC
May,2008 initial
03.07.2008
16.07.2008
04.08.2008
29.09.2008

(a)

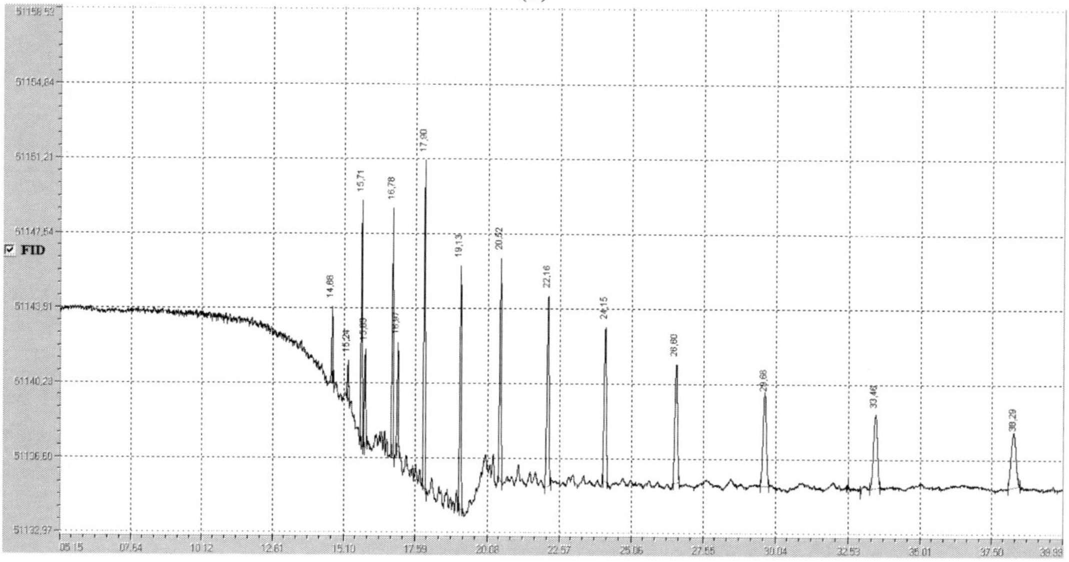
FID

(b)

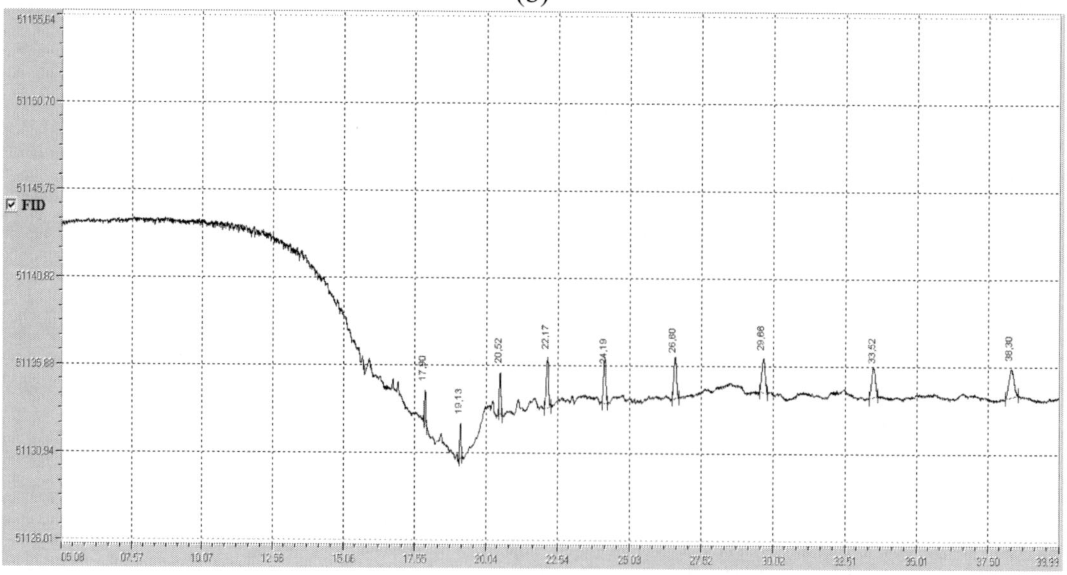
FID

(c)

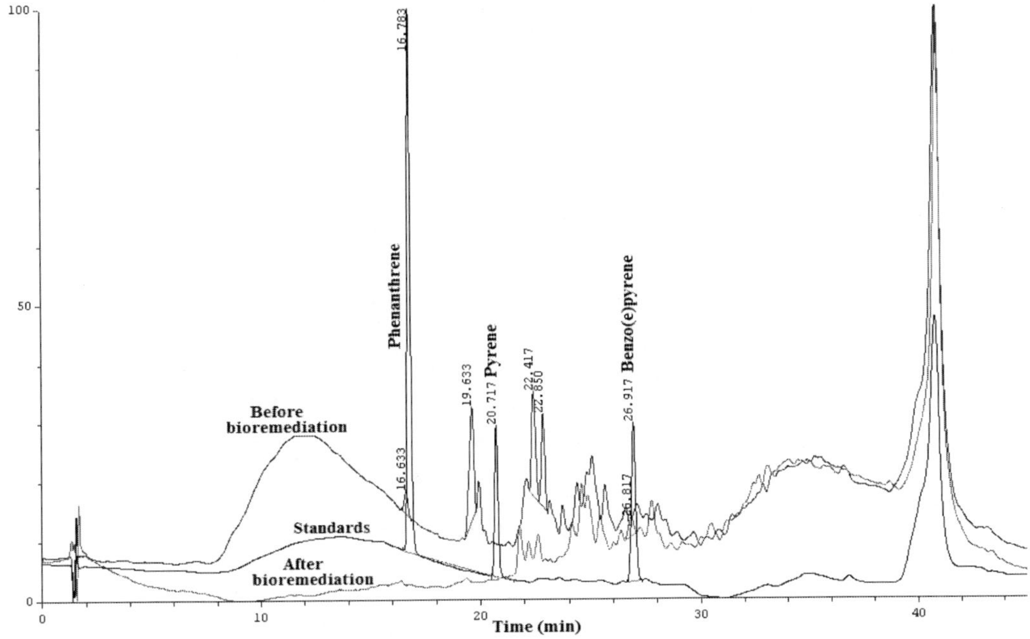

(d)

Figure 1. (a) Bioremediation of washed soil with the Rhoder before and during treatments, 2008; (b) GC analysis of HC in the washed soil before and (c) after bioremediation with the Rhoder; (d) HPLC analysis of PAH in the washed soil before and after bioremediation with the Rhoder.

4.1.4. Phytoremediation of the Washed off Soil with Residual Oil after Bioremediation

The phytoremediation of the soil after the previous bioremediation with the Rhoder in September, 2008, was not successful because of rain during all month. A lot of plantlets rotted through it. Salinity of the soil under these circumstances was decreased almost by 40% on this site (Table 2).

Thus, this technology, including the excavation of the soil polluted with oil out of the bog, and more complete extraction of the oil out of this soil and subsequent processing this oil and bringing it into a commodity form and then a realization of it, allow partially to offset outlays on the equipment service and energy expense. Subsequent the bioremediation of the washed off soil from residual oil (or the bioremediation + the phytoremediation) allows reducing a restoration period of oil polluted areas till 1-2 years and improving the quality of the remediation. This remediation technology of any soil contaminated by accidental oil spill in the Komi Republic can be really regarded as a comprehensive and non-waste one.

4.2. Bioremediation of the Impassible Bog in the Western Siberia, Muravlenko Town

The allocated object was very strongly polluted with oil (Figure 2), and it was difficult to expect a big success in such situation. Nevertheless, it was made a decision to test oil-oxidizing ability of the Rhoder in such extreme conditions. It was on the one hand; on the other hand it was necessary to be convinced that bioaugmentation with the Rhoder can initiate of self-restoration process though it may be not such effective as the ex-situ technology.

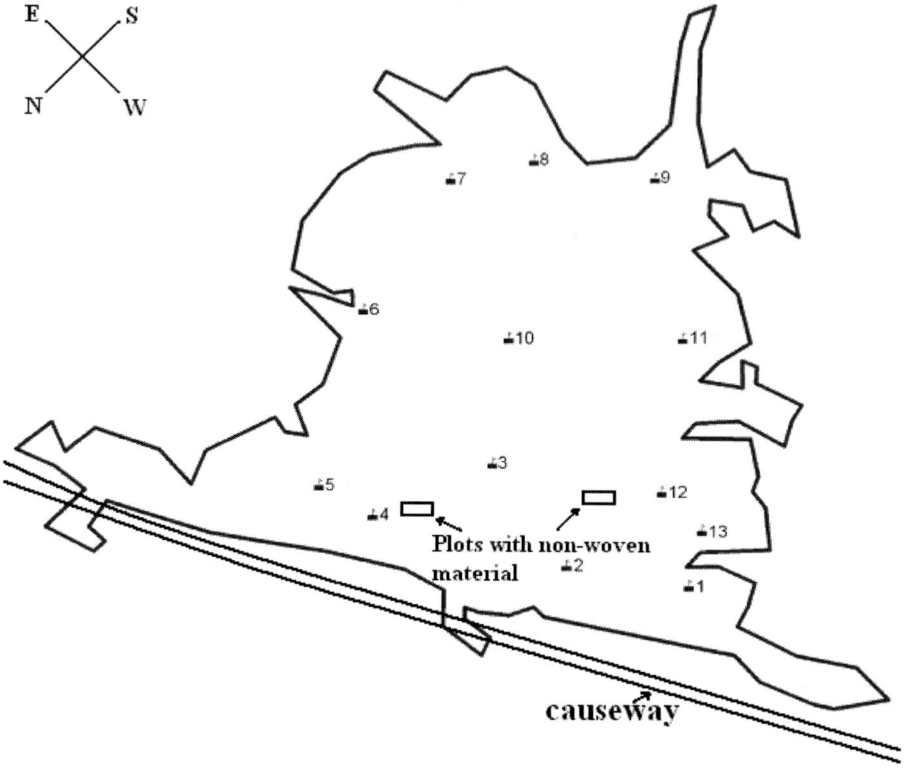

Figure 2. Scheme of the oil polluted bog with points of sampling, Western Siberia, Muravlenko, 2011.

4.2.1. Microbiological Monitoring

Preliminary microbiological analysis of soil samples showed (Table 3) that a lot of microorganisms were presented in layers of 0-10cm. In these upper layers of the soil the MPN of heterotrophic bacteria (HT) varied from $1.1*10^7$ to $6.1*10^8$ CFU/g of the soil. In these points the level of the oil contamination varied from 60.3 g/kg DM to 903.6 g/kg DM. The MPN of HCO bacteria varied from $1.2*10^6$ cells/g to $1.1*10^8$cells/g of the soil. In samples with a very high oil pollution the MPN of HCO cells was only $1.0*10^3$cells/g of the soil. In other samples taken from different depths of the bog the MPN of HT and HCO microorganisms was lower (Table 3). After three times introduction of the Rhoder the total number of HT microorganisms as a whole didn't decrease and even was increased in some samples by 1 order. The MPN of HCO bacteria increased by about 2 orders and more in the majority of the samples (Table 4). The negative influence of the oil-oxidizing preparation Rhoder on indigenous microorganisms wasn't observed.

4.2.2. Agrochemical Analysis

The boggy soil had initial pH from 4.9 to 5.4 and a content of nitrogen and phosphorus compounds in the soil was low (Table 3). Unfortunately, lime in amount of 400kg and fertilizers (600kg) were added manually on the oil polluted surface of the bog before the third application of the Rhoder due to circumstances beyond our control. Nearly one and a half months later it was observed that introduction of the lime had led to some increase of value pH in the soil and more favorable conditions for activity of the oil-oxidizing bacteria of the Rhoder (Table 4).

4.2.3. HC Degradation

Some samples of the moss selected for the preliminary examination of the bog were visually represented by oil slightly contaminated with moss. Several samples looked relatively non-polluted, the others were moderately polluted. 27 samples were selected from the different depth of the bog and analyzed before the bioremediation of this bog with the Rhoder.

Table 3. Microbiological and agrochemical characteristics of soil samples selected from different depths in the oil polluted area before the augmentation with the Rhoder

Samples	Depth of samples selection	Soil pH	HT CFU/g of soil	HCO, cells g of soil	Nitrogen N-NH_4^+ mg/kg of soil	Phosphorus PO_4^{3-}mg/kg of soil
1	(0-10)	-	-	-	-	-
	(10-25)					
2	(0-10)	5.2	$2.8*10^7$	$3.6*10^4$	5.08	-
	(10-25)	4.9	$2.5*10^7$	$4.3*10^4$	2.99	33.11
3	(0-10)	4.9	$6.1*10^8$	$8.1*10^7$	11.58	31.80
	(10-25)	5.0	$3.8*10^8$	$6.0*10^4$	9.18	22.28
4	(0-10)	5.4	$2.8*10^8$	$1.1*10^8$	17.40	33.53
	(0-25)	5.1	$6.1*10^7$	$3.8*10^7$	7.56	-
5	(0-10)	5.2	$1.1*10^7$	$4.9*10^7$	21.02	20.81
	(10-25)	5.0	$5.1*10^7$	$8.0*10^5$	15.91	-
	(25-40)	4.9	$1.8*10^7$	$7.9*10^5$	9.67	-
6	(0-10)	-	-	-	-	-
	(10-25)	5.	2.510^6	$6.0*10^4$	6.2	-
7	(0-10)	5.3	$1.9*10^7$	$8.4*10^4$	10.72	16.64
	(10-25)	4.9	$2.6*10^7$	$5.0*10^4$	16.01	-
8	(0-10)	4.9	$7.6*10^7$	$8.0*10^4$	11.14	19.10
	(10-25)	5.0	$6.4*10^6$	$7.7*10^3$	7.07	-
9	(0-10)	5.0	$1.1*10^8$	$7.1*10^3$	4.83	-
	(10-25)	4.9	$8.9*10^5$	$8.1*10^5$	7.56	-
10	(0-10)	5.1	$7.3*10^7$	$1.0*10^4$	6.35	-
	(10-25)	5.2	$2.8*10^6$	$1.0*10^7$	6.54	-
11	(0-10)	-	-	-	-	-
	(10-25)	5.0	$5.1*10^6$	$7.7*10^5$	3.30	-
12	(0-15)	-	-	-	-	-
	(15-30)	4.9	$5.9*10^7$	$1.2*10^6$	18.51*	-
13	(0-10)	-	-	-	-	-
	(10-25)	5.0	$6.1*10^7$	$9.6*10^4$	8.35	-

Note: not detected because samples were unable to determine due to their high oil content or an insufficient amount of it.

Table 4. Microbiological and agrochemical characteristics of soil samples selected from different depths on the oil polluted area after augmentation with the Rhoder

Samples	Depth Samples selection	Soil pH	MPN of HT CFU/ g of soil	MPN of HCO cells/g of soil	Nitrogen $N\text{-}NH_4^+$ mg/kg of soil	Phosphorus PO_4^{3-}mg/ kg of soil
2	(0-10)	5.2	$2.8*10^7$	$4.4*10^6$	172.5	5.17
	(10-25)	6.3	$2.5*10^7$	$4.5*10^6$	99.6	8.84
3	(0-10)	6.4	$2.2*10^7$	$4.3*10^6$	284.9	9.98
	(10-25)	6.3	$3.8*10^8$	$5.1*10^6$	66.0	9.46
4	(0-10)	5.7	$1.7*10^7$	$5.9*10^7$	275.8	12.46
	(0-25)	5.8	$7.1*10^6$	$8.9*10^6$	163.3	2.60
5	(0-10)	-	-	-	-	-
	(10-25)	6.0	$2.4*10^6$	$3.7*10^6$	88.8	3.55
7	(0-10)	-	-	-	-	-
	(10-25)	6.0	$3.2*10^7$	$1.1*10^6$	139.9	3.58
8	(0-10)	6.3	$2.4*10^7$	$6.6*10^6$	240.7	2.15
	(10-25)	4.9	$1.8*10^7$	$1.0*10^6$	71.0	1.83

Note: not analyzed.

On the right side of the bog in some places the preliminary concentration of the crude oil in the moss layers of 0-10cm was from 35.13 to 14.35kg/kg DM and residual concentration of HC in the same samples after extraction of the crude oil at the room temperature became from 290.6 to 66.9g/kg DM. The concentration of HC on the right side in two samples (0-10cm) varied from 543.1 to 522.99g/kg DM. In the soil layers of 10-25 cm the concentration of HC varied from 516.6 to 43.6g/kg DM. In soil layer of 15-30 cm the concentration of HC was about 300.0g/kg DM. This part of the bog was heavily polluted with the oil (Table 5, samples with a letter R). On the left side of the bog in one place in the moss layer of 0-10cm the crude oil concentration was 29.0kg/kg DM and after extraction of this crude oil under room temperature the residual HC concentration became 173.3g/kg DM. In the other samples the concentration of HC varied from 567.2 to 508.1g/kg DM. In the depth of 10-25 cm the concentration of HC varied from 9.3 to 82.3g/kg DM. In the soil layer of 25-40cm the concentration of HC was about 27g/kg DM. This part of the bog visually seemed a little bit purer than the right one (Table 5, samples with letter L).

The oil in the samples of the moss, which were severe contaminated of the real crude oil (35.1-14.5kg/kg DM), contained the saturated HC of 62.5±1.7%, the aromatic HC of 19.3±1.4%, resins and asphaltenes of 11.8±0.8% and from 5 to 7% of non HC (oxidized substances). Such composition of the oil is a typical for any high quality oil and such oil should be gathered and directed to a refinery plant. Oil contaminating samples of the moss with concentration HC of 850-460g/kg DM contained the saturated HC of 61.8±1.3%, the aromatic HC of 16.7±0.3%, resins and asphaltenes of 8.7±1.9%. Such contamination also represents the high oil quality and such oil should be gathered too. The oil in moss samples from the layers of 10-25 cm, 15-30 cm and 25-40 cm contained saturated HC of 49.4±1.12%, aromatic HC of 19.6±2.3%, resins and asphaltenes of 13.4±5.2% and 18% of non HC (oxidized substances). Such HC composition of the pollution indicated that the processes of the oil biodegradation with indigenous anaerobic microorganisms had begun inside of these layers. Obtained results showed that the initial huge amount of the crude oil in some places was decreased after bioaugmentation with the Rhoder (Table 5 and Figure 3), but the oil had appeared in other places, where previously it was absent. Content of the total saturated HC increased in these places. Probably, such changes in the amount of the crude oil and more impregnation of the top layers (0-10 cm) of the moss with the oil could be due to movement and displacement of the oil because of the small bias to a bulk of the sandy road.

Table 5. Content of crude oil and saturated hydrocarbons in soil samples before and after augmentation of oil polluted soil with the Rhoder

Samples	Depth of samples selection	Free crude oil, kg/kg DM	Saturated HC, g/kg**	Free crude oil, kg/kg DM	Saturated HC, g/kg**	Degradation, %
1_R	(0-10)	**15.24**	66.9	**5.39**	105.5	0
	(10-25)	**2.94**	51.3	**2.94**	59.8	0
2_R	(0-10)	**35.13**	73.7	**6.37**	234.1	0
	(10-25)	*	516.6	*	470.9	8.8
3_R	(0-10)	*	543.1	*	312.4	40.8
	(10-25)	*	84.7	*	45.3	31.2
4_L	(0-10)	*	567.2	*	567.8	0
	(10-25)	*	38.1	*	24.4	35.9
5_L	(0-10)	*	546.7	**15.64**	230.7	57.8
	(10-25)	*	11.8	*	5.1	56.8
	(25-40)	*	27.1	*	207.6	0
6_L	(0-10)	**29.03**	173.3	*	47.8	72.4
	(10-25)	*	82.3	*	433.9	0
7_L	(0-10)	*	515.0	*	11.3	97.8
	(10-25)	*	38.8	*	339.7	0
8_R	(0-10)	*	522.9	*	27.6	94.7
	(10-25)	*	43.6	**11.37**	217.1	0
9_R	(0-10)	**25.84**	77.8	*	26.5	65.9
	(10-25)	*	76.9	*	260.9	0
10_L	(0-10)	*	508.1	*	8.02	98.4
	(10-25)	*	9.3	**7.18**	330.2	0
11_R	(0-10)	**14.35**	280.0	*	190.7	32.1
	(10-25)	*	196.7	**4.96**	314.1	0
12_R	(0-10)	**25.04**	187.6	*	123.5	34.2
	(10-25)	*	53.5	**8.47**	301.5	0
13_R	(0-15)	**14.52**	290.6	*	318.0	0
	(15-30)	*	308.0	*	384.1	0

Note: **R** – right side of area, **L** – left side of area, * - free oil is absent; ** - residual saturated HC in the samples after separated the crude oil.

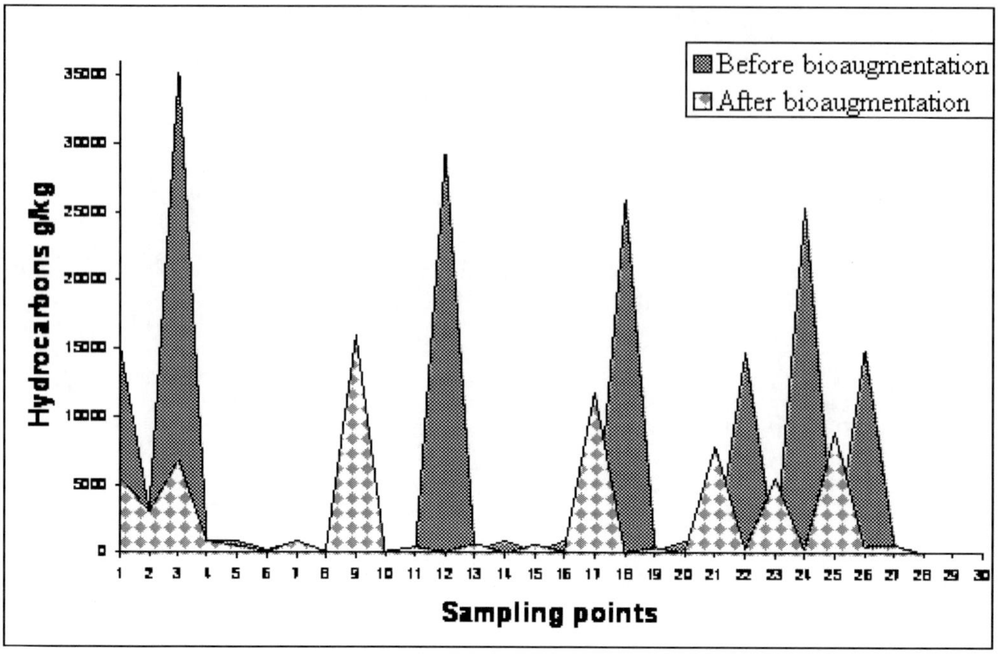

Figure 3. Bioremediation of the oil polluted bog with the Rhoder, Muravlenko, 2011.

Chromatograms of hydrocarbons of the contaminated moss with extremely high and medium levels of the oil pollution before and after bioaugmentation are presented on Figure 4A-C and 4A1-4C1 and confirm results that are described above and below.

After ending of the bioaugmentation with the Rhoder on the right side of the bog the saturated HC of 60.5±0.7% and the aromatic HC of 21.5±0.7%, and resins and asphaltenes of 10.0±0.01% and about 8% of non HC (oxidized substances) were found in samples of moss from the depth of 0-10 cm, which initially contained a lot of crude oil. The oil contained saturated HC of 54.0±0.01% and aromatic HC of 19.5±8.5% and resins and asphaltenes of 6.8±0.4% and about 20% of non HC in samples of the soil from layers of 10-25cm. It is interesting, in the depth of 10-25 cm (anaerobic conditions) degradation process often was more intensive than on the surface of soil. Oil contained of 53.5±0.01% of the saturated HC, the aromatic HC of 23.5±0.01%, resins and asphaltenes of 11.5±0.01% and about 13% of oxidized substances in the samples from the moss layer of 25-40cm. The composition of the oil pollution changed and became the worse if the layer of soil was lower.

Another situation was observed in oil samples from the soil on the left side after ending the bioaugmentation with the Rhoder. The saturated HCs were found of 32.9 ± 5.8% and the aromatic HCs were found of 23.3 ± 1.8% and resins and asphaltenes were found of 29.3 ± 5.9% and oxidized substances were more than 14% in the depth of the moss layers of 0-10cm, that indicated on significant oil oxidizing processes which caused by using of the Rhoder.

Oil contained 60.0 ± 1.6% of the saturated HC and 21.0 ± 0.9% of the aromatic HC and 10.8 ± 1.3% of resins and asphaltenes and about 8% of oxygenated compounds in the samples from the moss layers of 10-25cm (the left side). The quality of oil in 10-25cm of soil layers was better than in the upper layers.

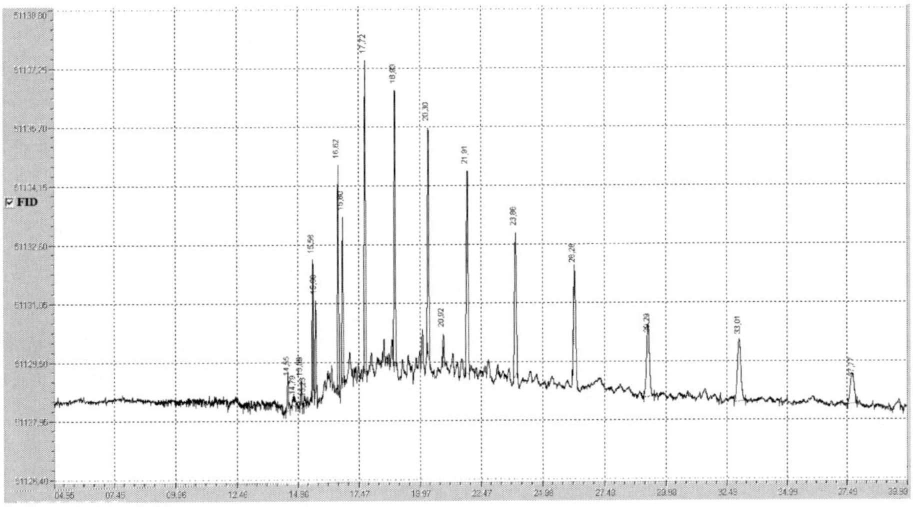

(a)

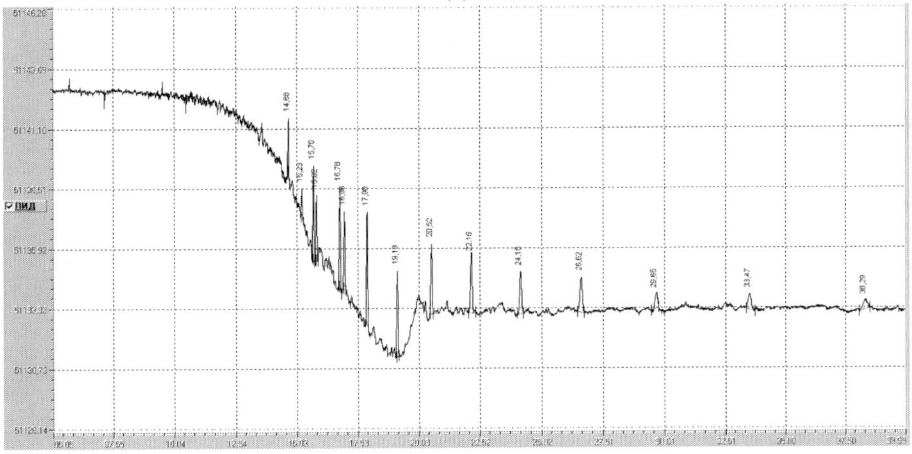

(a1)

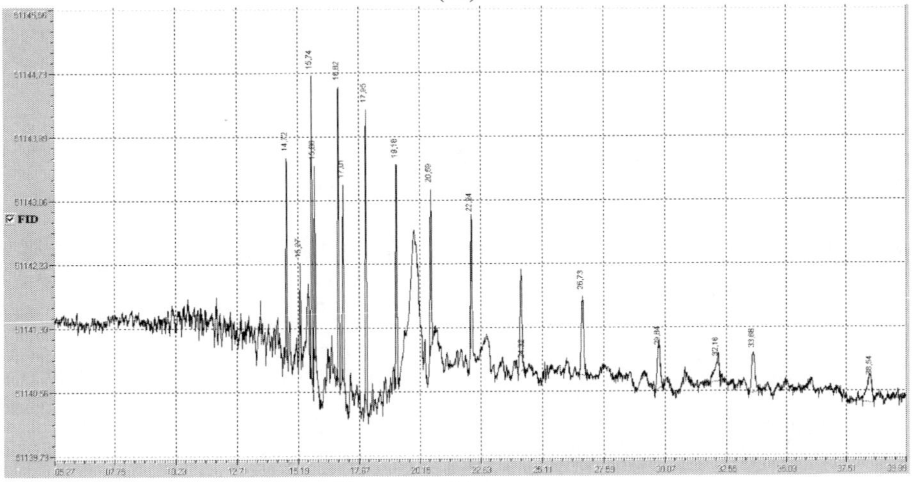

(b)

Figure 4. (Continued)

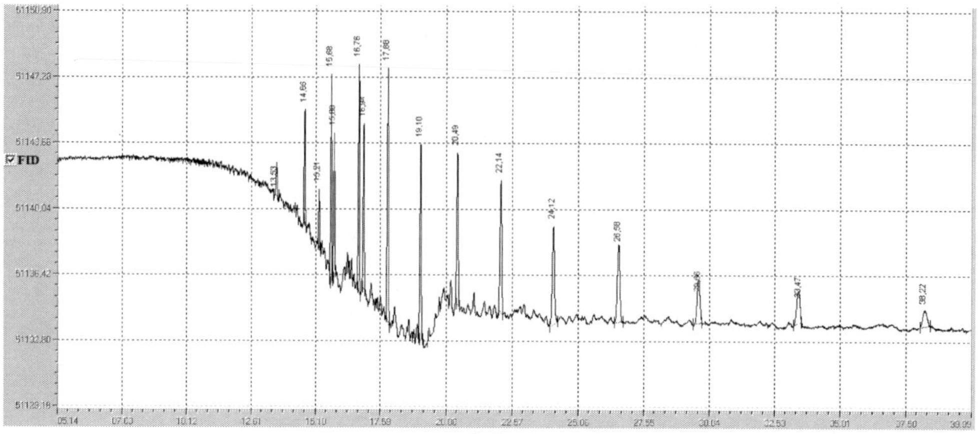

(b1)

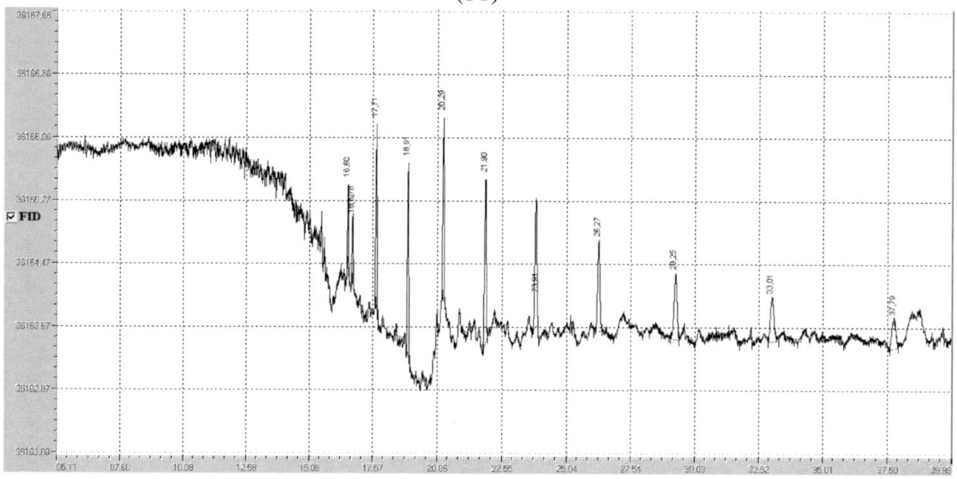

(c)

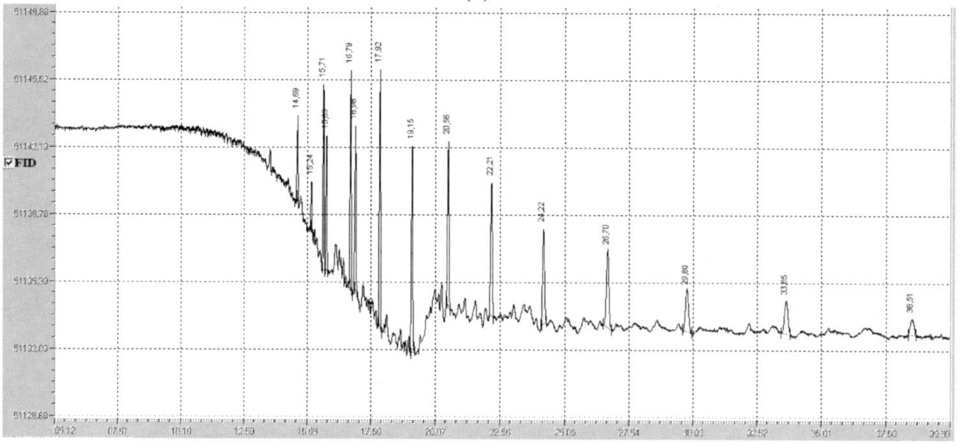

(c1)

Figure 4. (a) GC analysis of HC in the soil with extremely high oil pollution, selected from the depth of 0-10cm before and (a1) after augmentation with the Rhoder; (b) selected from the depth of 10-25cm before and (b1) after augmentation with the Rhoder; (c) GC analysis of HC in the soil samples with an average level of oil pollution, selected from the depth of 25-40cm before and (c1) after augmentation with the Rhoder.

Analysis of the residual HC contamination by HPLC method in the moss after bioaugmentation with the Rhoder showed the oil degradation (Figure 5A-C) in the layers of 0-10 cm and 10-25 cm of the moss (except the layer of 25-40 cm) and confirmed that the degradation of the aromatic HC was observed in these layers of soil. Tentatively the average efficiency of the Rhoder application can be estimated as 55.2±26.2% for not so favorable weather conditions if an average percentage of oil degradation be calculated (Table 5). It is significant that the oil spill on the bog was the fresh (in spring), and the Rhoder was prepared as a liquid concentrate of cells with a high hydrocarbon oxidizing activity ($1.0*10^{11}$ cells per 1mL of the concentrated product).

Thus, the obtained results have shown, on the one hand, that the Rhoder is able to operate in extreme conditions, such as a super high level of the oil pollution under unfavorable weather conditions without milling of moss that useful for the bioremediation at all. On the other hand, despite of the results described above, there was still a lot of oil on the surface of the bog. Multiple repetition of the bioaugmentation with the Rhoder on the bog heavily polluted with oil will be required for several years to fully restore this bog. The bioaugmentation technology described above cannot be considered as an effective one for the restoration of bogs polluted with oil in severe climatic conditions in the northern part of the Western Siberia. It is necessary to develop a new bioremediation technology, may be with using aerobic-anaerobic process of oil biodegradation. Besides, the valuable energy feedstock has been irretrievably lost, and the negative influence on the environment will keep for a long time because of spreading the oil contamination far from the oil polluted sites and inpour into the groundwater.

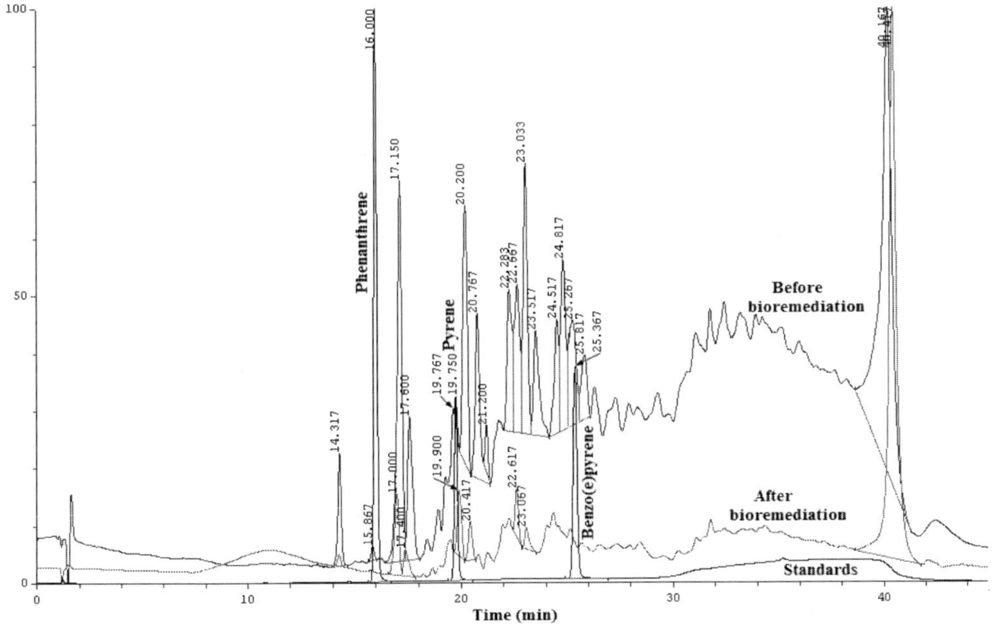

(a)

Figure 5. (Continued)

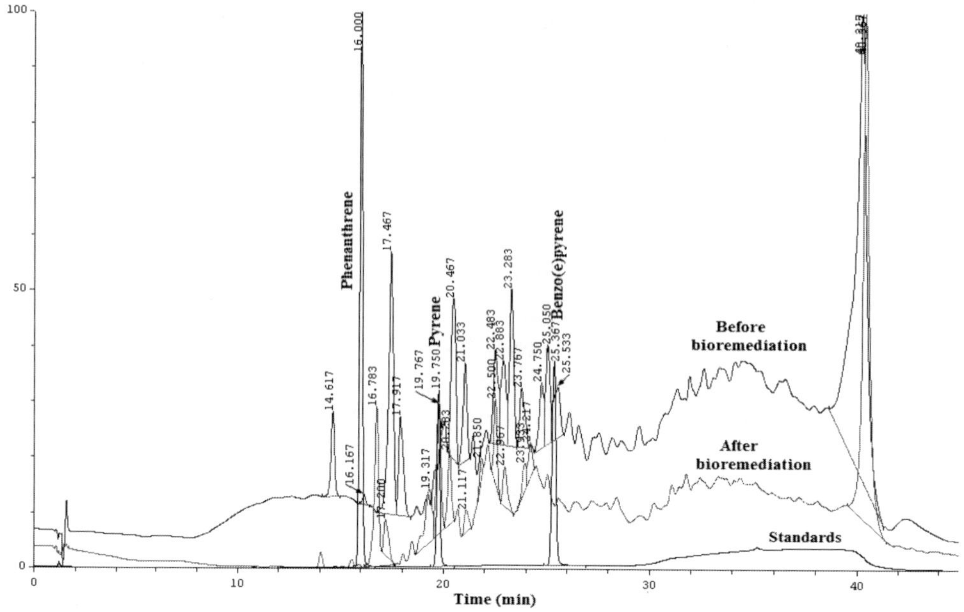

(b)

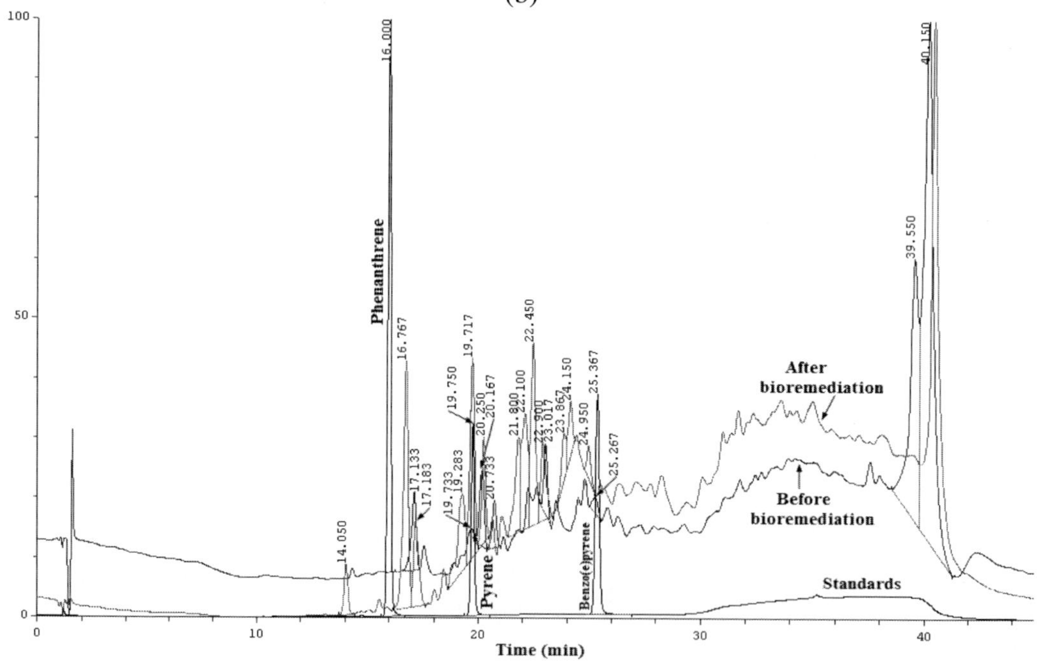

(c)

Figure 5. (a) GC analysis of HC in the soil with extremely high oil pollution, selected from the depth of 0-10cm before and (a1) after augmentation with the Rhoder; (b) selected from the depth of 10-25cm before and (b1) after augmentation with the Rhoder; (c) GC analysis of HC in the soil samples with an average level of oil pollution, selected from the depth of 25-40cm before and (c1) after augmentation with the Rhoder.

4.3. Analysis of the Moss Toxicity after the Bioaugmentation

After the third application of the Rhoder the seeds of oats and a mixture of perennial grasses were sown on the bog polluted with oil to test the phytotoxicity of the moss and possibly for the phytoremediation. Half of the seeds had been watered with the working solution of the humate "Extra". The seeds watered with the humate were sowed on the right side of the bog more polluted with oil. On the left side of the bog the seeds were sowed without watering with the humate. Two small plots were isolated on the right and on the left sides of the bog after ending application of the Rhoder (Figure 2) and covered these plots with non-woven material to protect seedlings against adverse weather conditions. This material passes oxygen and rain moisture to the soil.

Seeds watered with the humate and without it did not grow up at all after six weeks in places with the very high level of the oil pollution on the whole bog. In the left side of the bog where the concentration of oil previously was below 800-900g/kg DM seeds of oats without humate grew up (length of seedlings was about 10 cm) but seeds of perennial grasses mixtures did not grow up at all. On the right side of the bog where seeds watered with the humate were sown, the seeds of oats and especially perennial grasses germinated. Grown seedlings mixtures of perennial grasses were about 7cm in length and possessed strong roots. In plots under non-woven material it was observed the same situation: on the left half of the oil contaminated bog (seeds without the humate) mainly oats seedlings grew. On the right side the grass mixture and a little bit seedlings of oat grew too. And oat, and grass mixture well grew in places of both plots where the level of the oil contamination was below 100g/kg DM.

These results showed that the humate, containing humic and fulvic acids, had a positive effect on the germination and growth of roots of the perennial grasses mixture and practically had no effect on the germination and growth of oats (Avena sativa), which was more resistant to the oil pollution and used for the phytoremediation mainly in the northern part of Russia. Similar results were obtained earlier in laboratory experiments on bioremediation of oil polluted soil with the Rhoder and addition of Pawhumus (Germany) [17].

CONCLUSION

Comparison of two remediation technologies for oil-polluted bogs in the Northern part of Russia has shown that the most effective technology is ex-situ. It allows to do full remediation of soil polluted with oil in a short time, but a cost of this technology is significantly high. In addition ex-situ bioremediation technology allows refining gathered oil and partially offsetting outlays on equipment service and energy expense. The second bioremediation technology (in-situ) cannot be considered as the effective one for impassable bogs polluted with oil behind the Polar Circle in the Western Siberia, because it will really require of 3-4 years or more to restore bogs in severe climatic conditions there. Obtained results showed that the processes of oil biodegradation had begun inside of the bottommost layers of the bog due to indigenous anaerobic microorganisms. So it is necessary to develop in future a new option of bioremediation technology with using aerobic-anaerobic biodegradation of oil for such contaminated bogs which would be more favorable for environment and attractive for economic. Nevertheless, the oil-oxidizing preparation Rhoder during in-situ bioremediation is able to degrade oil (55.2±26.2%) in extreme conditions: a super high level of oil pollution (from 14.4-35.1 kg of oil per kg of dry moss to 516.6 - 43.6g/kg of dry moss) for unfavorable weather conditions without milling which is favorable for any bioremediation.

The humate "Extra" (Russia) containing humic and fulvic acids had a positive effect on the germination and roots growth of perennial grasses mixture and practically had no effect on the germination and growth of oats (Avena sativa) on the bog contaminated with oil.

ACKNOWLEDGMENTS

The authors express their gratitude to Mr. A.B. Kurchenko (Director-General of Joint Stock Co SPASF "Priroda", the Komi Republic) and Mr. I.I. Zhukov (Deputy Director-General of the AIE "Ecoterra", Moscow) for financial support and opportunity to use the oil-oxidizing preparation Rhoder for the bioremediation of natural objects polluted with oil.

REFERENCES

[1] Kurchenko, A. In Proc. 496 Intren. Oil Spill Confer, *Seattle*, 231 (1999).

[2] Kurchenko, A. B. In Proc. Of The Fifth Scientific And Practical Conference: Ecology Works On Oil Fields Of The Timano-Pechorsky Province. Current State And Prospects, *Syktyvkar*, 132 (2008).

[3] Murygina, VP; Arinbasarov, MU; Kalyuzhnyi, SV. *Ecology And Industry Of Russia*, (8), 16 (1999). (In Russian)

[4] Valentina, P; Murygina; Maria, Y. Markarova; Sergey, V. Kalyuzhnyi. *Environmental International*, 31 (2), 163 (2005).

[5] Valentina, Murygina; Maria, Markarova; Sergey, Kalyuzhnyi. In Proc. Of IPY-OSC Symp., Norway, Oslo, (2010). Http://Www.Ipy-Osc.No/

[6] Murygina, V; Gaidamaka, S; Iankevich, M; Tumasyanz^, A. *Progress In Environmental Science And Technology*, III, 791 (2011).

[7] Gally, P; Murygina, VP; Kalyuzhnyi, SV. Proc. Of The 8th International In Situ And On-Site Bioremediation Symposium, Baltimore, MD. 421 (2005).

[8] Kamentschikov, FA; Chernikh, LN; Murygina^ , VP. *Oil Economy*, No. 3, 80 (1998). (In Russian).

[9] Ouyang, W; Yu, Y; Liu, H; Murygina, V; Kalyuzhnyi, S; Xiu, Z. *Process Biochemistry*, 40 (12), 3763 (2005).

[10] Wei, Ouyang; Hong, Liu; Yong-Yong, Yu; Murygina, V; Kalyuzhnyi, S; Zeng-De, Xiu. *Huanjing Kexue/Environmental Science*., 27 (1), 160 (2006).

[11] De-Qing. S; Jian, Z; Zhao-Long, G; Jian, D; Tian-Li, W;Murygina, V; Kalyuzhnyi, S. *Water, Air, And Soil Pollution*, 185 (1-4), 177 (2007).

[12] Yu.S.Drugov; Zenkevich, IG; Rodin, AA. *Gaschromathography Identification Of Air, Water And Soil And Bio-Nutrients Pollutants*. Binom, Moscow 2005, 752 P (In Russian).

[13] Aksenyuk, DA; Gerasimova, CA; Yu.V. Mikhailik. (2009). Lab@Ecocity.Ru
[14] Mineev, VG. (Ed.) *Practical Handbook On Agro Chemistry*. Moscow State University, Moscow, Russia 2001. 688 P (In Russian).
[15] Netrusov, AI; (Ed.) 2005. *Practical Handbook On Microbiology*. Academia, Moscow, Russia (In Russian).
[16] Nazina, T; Rozanova, Ye; Belyayev, S; Ivanov, M. *Chemical And Microbiological Research Methods For Reservoir Liquids And Cores Of Oil Fields*. Preprint Biological Centre Press, Pushchino, 1988, 35 P (In Russian).
[17] Salem, KM; Perminova, IV; Yu Grechischeva, N; Murygina, VP; Mescheryakov, SV. *Ecology And Industry Of Russia*. (4), 19 (2003), (In Russian).

In: News in Chemistry, Biochemistry and Biotechnology ISBN: 978-1-63117-273-1
Editors: G. E. Zaikov, G. Nyszko, L. P. Krylova et al.

Chapter 16

STRONG POLYELECTROLYTE-INDUCING DEMIXING OF SEMIDILUTE AND HIGHLY COMPATIBLE BIOPOLYMER MIXTURES

***Y. A. Antonov*[1] *and Paula Moldenaers*[2]**
[1]N.M. Emanuel Institute of Biochemical Physics, Russian Academy of Sciences, Moscow, Russia
[2]K.U. of Leuven, Department Chemical Engineering, Leuven, Belgium

ABSTRACT

The weak intermacromolecular interactions caused by the presence of a complexing agent in two phase biopolymer mixture can affect its phase equilibrium and morphology. In this communication, the attempt was performed to induce demixing in semidilute and highly compatible sodium caseinate/sodium alginate system (SC-SA) mixtures in the presence of sodium salt of dextran sulfate (DSS) at pH 7.0, (above the isoelectrical point of caseins), and to characterize phase equilibrium, intermacromolecular interactions, and structure of such systems by rheo-small angle light scattering (SALS), optical microscopy (OM), phase analysis, dynamic light scattering (DLS), fast protein liquid chromatography (FPLC), ESEM, and rheology. Addition of dextran sulfate sodium salt (DSS) to the semidilute single phase SC-SA system, even in trace concentrations (10^{-3} wt %), leads to segregative liquid-liquid phase separation, and a substantial increase in storage and loss moduli of the system. The degree of the protein conversion in the complex grows, when the concentration of SC in the system increases from 1 to 2 wt%. It is also established here that demixing of semidilute biopolymer mixtures, induced by the minor presence of DSS is a rather common phenomenon, because its also was observed here for other biopolymer pairs. At high shear rates SC becomes even less compatible with SA in the presence of DSS than at rest. Experimental observations suggest that the approach for inducing demixing of semidilute and highly compatible biopolymer mixtures by physical interactions of the constituents is a promising tool for regulation of biopolymer compatibility and achieving better predictions of phase behavior of aqueous protein-charged polysaccharide systems.

Keywords: Biopolymer mixture, demixing, complex formation, structure formation, rheo-optics

INTRODUCTION

The importance of the phase behavior in biopolymer mixtures is evident in many technological processes, such as isolation and fractionation of proteins (see, for example, [1-6]), and enzymes [7], enzyme immobilization [8,9], encapsulation [10] and drug delivery [9,11]. Aqueous, two phase systems are used in modern technological processes where clarification, concentration, and partial purification are integrated in one step [12]. Thermodynamic incompatibility, or, in other words, segregative phase separation, determines the structure and physical properties of biopolymers mixtures in quiescent state [13-15] and under flow [16-18] and plays an important role in protein processing in food products [14].

From a technological point of view, especially important are biopolymer systems which undergo liquid-liquid phase separation in a wide concentration range, starting from low concentrations [19]. But whether phase separation is desired or not, it is important for practical applications to understand the underlying mechanisms and molecular interactions governing the phase behavior of a given system [20]. Despite the considerable amount of research in the field of segregating polymer mixtures, the molecular interactions in the systems are inadequately understood, although theoretical models have been proposed. [21-28]. There have, as of yet, been comparatively few studies on phase separation in mixtures of similarly charged polyelectrolytes[29,30]. Such systems may have advantages over uncharged systems in the separation of proteins due to the tunable charge in the system arising from the dissociated counter ions of the polyelectrolytes.[29,30]. Although the majority of biopolymer mixtures show phase separation [14,32], in most cases the phase separation takes places at critical total concentrations, which are much higher (7-12 wt%) [31,32] compared with those of synthetic polymers (less than 1-2 wt%). Unlike synthetic polymers with flexible chains, many proteins are known to be relatively symmetric compact molecules and are usually able to form solutions that can still be considered dilute for concentrations 10-fold higher than for synthetic polymers of the same molecular weight [33].

The aims of this study to induce demixing in semidilute and highly compatible sodium caseinate/sodium alginate system (SC-SA) mixtures in the presence of sodium salt of dextran sulfate (DSS) at pH 7.0, (above the isoelectrical point of caseins), and to characterize phase equilibrium, intermacromolecular interactions, and structure of such systems by rheo-small angle light scattering (SALS), optical microscopy (OM), phase analysis, dynamic light scattering (DLS), fast protein liquid chromatography (FPLC), ESEM, and rheology. The molecular weight, charge, and topography of the accessible surface of water soluble complexes of proteins with anionic polysaccharides are differ markedly from the "free" proteins. Therefore it can be assumed that all these factors may affect the phase separation. In the present work, we focus our study on the phase transitions in aqueous semidilute homogeneous sodium caseinate/sodium alginate systems (SC-SA) with the total concentration of biopolymers 1,5 wt%-2.5 wt%, i.e., much below the critical concentrations for phase separation [17]. The phase state of the SC-SA mixtures is not sensitive to changes in pH, ionic strength and temperature in the quiescent state [31,32] and under of shear flow [17].

Therefore, the effect of demixing that can be reached for this system can be easily reproduced for other emulsions in which the phase equilibrium is more sensitive to physicochemical parameters. Here it will be explored how far this strategy of demixing can be extended to other biopolymer pairs. For this reason gelatin-SA and gelatin –SC systems will be investigated to assess the generality of our observations. In addition, the shear induced behavior of the decompatibilized semidilute SC-SA system will be presented and compared with that of the "native" SC-SA system.

Alginate is an anionic polysaccharide consisting of linear chains of (1–4)-linked ß-D-mannuronic and-α-L-guluronic acid residues. These residues are arranged in blocks of mannuronic or guluronic acid residues linked by blocks in which the sequence of the two acid residues is predominantly alternating [33,34]. Casein is a protein composed of a heterogeneous group of phosphoproteins organized in micelles. These biopolymers are well known, widely used in industry for their textural and structuring properties [14,31,32,33,35], and the thermodynamic behavior of the ternary water–caseinate–alginate systems is known from literature [17,31,32,35].

II. Materials and Methods

The caseinate at neutral pH is negatively charged, like alginate, and DSS. The sodium caseinate sample (90% protein, 5.5% water content, 3.8% ash, 0.02% calcium) was purchased from Sigma Chemical Co. The isoelectric point is around pH = 4.7–5.2 [36].The weight average molecular mass of the sodium caseinate in 0.15 M NaCl solutions is 320 kDa. The medium viscosity sodium alginate, extracted from brown seaweed (*Macrocystis pirifera*), was purchased from Sigma. The weight average molecular weight of the sample, M_w was 390 kDa [16]. Dextran sulfate, DSS (M_W = 500 kDa, M_n = 166 kDa, η (in 0.01 M NaCl) = 50 mL/g, 17% sulfate content, free SO_4 less than 0.5%) was produced by Fluka, Sweden (Reg. No. 61708061 A, Lot No. 438892/1). The gelatin sample used is an ossein gelatin type A 200 Bloom produced by SBW Biosystems, France. The Bloom number, weight average molecular mass and the isoelectric point of the sample, reported by the manufacturer are respectively 207, 99.3 kDa, and 8-9.

Preparation of the Protein/Polysaccharide Mixtures

Most experiments were performed in the much diluted phosphate buffer (ionic strength, I= 0.002). To prepare molecularly dispersed solutions of SC, SA, gelatin, or DSS with the required concentrations, phosphate buffer (Na_2HPO_4/NaH_2PO_4, pH 7.0, I=0.002, and 0.015) was gradually added to the weighed amount of biopolymer sample at 298 K, and stirred, first for 1 h at this temperature and then for 1 h at 318 K. The solutions of SC, SA, and DSS were then cooled to 296K and stirred again for 1 h. The required pH value (7.0) was adjusted by addition of 0.1–0.5M NaOH or HCl. The resulting solutions were centrifuged at 60,000 *g* for 1 h at 296K, or 313 K (gelatin solutions) to remove insoluble particles. Concentrations of the solutions are determined by drying at 373K up to constant weight. The ternary water–SC–SA systems with required compositions were prepared by mixing solutions of each biopolymer at

296 K. After mixing for 1 h, the systems were centrifuged at 60,000 *g* for 1 h at 296K to separate the phases using a temperature-controlled rotor.

Determination of the Phase Diagram

The effect of the presence of DSS on the isothermal phase diagrams of the SC-SA system was investigated using a methodology described elsewhere [36]. The procedure is adapted from Koningsveld and Staverman [37] and Polyakov et al. [38]. The weight DSS/SC ratio in the system, (q) was kept at 0.14. The threshold point was determined from the plot as the point where the line with the slope −1 is tangent to the binodal. The critical point of the system was defined as the point where the binodal intersects the rectilinear diameter, which is the line joining the centre of the tie lines.

Rheo-Optical Study

A rheo-optical methodology based on small angle light scattering (SALS) during flow, is applied to study in-situ and on a time-resolved basis the structure evolution. Light scattering experiments were conducted using a Linkam CSS450 flow cell with a parallel-plate geometry. A 5 mW He-Ne laser (wavelength 633 nm) was used as light source. The 2D scattering patterns were collected on a screen by semi transparent paper with a beam stop and recorded with a 10-bit progressive scan digital camera (Pulnix TM-1300). Images were stored on a computer with the help of a digital frame grabber (Coreco Tci-Digital SE). The optical acquisition set-up has been validated for scattering angles up to 18°. The gap between the plates has been set at 1 mm and the temperature was kept constant by means of a thermostatised water bath. In house developed software was used to obtain intensity profiles and contour plots of the images (New SALS SOFT-WARE-K.U.L.). Turbidity measurements have been performed by means of a photo diode. *Microscopy observations* during flow have been performed on a Linkham shearing cell mounted on a Leitz Laborlux 12 PolS optical microscope using different magnifications.

Rheological measurements were performed using a Physica Rheometer, type CSL2 500 A/G H/R, with a cone-plate geometry CP50-1/Ti ~diameter 5 cm, angle 0,993°, Anton Paar. The temperature was controlled at 23 °C by using a Peltier element. For each sample, flow curves were measured at increasing shear rate ~from 0.1 to 150 s^{-1}. The ramp mode was logarithmic and the time between two measurements was 30 s. Frequency sweeps ~0.1–200 rad/s were carried out as well for a strain of 3.0%, which was in the linear response regime. During the rheological measurements, all samples were covered with paraffin oil to avoid drying.

Dynamic Light Scattering

Determination of Intensity- weighted distribution of hydrodynamic radii (R_H) of SC, SA, and DSS solutions and their mixtures was performed, using the Malvern ALV/CGS-3 goniometer. Concentration of the protein in protein-dextran sulfate mixtures was kept at 0.1

(w/w). For each sample the measurement was repeated 3 times. The samples were filtered before measurement through DISMIC-25cs (cellulose acetate) filters (sizes hole of 0.22 μm for the binary water-casein and water-dextran sulfate solutions and 0.80 μm for the protein-polysaccharide mixtures). Subsequently the samples were centrifuged for 30 seconds at 4000 g to remove air bubbles, and placed in the cuvette housing which was kept at 23°C in a toluene bath. The detected scattering light intensity was processed by digital ALV-5000 Correlator software. The second order cumulant fit was used for the determination of the hydrodynamic radii. The asymmetry coefficient (Z) of the complex particles was estimated by Debye method based on determining the scattering intensity at two angles 45° and 135°, symmetrical to the angle 90°.

Zeta Potential Measurement

The ζ–potential measurements of SC and DSS solutions and their mixtures at different q values were performed at 23°C with a Malvern-Zetamaster S, model ZEM 5002 (England), using a rectangular quartz capillary cell. The concentration of the protein in solutions was 0.1 wt%, and the concentrations of DSS in the protein-polysaccharide solutions were variable. All solutions were prepared in phosphate buffer (Na_2HPO_4/NaH_2PO_4, pH 7.0, I=0.002).The zeta potential was determined at least three times for each sample. The zeta potential was calculated automatically from the measured electrophoretic mobility, by using the Henry equation:

$$U_e = \varepsilon z\rho f/6\pi\eta, \quad (1)$$

where U_e is electrophoretic mobility, ε is the dielectric constant, η is the viscosity and $z\rho$ is the zeta potential. The Smoluchowski factor, f =1.5 was used for the conversion of mobility into zeta potential.

Environment Scanning Electron Microscopy (ESEM)

Micro structural investigation was performed with the environment scanning electron microscope Philips XL30 ESEM FEG. The instrument has the performance of a conventional SEM but has the additional advantage that practically any material can be examined in its natural state. The samples were freeze-fractured in freon and immediately placed in the ESEM. Relative humidity in the ESEM chamber (100%) was maintained using a Peltier stage. Such conditions were applied to minimize solvent loss and condensation, and control etching of the sample. Images were obtained within less than 5 minutes of the sample reaching the chamber. The ESEM images were recorded multiple times and on multiple samples in order to ensure reproducibility.

Fast Protein Liquid Chromatography (FPLC)

Solutions of sodium caseinate, (0.5 wt %), dextran sulfate (0.5 wt%) and their mixtures, containing 0.5 wt% of the protein and variable amount of dextran sulfate were applied on a Superose 6 column (HR 10/30), Amersham Biosciences mounted on an FPLC apparatus (Pharmacia, Uppsala, Sweden). Elution was performed at room temperature with phosphate buffer (5mM Na_2HPO_4/NaH_2PO_4, pH 7.0) 2% (v/v) n-propanol (Riedel-de Haen, Seelze, Germany)) and 0.015 M NaCI. The samples and the elution buffer were filtered through a 0.22 um sterile filter. The flow rate was 0.2 mL min-1 and the column was monitored by UV detection at 214 nm.

Determination of Dextran Sulfate Content

The phenol-sulphuric acid method of Dubois et al. [39] was applied. 50 uL. 80% (w/w) phenol in water and 5mL sulphuric acid were added to the measured samples of 0.5 mL. After 30 min at room temperature the absorbance at 485 nm was measured. A calibration plot was constructed with D-glucose (Riedel-de Haen).

III. Results and Discussion

A. DSS-Induced Demixing

The experimental results shown in this section have been obtained on water (97.5 wt %)-SC (2.00 wt %)-SA (0.5 wt %) semidilute systems. This system is located in the one-phase region far from the binodal line. To study the effect DSS on the phase behavior, a flow history consisting of two shear zones is used. First, a preshear of 0.5 s^{-1} is applied for 1000 s (500 strain units) to ensure a reproducible initial morphology. Subsequently, this preshear is stopped, and the sample is allowed to relax for 30 s leaving enough time for full relaxation of deformed droplets. Then SALS patterns are monitored.

The SALS patterns and the scattering intensity upon adding different amount of DSS are shown in Figures 1 (a-f), and 2, starting from a concentration of DSS as low as 2.08 10^{-3} wt%. In the absence of DSS no scattered light is observed (data are not presented). The presence of even only 2.08 10^{-3} wt% DSS in the homogeneous system led to appreciable increase the SALS pattern (Figure 1) and accordingly the light scattering intensity (Figure 2). It is important to note that that the SC-DSS system remains homogeneous in the DSS concentration range studied here. Centrifugation of the SC-DSS systems (120 min. 60.000 g, 296 K) prepared at the same conditions did not show phase separation. When the DSS concentration in the SC-SA system increases the SALS pattern (figure 1) and the scattering intensity (figure 2) of the system sharply grows. This indicates that the position of the system on the phase diagram changes deeply into the two phase range. The corresponding microscopy images for the same concentrations of DSS and the same flow conditions are shown in Figure 3. One can see that the phase separation led to formation of liquid-liquid emulsions. At the lowest DSS concentration (2.08 10-3 wt %), the system contains ultra small droplets of the dispersed phase having a size of 2-3 μm. At higher DSS concentrations the size of the droplets increases significantly in agreement with SALS data achieving more than 50 μm. in diameter.

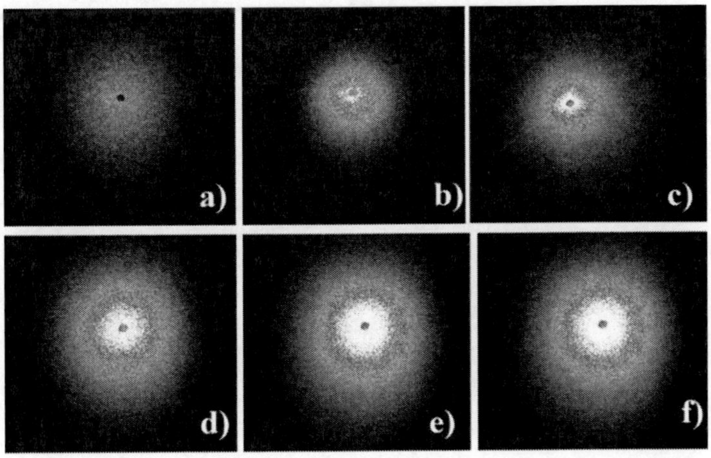

Figure 1. Effect of the concentration of DSS on the SALS patterns of water (97.5 wt %)-SC (2.00 wt %)-SA (0.5 wt %) single-phase systems. pH 7.0. I=0.002 (phosphate buffer). Temperature 296 K. Concentrations of DSS in mixture, wt%: (a) 2.08×10^{-3}, (b) 4.10×10^{-3}, (c) 1.61×10^{-2}, (d)$7{,}50\times10^{-2}$, (e) 0.15, (f) 0.29, and resulting DSS/SC ratio: (a) 0.001, (b) 0.002, (c) 0.008, (d) 0.0375, (e) 0.075, (f) 0.145.

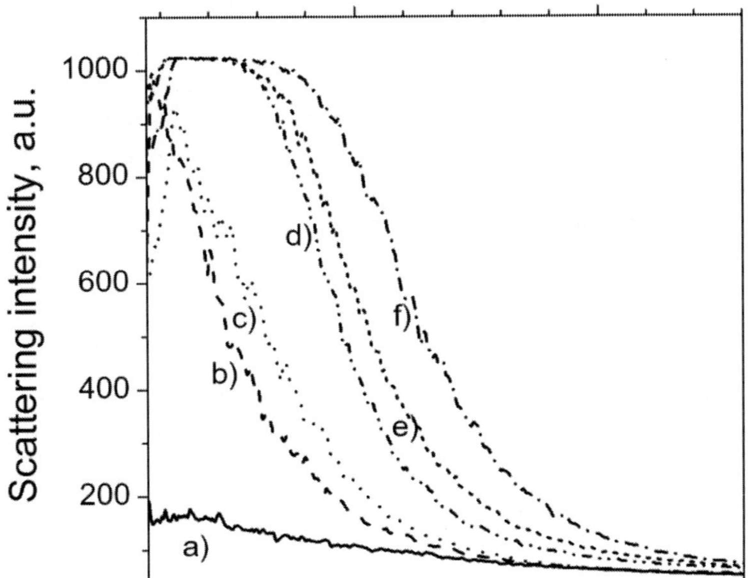

Figure 2. Effect of the concentration of DSS on the scattering intensity of water (97.5 wt %)-SC (2.00 wt %)-SA (0.5 wt %) single-phase systems as a function of the distance from the bean stop. The other parameters are the same as in Figure1.

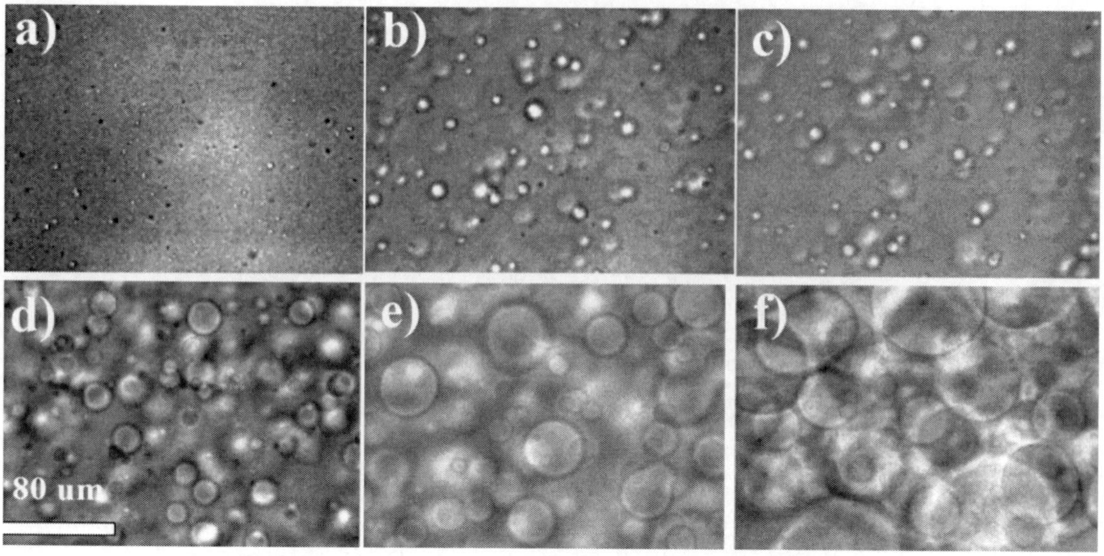

Figure 3. Microscopy images of water (97.5 wt %)-SC (2.00 wt %)-SA (0.5 wt %) system after addition of different amounts of DSS. pH 7.0, *I*=0.002 (phosphate buffer). Temperature 296 K. The other parameters are the same as in Figure1.

In order to quantify the effect of DSS on phase equilibrium in semidiluted SC-SA system, the isothermal phase diagram of the system was determined in the presence of DSS, at DSS/SC weight ratio (q) =0.14, plotted in the classical triangular representation, and compared with that obtained in the absence of DSS (Figure 4). The phase separation in the presence of DSS has a segregative character with preferential concentrating of SC and SA in different phases. The phase diagram of the initial system, without DSS, is characterized by a high total concentration of biopolymers at the critical point (C_c^t = 62.9 g/L), and a strong asymmetry (*K*s = 15.5). The presence of DSS affects dramatically the phase separation, significantly increasing the concentration range corresponding to two phase state of the system. The total concentrations of biopolymers at the critical point decreases to 10.6 g/L. The phase separation is observed at total concentrations of biopolymers just above 1 wt%, i.e., level of compatibility of the biopolymers after an addition of DSS seems to be one of the smallest known for biopolymer mixtures (see, for example, [40,41]). The decrease in compatibility of casein and alginate is especially surprising when taking into account that the phase composition of this system is weakly dependent on many physicochemical factors, such as pH (in the pH range from 7 to 10), ionic strength and temperature (from 5 to 60°C) [17,32,34].

B. Rheological Behavior of the Demixed Systems

For the rheological investigations, the homogeneous W-SC (2.0 wt%)-SA (0.5 wt %) system (point A on the phase diagram, Figure 4) was characterized before, and after addition of DSS at the DSS/SC weight ratio, q=0.045, and q=0.15 respectively. The latter two systems were two phase ones with the content of the casein enriched phase 15 w/w, and 55% w/w accordingly. The experimental flow protocol applied was the same as the one used for rheo-

SALS. The mechanical spectrum and flow curve were determined in order to characterize the state of the systems through their viscoelastic behaviors. It has been shown [42] that at moderately low shear rates, the biopolymer emulsions can be regarded as conventional emulsions and various structural models that are available in the literature for prediction of the morphology in these emulsions can also be used for prediction of the structure in aqueous biopolymer emulsions.

The evolution of the mechanical spectrum was investigated as a function of DSS concentration. These viscoelastic behaviors were monitored and compared with the behavior of the W-SC-SA system without DSS. The dynamic modulus *G*' (elastic) and *G*" (viscous) were measured with frequency sweep experiments at a constant strain of 3%, which was checked as being in the linear regime. The obtained data are presented in Figure 5.

For the single phase system, and the system containing 0.09 wt% DSS, G' was too low to be measured accurately. Under these conditions the system behaves as purely viscous liquid with the curve of G" versus frequency displaying a slope of one on a double logarithmic graph. In the present of DSS the system undergoes phase separation, and this transition leads to an appreciable increase of the moduli. The elastic properties of the decompatibilized W-SC-SA system were mainly induced by the presence of the DSS. In the presence of high ionic strength (0.25, NaCl), when electrostatic interactions were suppressed the mechanical spectrum of the system (q=0.14) becomes insensitive to the presence of DSS (data are not presented). Flow curves determined at the same concentrations show an increase in viscosity for the demixed systems, especially remarkable at a low shear rates (Figure 6).

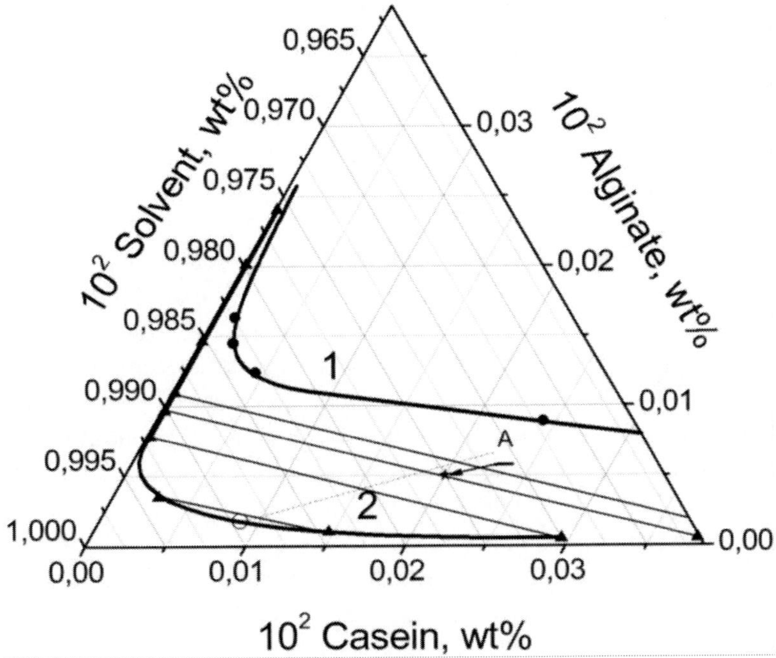

Figure 4. Isothermal phase diagrams of the W-SC-SA system. pH 7.0, I=0.002 (phosphate buffer), 296 K. 1. In the absence of DSS. 2. In the presence of DSS, at DSS/SC weight ratio (q) =0.14.

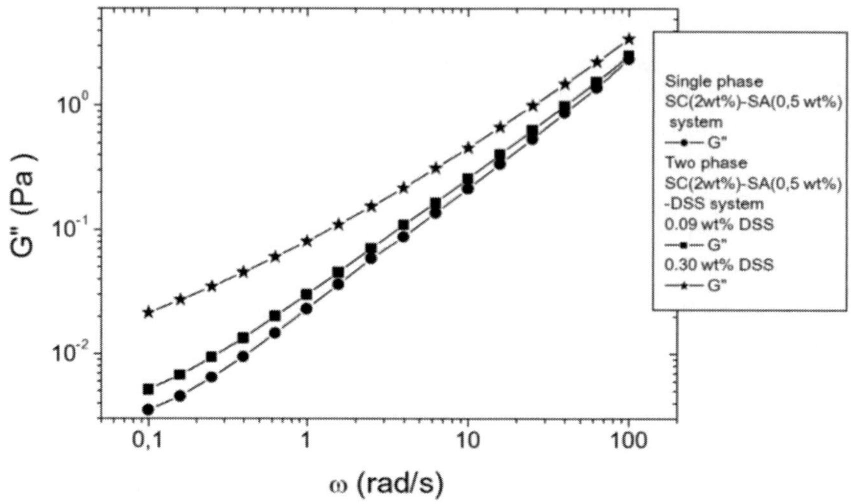

Figure 5. Dynamic spectra of single phase W-CS (2 wt%)-SA (0.5 wt %) system, and two phase W-CS (2 wt%)-SA (0.5 wt%)-DSS systems. pH 7.0, *I*=0.002 (phosphate buffer), 296 K.

More detailed experiments were then carried out on the single phase W-SC (4 wt%)-DSS (variable), and W-SA (0.5 wt%)-DSS (variable) systems to understand how DSS affects the mechanical spectrum of the casein and alginate solutions, and accordingly the coexisting phases. The behavior of these solutions in the presence of sulfated polysaccharide is clearly different (Figures. 7 and 8); the casein-enriched phase is sensitive to the presence of DSS, while the viscoelastic properties of the alginate-enriched phase in the presence of DSS remain almost unaltered. As reported in Figure 7 a, the dependence of the G" on the DSS/casein ratio has an extreme character, with a maximum at a DSS/casein ratio around 0.14. In the presence of even small amounts of DSS (0.01-0.05 wt %), a dramatic increase of the G" of the emulsion takes place. Thus, in the presence of 0.5 wt% of DSS (at q=0.14) and at a frequency 1 rad/s, G" values is more than 1400 times, higher compared with those of the single phase system with almost the same composition. From theory we know that such dependences are typical for the formation of inter-polymer complexes [42]. Similar changes were observed for the viscosity (Figure 8 a,b). At q=0.14 and a shear rate of 10 s^{-1} the viscosity is more than 940 times higher compared with those of the single phase system with almost the same composition (Figure8 b). It is important to note that in the shear rate range from 0.1 to 150 s^{-1} we did not find any difference in the flow curves obtained in conditions with increasing versus decreasing shear rate (data are not presented). It can be assumed that the dramatic changes in rheological behavior of the casein-alginate system in the presence of DSS are due to interactions of the casein molecules with the DSS molecule. It can be suggested that casein interacts with DSS, and this interaction may have an effect on the phase separation. Note, that the viscosity of the demixed system in figure 8 decreased from 5.72 to 1.74 Pa s with increasing shear rate from 0.1 to150 s^{-1}, which highlight the shear thinning behavior of the demixed system, indicating a structural change. The result was striking since most concentrated protein-polysaccharide mixtures can be shear thinning only due to the polysaccharide relaxations. In the absence of structure-induced formation the rheological behavior of concentrated polysaccharide solutions is monotonically shear thinning; the viscosity varies between two extremes η_o and η_∞. A possible additional mechanism would be the breakdown of structures due to the breakup of physical bonds at high shear. This structure was most probably due to the electrostatic interactions between SC and DSS. Indeed, in the presence of 0.25 M NaCl, when no attractive interaction took place, no shear thinning behavior was observed (data are not shown). More detailed experiments were then carried out to understand the mechanism of demixing.

C. Intermacromolecular Interactions and the Mechanism of Demixing in SC-SA-DSS System

An important property of the demixed semidilute SC-SA systems described above is their high stability against homogenization and low sensitivity to change in temperature. Thus, for the mixtures with different composition we observed constancy of absorption values at 500 nm during 6 h storage, as well as in processes of their heating from +5°C to 70°C. The results obtained (Figs. 7 and 8) show the presence the intermacromolecular interactions between SC and DSS. Usually coulomb protein-polysaccharide complexes are formed only in the vicinity of the isoelectric point of the protein [44], but for several systems formation of soluble protein – polysaccharide complexes has been registered even at pH 6.-8.0 [45-47]. A beneficial consequence of complexation of sulfated polysaccharide with caseins at pH values above IEP is the protection afforded against loss of solubility as a result of protein aggregation during heating or following high-pressure treatment [48,49]. The mechanism of this protection has been unclear until now. Snoeren, Payens, Jevnink, and Both, assumed [50] that there is a nonstatistical distribution of positively charged amino acid residues along the polypeptide chain of kappa casein molecules and, as a consequence, the existence of a dipole interacting by its positive pole with sulfur polysaccharide is responsible for complex formation in such systems.

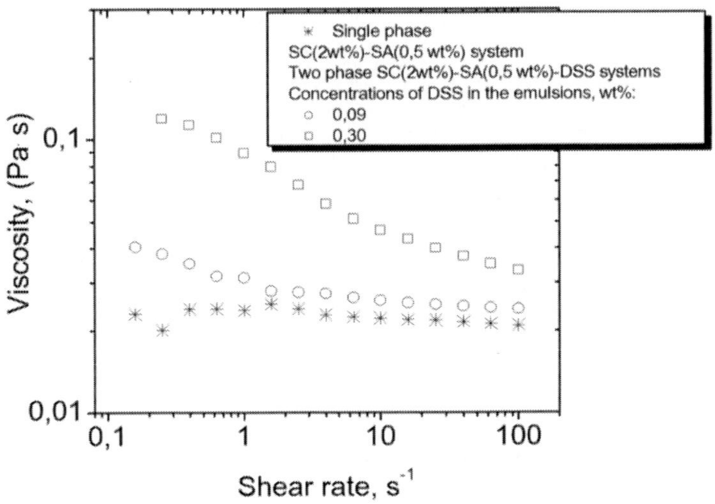

Figure 6. Flow viscosity of single phase W-CS (2 wt%)-SA (0.5 wt %) system, and two phase W-CS (2 wt%)-SA (0.5 wt%)-DSS systems, after application of increasing shear rates. pH 7.0, I=0.002 (phosphate buffer), 296 K.

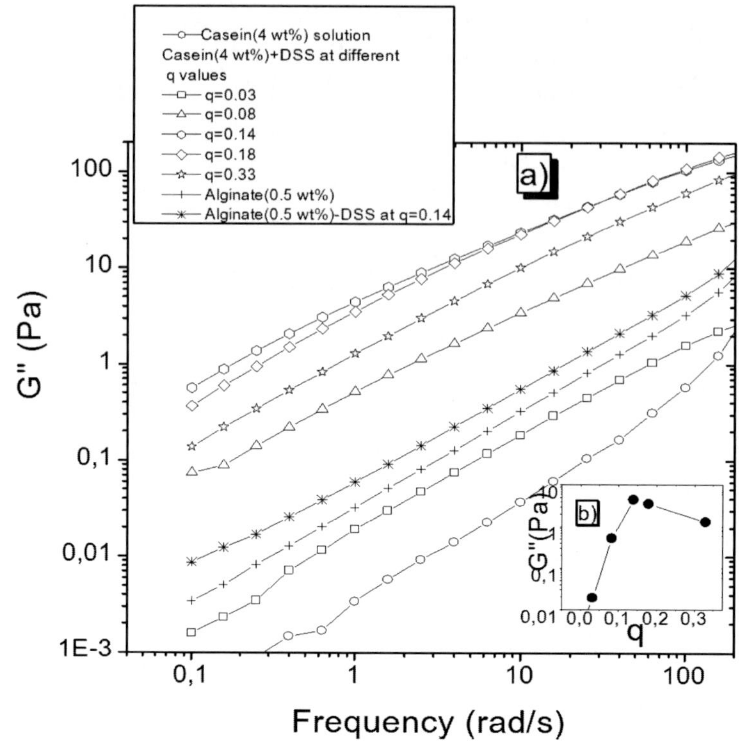

Figure 7. (a), G" of W-SA(0.5 wt%),W-SA(0.5 wt%)-DSS ,and W-SC(4 wt%) –DSS, systems at different q values ,(b) the dependence of G" on q values for W-SC(4 wt%) –DSS, system at frequency 1.0 rad/s. pH 7.0, *I*=0.002 (phosphate buffer), 296 K.

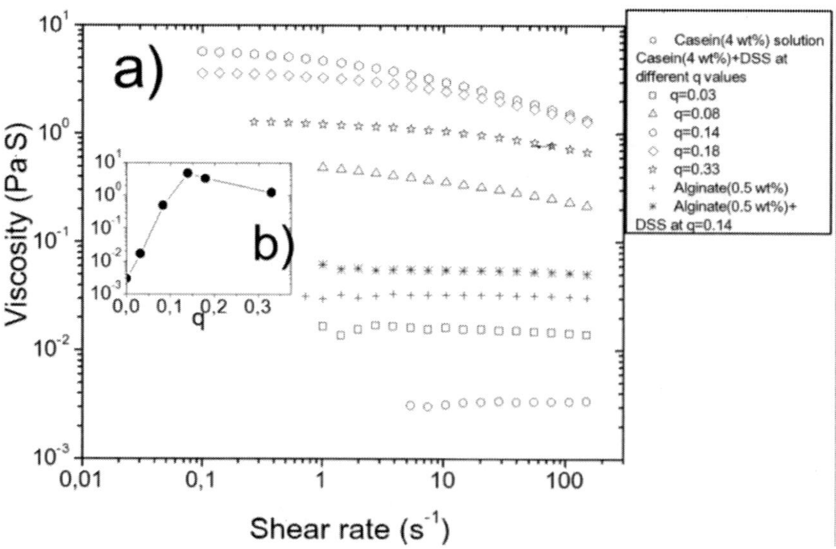

Figure 8. (a) Dependences of flow viscosity of W-SC(4 wt%), W-SA(0.5 wt%), and W-SC (4 wt%) – DSS (var) systems (b) and the dependence of flow viscosity of the W-SC-DSS system on q values at shear rate 1.0 s^{-1}. pH 7.0, *I*=0.002 (phosphate buffer), 296 K.

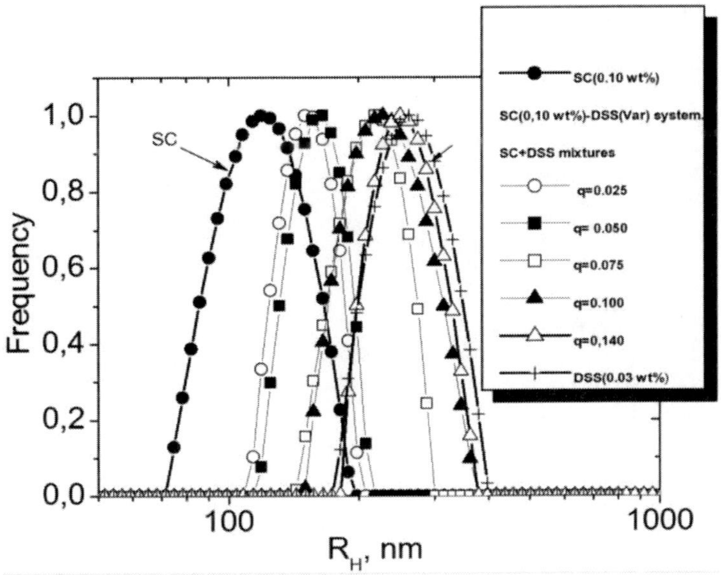

Figure 9. The intensity- weighted distribution of hydrodynamic radii (R_H) of solutions of sodium caseinate, dextran sulfate and their mixtures. Concentration of SC is equal to 0.1 (w/w). pH 7.0, *I*=0.002 (phosphate buffer), 296 K.

Many scientists suppose [51,52] that nonelectrostatic forces, hydrophobic and (or) hydrogen bonds, play a determinant role in this process. In the case of sulfated polysaccharides this assumption is confirmed by experimental data showing the capacity of the sulfate groups to form hydrogen bonds with the protein cationic groups [53].

Introduction of NaCl in the initial buffer results in full insensitivity of the viscosity and the phase diagram of the SC-SA system to the presence of DSS in all the *q* range studied. On the other hand, an addition of 0.2 M NaCl in the SC-SA-DSS system at q=0.14 after a 24 h storage results in a sharp increase in the level of compatibility of SC with SA to that of SC-SA solution alone. This shows that the complexes are formed and stabilized via electrostatic interaction, rather than through hydrogen bonds formation or hydrophobic interaction. The role of salt is to "soften" the interactions, which is equivalent to making the electrostatic binding constant smaller.

To study intermacromolecular interactions in the process of demixing of the SC-SA system, at first, we focus our attention to the interaction between SC and DSS in aqueous solutions within the region of pair interaction. To this aim, we have chosen SC and DSS concentrations low enough to exclude or considerably diminish effects of possible aggregation. This allows us to single out information on interaction processes between the two types of macromolecules, well separated from the subsequent aggregation process. DLS can provide information about the hydrodynamic radius of proteins and polysaccharides and about the binding of ligands to these types of macromolecules. Figure 9 shows the intensity-weighted distribution of hydrodynamic radii (R_H) of solutions of sodium caseinate, dextran sulfate and their mixtures with the concentration of the protein equal to 0.1 (w/w),i.e., at the total concentrations below the critical concentration of phase separation of SC-SA system (see Figure 4). At 296 K, molecules of SC and DSS have R_H values 119 nm and 250 nm respectively.

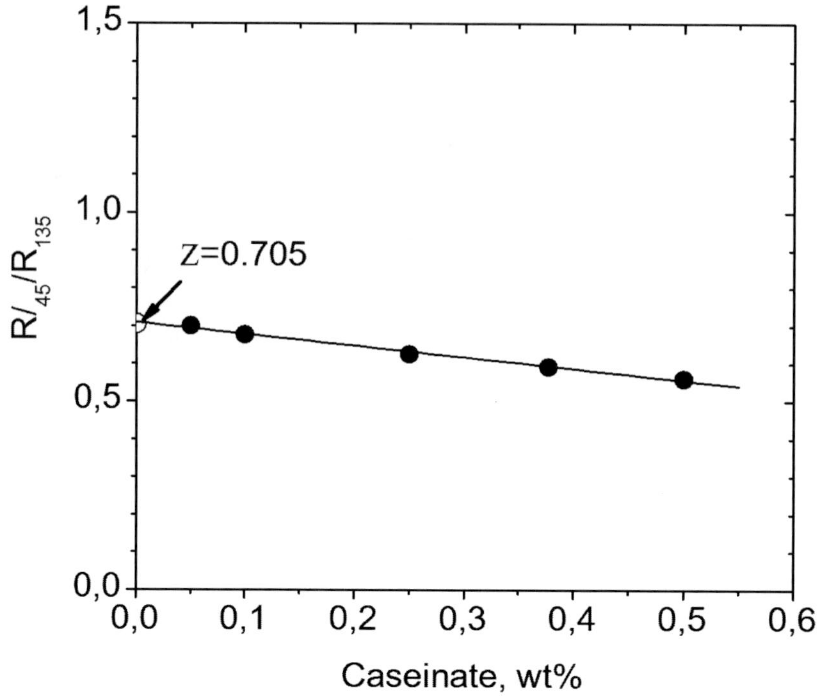

Figure 10. Dependence of the ratio of the scattering intensity, R at 45° and 135° on the concentration of casein in the SC-DSS mixture at q=0.14. pH 7.0, *I*=0.002 (phosphate buffer), 296 K.

An addition of DSS to SC solution at DSS/SC weight ratios ranging (q) from 0.025 to 0.05 leads to significant increase in the R_H toward the values R_H for DSS solution. At higher q values = 0.14, R_H of the mixed associates achieve the values R_H for DSS, and their size does not change with the further increase of q values. This is an indication of intermacromolecular interaction of the casein molecules with DSS and formation of complexes. At q=0.14, function of the intensity-weighted distribution of hydrodynamic radii (R_H) is placed completely outside that describing free SC.

The asymmetry coefficient (Z) of the complex associates was estimated by Debye method based on determination of the scattering intensity,(R) at two angles 45° and 135°, symmetrical to the angle 90° and subsequent extrapolation of the R45°/R/135° to zero concentration. The results obtained are presented in Figure 10. The complex associates are asymmetric with Z values equal to 0.7.

Figure 11 presents Zeta potential values and the total concentration of the biopolymer at the critical point C^t_{cr} as a function of the DSS/SC ratio, q. After an addition of DSS the negative value of the zeta potential increases and C^t_{cr} decreases achieving correspondingly the maximal and minimal values at q =0.14.

Once the negative charge of a protein becomes higher in the presence of DSS, interactions between casein molecules could be hindered by an overall effect of electrostatic repulsion. Thus, an increase in the net charge of casein due to DSS binding could lead to an enhancement in the extent of such repulsions, contributing to the suppression of the further association and aggregation. Obviously, at pH = 7.0, the total charge of the high molecular weight DSS molecule is higher than the total positive charge of the relatively small SC molecule. This gives the possibility to regard complex formation between these biopolymers (similarly to other weak-polyelectrolyte– strong-polyelectrolyte interactions [54,55]) as a mononuclear association in which the DSS molecule is the nucleus and the casein molecule is a ligand. Therefore, the formation of casein–DSS complex can be regarded as the reaction of few casein molecules successively joining one molecule of DSS nucleus. Note that at pH = 7.0, (experimental conditions) all the cationic groups of casein, as well as all the sulfate groups of DSS are ionized. It easy to show that at $q^o = q^*$ (0.135), the ratio of the total amount of sulfate groups in DSS molecule and cationic groups in casein molecule ($\frac{[S^-]}{[Cat^+]}$) is close to unity. Actually, the total amount of cationic groups in casein molecule is 0.76 mmol/g [50,56] and the content of sulfur groups in DSS molecule is equal to 5.43 mmol/g [57]. Therefore $\frac{[S^-]}{[Cat^+]} = q\,\frac{5.43}{0.76}$. At $q^* = 0.135$ one can obtain $\frac{[S^-]}{[Cat^+]} = 0.964$.

Figure 12 presents the chromatograms of the initial solutions of SC (0.25 wt %) and DSS (0.25 wt %), and the SC- DSS system (q=0.14, concentrations of SC =0.25 wt% and 1.0 wt %), showing distribution of the protein, polysaccharide, and complex associates in the chromatographic fractions.

Free SC exhibited at pH 7.0 two unequal peaks. The first peak (83% from the total square) presents SC molecules, and the second one (17% from the total square) corresponds to the SC associates. Estimation of the molecular weights of these components on the basis of known molecular weights of alpha, beta, and gamma gelatins gave 260 kDa and 380 kDa accordingly. The weight average molecular weight of both fractions was about 300 kDa.

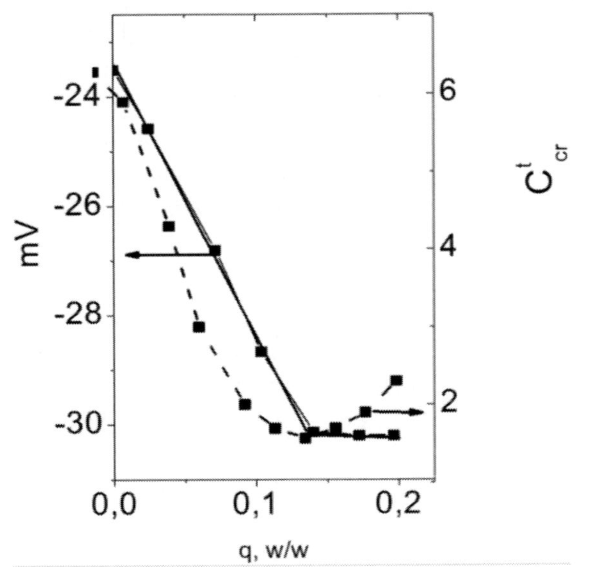

Figure 11. Dependence of Zeta potential, and the total critical concentration of biopolymers corresponding to phase separation of W-SC-SA-DSS system on q. SC/SA weight ratio is 4. pH 7.0, *I*=0.002 (phosphate buffer), 296 K.

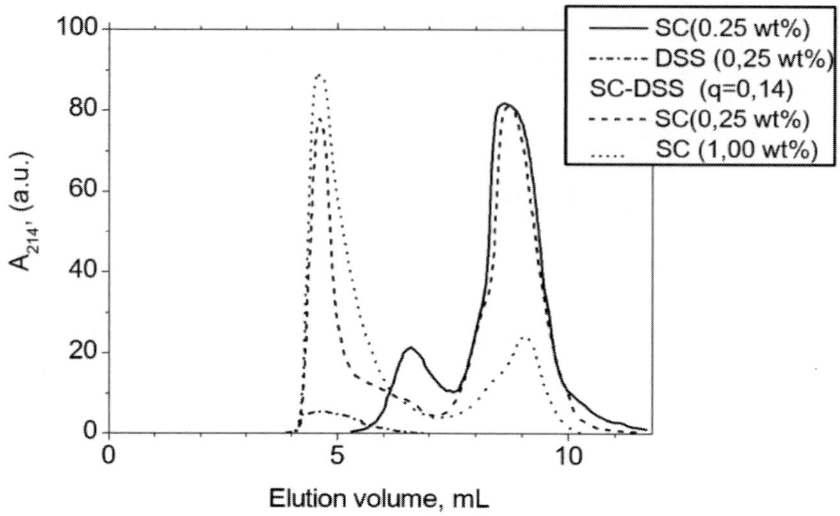

Figure 12. Chromatograms of the initial solutions of SC and DSS, and the SC- DSS system (q=0.14), showing distribution of the protein, polysaccharide, and complex associates in the chromatographic fractions at concentrations of the protein 0.1 wt% and 1.0 wt%. pH 7.0, *I*=0.015 (phosphate buffer), 296 K.

DSS exhibited at the same conditions a weak wide signal in the excluded volume. The chromatograms of the SC (0.25 wt %) -DSS systems at q=0.14 gave a new high molecular weight component corresponding to excluded volume and the peak corresponding to the elution volume of the free (unbounded SC). It is interesting to note that at concentration of SC below the critical concentration of the phase separation, the degree of conversion of the SC in water soluble complex with DSS is low (30%), and mainly the high molecular fraction of SC interact with DSS. The interaction becomes stronger when the concentration of the SC in the mixture increases up to 1.0 wt % (inside two-phase range of SC-SA system in the presence of DSS (q=0.14).In such conditions 83 % of SC form complex with DSS. Taking into account that the maximal yield of the complex takes place at q=0.14, knowing the weight-average molecular weights of SC and DSS and the degree of the protein conversion in protein-polysaccharide complex, we can roughly evaluate the SC/DSS molar ratio in the complex in the selected conditions corresponding to demixing of the mixed solutions of SC (2 wt%)-SA (0.5 wt%) in the presence of DSS (q=0.14). Simple calculation showed that about 10 molecules of SC join to 1 DSS molecule, forming large associates with high molecular weight. Systematic experimental data concerning dependence of C^*_t upon the radius, or molecular weight of synthetic or natural polymers are unknown untill now, although it is generally accepted that thermodynamic compatibility of polymers decreases with increase in molecular weights. It has been shown recently[58] that the total concentrations of biopolymers at the threshold point (C^*_t) for casein-guar gum system changes in accordance to $C^*_t \propto {M^{cas}_w}^{-0.27}$, where M^{cas}_w is molecular weight of caseins. This dependence has been established in a wide range of M^{cas}_w (from 25 kDa to 160.000 kDa). In that way, formation of large SC-DSS associates should decrease considerably compatibility of SA with bonded SC compared with that of "free" casein molecules that was observed in present work (Figure 4).

D. Commonality of the DDS-Induced Demixing at Rest

The other question arising from the demixing phenomenon in diluted biopolymer systems in the presence of DSS is, what is the key factor determining complex formation between DSS and caseins at pH 7.0 (far from the pH value corresponding to iep of caseins)? Is the high local charge density of the positively charged kappa casein responsible for that, or its mainly determined by the structural features of DSS, such as the concentration of sulfate groups, charge density, and conformation of the polysaccharide. Specific interaction between k-casein and carrageenan has been ascribed by Snoeren et al. [50] to an attraction between the negatively charged sulfate groups of carrageenan and a positively charged region of κ-casein, located between residues 97 and 112. It does not occur with the other casein types. Since the positive patch on κ-casein is believed to have a size of about 1.2 nm and is surrounded by predominantly negatively charged regions, the importance of the inter sulfate distances is unmistakable. To extend the Snoeren suggestion to our system, containing more stronger polyelectrolyte than carrageenan, or to reject it, we investigated the effect of DSS on the phase equilibrium in semidilute single phase biopolymer systems containing the protein (gelatin) with the statistical distribution of the positively charged functional groups. Two systems were under consideration; gelatin type A-SA, and gelatin typeA-SC. The former is a single phase one in water over a wide concentration range, and it undergoes phase separation at ionic strength above 0.2 [59]. The latter system undergoes phase separation only at a very high ionic strength (above 0.5) [60] and is characterized by a very high total concentration of the biopolymer (>15-20 wt%) at the critical point [61].

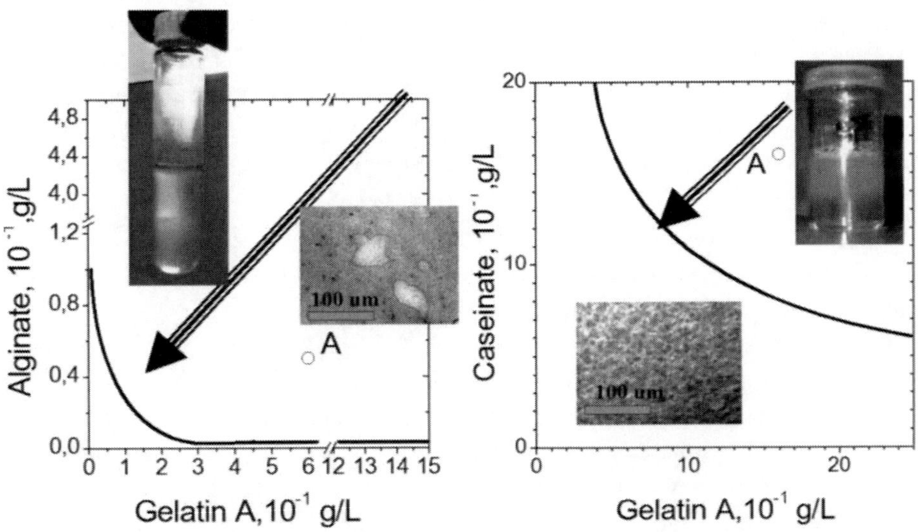

Figure 13. Shift of the bimodal line of the W-gelatin type A-SA, and W-SC-gelatin type A systems in the presence of DSS at DSS/protein weight ratio 0.14; photo images and microscopy images of the demixed W-gelatin type A(6 wt%) –SA (0.5 wt%), and W-SC (16 wt%)-gelatin type A (16 wt%) systems (points A on phase diagrams). W-gelatin type A-SA system was prepared at pH 5.0, and W-SC-gelatin type A system was prepared at pH 7.0.

The compatibility of these biopolymer pairs in water in the presence of DSS (at q=0.14) was studies. The phase separation of both systems in the presence of DSS was established, and the binodal lines for them were determined (Figure 13).The binodals for the systems without DSS are placed outside the concentration range studied. In both systems the phase separation leads to formation of water in water emulsions with liquid coexisting phases (Figure 13).Two important conclusions can be made from these data. First, the DSS induced phase separation in semidilute biopolymer solutions at rest is a rather general phenomenon not an exceptional case. Second, the structural features of DSS molecules is the more important factor determining complex formation of SC with DSS and subsequent demixing of the single phase semidilute systems, rather than the characteristics of the distribution of the positively charged groups in the protein molecules. The last conclusion is in agreement with the FPLC data (figure 12). As can be seen the degree of the protein conversion in complex achieves 80% whereas the content of kappa casein in SA is only 12-14 % [62]. It is known that the sulfate groups of DSS are more closely packed than that of κ-carrageenan (0.5 nm for DSS and 1.2 nm for carrageenan [62,63]. The later can allow for the attractive forces to overcome the repulsive forces acting outside the positive patch. Bowman, Rubinstein, and Tan, characterizing complex formation between negatively charged polyelectrolytes and a net negatively charged gelatin by light scattering, suggested[64] that the protein is polarized in the presence of strong polyelectrolyte. Junhwan and Dobrynin have recently presented the results of molecular dynamics simulations of complexation between protein and polyelectrolyte chains in solution[65]. They found that protein placed near polyelectrolyte chains is polarized in such a way that the oppositely charged groups on the protein are close to the polyelectrolyte, maximizing effective electrostatic attraction between the two while the similarly charged groups on the protein far away from the polyelectrolyte minimize effective electrostatic repulsion. In dilute and semidilute solutions, which are subjects of our study, polyampholyte chains usually form a complex at the end of polyelectrolyte chains resulting from the above polarization effect by polyelectrolyte. We believe that polarization-induced attraction is the main mechanism of complexation SC and DSS.

E. Discussion on the Structure of the SC-DSS Complexes and SC Enriched Phase of the Demixed SC-SA System

From study of polyelectrolyte complexes we know that interaction between oppositely charged polyelectrolyte's leads to partial or complete neutralization of charges, complexes remain soluble or precipitate, and in some cases gel-like networks are formed. If neutralization of charges is significant, the so called "scrambled egg" compact structure will be formed. When neutralization of charges is far from complete, a "ladder" structure of complex can be formed [66].

The results of the Zeta potential measurements, DLS and flow experiments shown that the negative charge of the SC increases during interaction with DSS, and the maximal binding takes place at approx 0.14 DSS/SC weight ratio. Such features of the intermacromolecular interactions do not promote formation of the "scrambled egg" structure, because DSS molecule having many combined SC molecules and considerable negative charge can not be fold. Therefore the ladder structure is more preferable for the system (Figure 14 a). The overage size of the SC-DSS complex associates established from the DLS experiments is 0.2

um. Such a length scale would be in line with the fact that the SC/DSS solution is slightly turbid. This turbidity arises from a length scale in the micrometer range. Obviously, heterogeneities on a micrometer scale were formed. If SC/DSS solution was made of a homogeneous structure of polymers on the nanometer scale, it would be transparent. In the presence of free polymer-SA, complex associates of SC and DSS undergo further association and the system becomes two phasic. This suggestion finds confirmation in the flow experiments; viscosity of the demixed SC-SA system is considerably higher than that of undemixed SC-SA system having the same concentrations (Figure 6). This difference is even much higher in the case of higher protein concentration in the single phase SC-DSS system (Figure 8) this is a clear indication of association of the "ladder " structure of the complex associates, and formation of network (Figure 14b). Figure 15 presents ESEM images obtained for the SC-DSS system at q=0.14 (at maximal binding) and different concentrations of SC in the system. One can see that at concentration of SC equal to 2 wt% the formation of the regular structure is observed, which transfer to some "network" structure at higher concentration of SC (6 wt%) in the system.

F. Shear-Induced Behavior of the SC-SA System in the Presence of DSS

The experimental results shown in this section have been obtained on a water (97.5 wt%)-SC (2.0 wt%)-SA (0.5 wt%)-DSS ($2\odot10^{-3}$ wt %) system. It contains 99 wt % of the SC enriched phase and 1 wt % of the SA enriched phase which have been mixed by hand, typically resulting in a very fine morphology. This emulsion is located in the two-phase region not far from the binodal line. The coexisting phases have Newtonian viscosities at 296 K, of 0.03 Pa· s and 0.02 Pa· s for the SC enriched and the SA enriched phase, respectively.

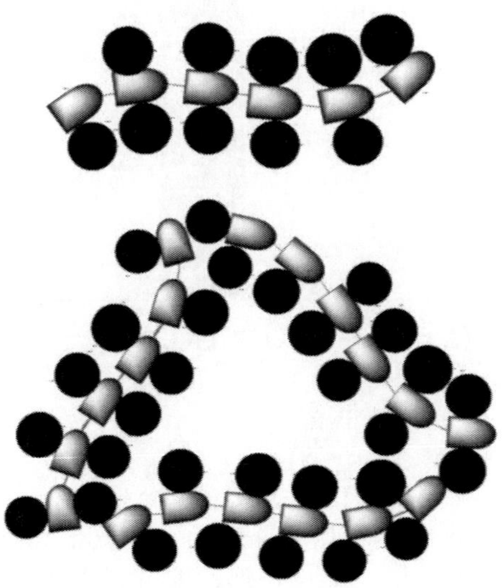

a)

b)

Figure 14. Schematic representation of the possible structures of (a) ladder-like and (b) gel-like. The long chain represent DSS molecule and the balls represent casein chains.

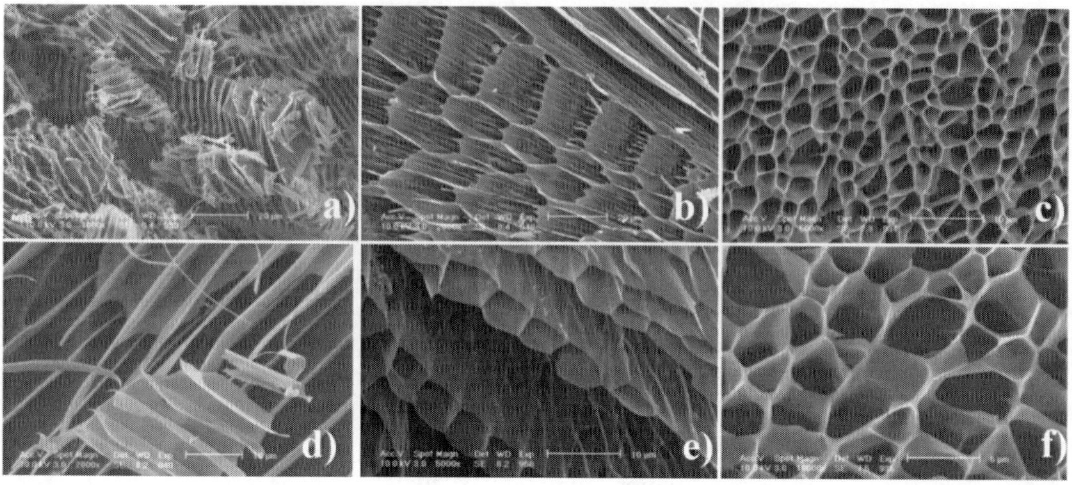

Figure 15. ESEM images of the SC-DSS systems at q=0.14. pH 7.0. Concentration of SC in the system, wt%; a,d- 1.0, b,e-2.0, c,f- 6.0.

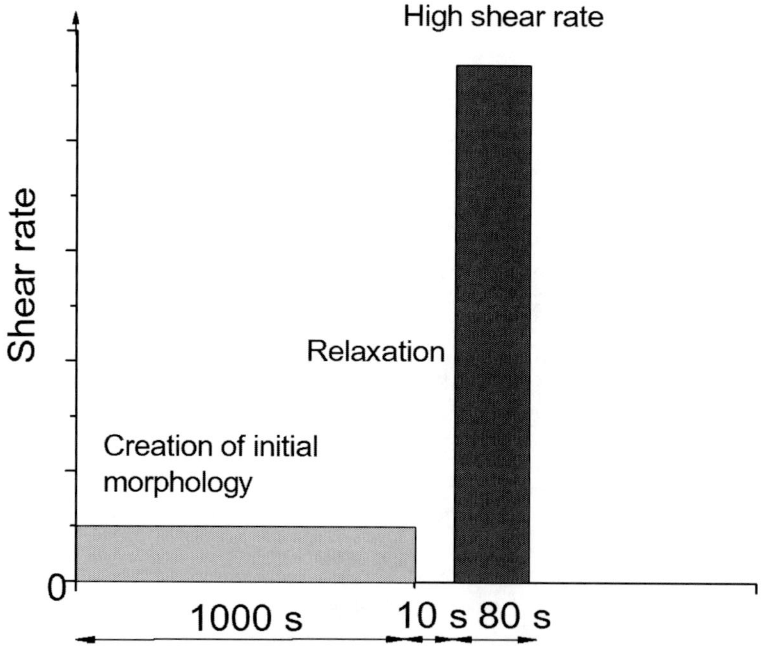

Figure 16. Schematic representation of the shear history.

To study the effect of flow on the phase behavior, a flow history consisting of three shear zones is used (Figure 16). First, a preshear of 0.5 s^{-1} is applied for 1000 s (500 strain units). It has been verified that this procedure leads to a reproducible initial morphology. Subsequently, this preshear is stopped and the slightly deformed droplets are allowed to retract to a spherical shape. The resulting droplet radius is of the order of 5 micron. Finally, the shear rate is suddenly increased to a high value for 80 s, and after stopping flow the evolution of the SALS patterns are monitored.

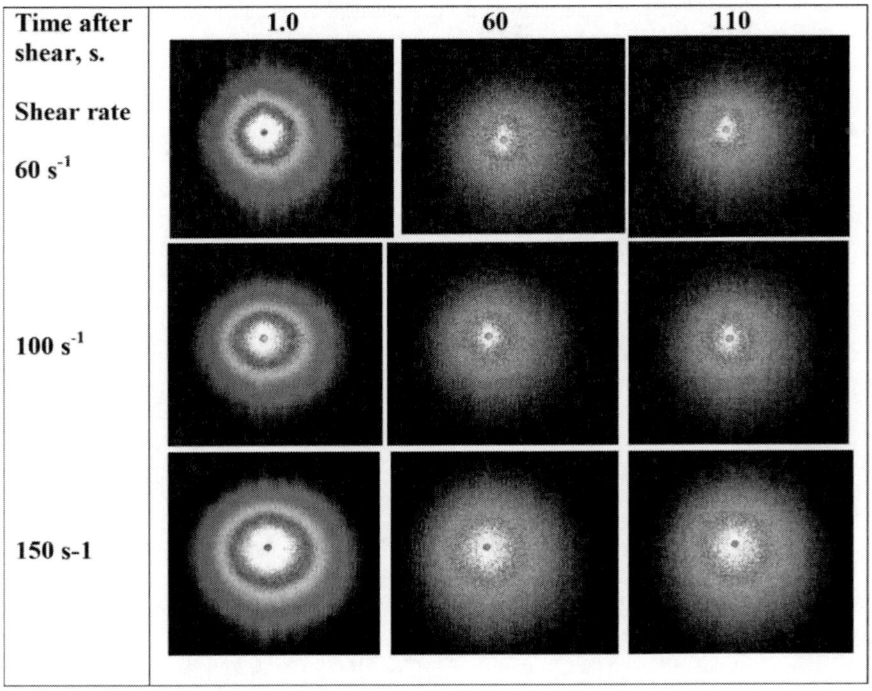

Figure 17. Evolution of the SALS patterns of water (97.5 wt %)-SC (2.00 wt %)-SA (0.5 wt %)-DSS (2 10^{-3} wt%) after cessation of a high shear rate flow.Shear rates and times of the shear as indicated on the figure. pH 7.0. I=0.002 (phosphate buffer). Temperature 296 K. The SALS pattern of water (97.5 wt %)-SC (2.00 wt %)-SA (0.5 wt %)-DSS ($2 \cdot 10^{-3}$ wt%) system before high shear rate is shown in Figure1a.

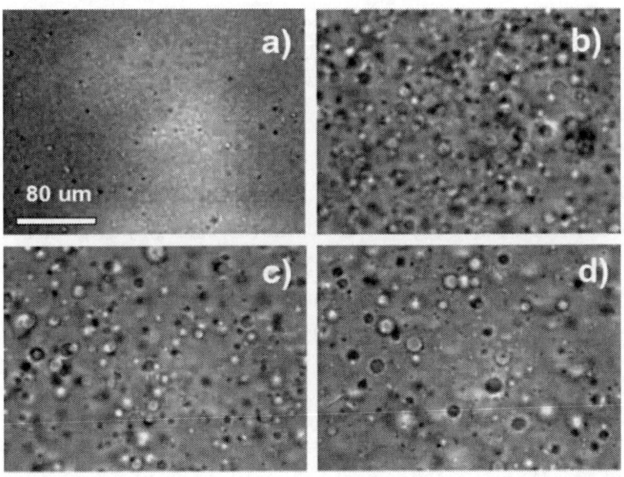

Figure 18. The evolution of microscopy images of of water (97.5 wt %)-SC (2.00 wt %)-SA (0.5 wt %)-DSS (2 10^{-3} wt %) system before high shear rate flow (a) and just after cessation of a high shear rate flow. Shear rate: b) 60 s^{-1}, c) 100 s^{-1}, d) 150 s^{-1}. pH 7.0. I=0.002 (phosphate buffer). Temperature 296 K.

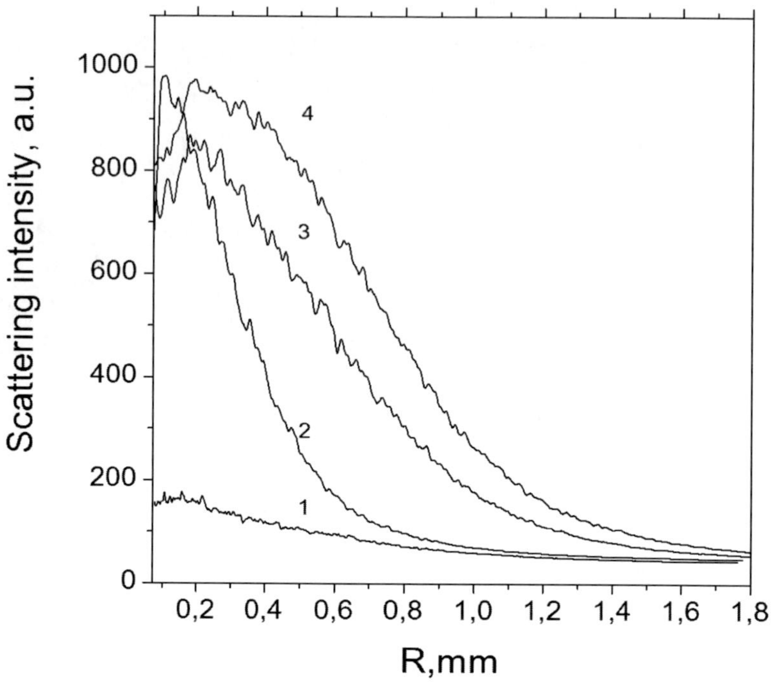

Figure 19. Dependence of the scattering intensity of the demixed water (97.5 wt %)-SC (2.00 wt %)-SA (0.5 wt %) - DSS ($2\odot10^{-3}$ wt %) system after preshear (curve 1), and just after cessation of flow at 60 s^{-1} (curve 2) and water (87.8 wt%)-SC (12.2 wt%)-SA (0.1 wt %) system after preshear (curves 3), and just after cessation of flow at 60 s^{-1}, on the distance from the bean stop. Both systems contain 1.0 wt% SA enriched dispersed phase.

The evolution of the SALS patterns after cessation of steady-state shear flow at 60 s^{-1}, 100 s^{-1} and 150 s^{-1} is shown in Figure 17. In each experiment, a freshly loaded sample has been used. As can be seen at all shear rates selected, the SALS patterns become more intensive just after cessation of flow. The higher the shear rate applied the more intensive the SALS pattern becomes. This is a clear indication of shear induced demixing in SC-SA system in the presence of DSS. After cessation of shear flow the light intensity is slowly decreasing (Figure17), but the complete recovery of the initial SALS pattern takes place only after 1-2 hours (data are not presented). In Figure 18 microscopy images corresponding to the same emulsion as in SALS experiments are presented first, after preshear of the emulsion at 0.5 s^{-1} for 1000 s. with subsequent cessation of steady-state shear flow at 60 s^{-1}(a), 100 s^{-1} (b), and 150 s^{-1}. One can see an appreciable increase of the droplet size after cessation of high shear rate flow, in accordance with SALS data.

In Figure 19 the light scattering intensity of semidilute demixed water (97.5 wt%)-SC (2.0 wt%)-SA (0.5 wt%)-DSS ($2\odot10^{-3}$ wt %) system after preshear (curve 1) and just after cessation of flow at 60 s^{-1} (curve 2) is compared with that of water (87.8 wt%)-SC (12.2 wt%)-SA (0.1 wt %) system, containing 1 wt% SA enriched phase at the same shear history (curves 3 and 4). It is seen that the increase in the light intensity after cessation of flow takes place for both systems, however for the former system the light intensity increased much higher that for the latter one. These observations can be explained on the basis of a comparison of the molecular weights of the "free" SC and SC, combined with DSS (see

Figure 12). The molecular weight of the latter one is much higher than that of the former one. Note, that the second virial coefficients on the molar scale, related to pair interactions of similar SC macromolecules, A_{22} depends on the molecular weight inversely [67]. Therefore, according to conditions of the phase separation in biopolymer systems in flow [68]:

$$A_{23} > \sqrt{A_{22}A_{33}} \quad (2)$$

in which A_{ij} are the second virial coefficients on the molar scale, related to pair interactions of similar (2-protein, 3-polysaccharide) and dissimilar macromolecules, the protein-polysaccharide mixture containing macromolecules with lower values of A_{22} will be more predisposed to shear induced demixing.

CONCLUSION

It well known that phase equilibrium in aqueous system containing casein and linear acid polysaccharide is weakly sensitive to changes of the main physico-chemical parameters, such as pH, ionic strength, and temperature. This is the case both at rest [17,31,32,34] and under shear flow [69]. It has been shown in many studies that proteins interact with acid polysaccharides forming intermacromolecular coulomb complexes mainly at pH values below the isoelectrical point of the protein (iep) when both biopolymers are oppositely charged, or at pH values slightly above iep.

In this work the attempt was performed to induce demixing in semidilute and highly compatible sodium caseinate/sodium alginate system (SC-SA) mixtures in the presence of sodium salt of dextran sulfate (DSS) at pH 7.0, (above the isoelectrical point of caseins) and to characterize phase equilibrium, intermacromolecular interactions, and structure of such systems by rheo-small angle light scattering (SALS), optical microscopy (OM), phase analysis, dynamic light scattering (DLS), fast protein liquid chromatography (FPLC), ESEM, and rheology.The results obtained in the present study can be summarized as follows:

DSS is able to induce a deep segregative phase separation in semidilute SC-SA systems (at a SC concentration as low as 1 wt %) at a trace concentrations (10^{-3} wt %); DDS significantly increases the phase separation range, as well as viscosity and mechanical moduli of the system. The phase separation observed is the result of formation at pH 7.0 (i.e., far away from the iep of the caseins /4.4-4.6/) of DSS/SC water soluble charged associates (1:10 mol/mol), having R_H=0.26 um and electrostatic nature. The minimal compatibility of SC and SA was observed at the DSS/SC weight ratio of 0.14, which corresponds to an equality of the cationic groups in the protein molecules and sulfate groups in DSS. At a higher SC concentration (4 wt %) SC-DSS associates forms some kind of network. Data of chromatography indicate that DSS interacts first with SC associates having higher molecular weight. The degree of the protein conversion in the complex increases from approx. 30 % to 80% when the concentration of SC in the system grows from 1 to 2 wt %. Phase separation of semidilute ternary water-biopolymer 1-biopolymer 2 systems in the presence of DSS is observed here to be a rather common phenomenon, observed for different types of biopolymers, e.g., SC-gelatin type-A, gelatin type A-SA. Therefore the use of DSS as a decompatibilizer for semidilute biopolymer systems can find applications in processes for concentrating biological materials in two phase systems, because DDS induced demixing allows decreasing the critical concentration of the phase separation significantly. Moreover, at a high shear rate flow ($60s^{-1}$-$150\ s^{-1}$), semidilute phase separated SC-SA-DSS systems undergo further segregative separation, decreasing the critical concentration of phase separation into the range of dilute solutions. Such peculiarities of thermodynamic and rheological behaviors allow us to consider sulfate polysaccharide interacting with protein in aqueous protein-polysaccharide mixture as a new type of decompatibilizer for biopolymer emulsions. Therefore the results obtained promote more thorough understanding of the relationships between intermacromolecular interactions in aqueous biopolymer systems and their thermodynamical properties.

REFERENCES

[1] Hidalgo, J; Hansen, PMT. *J. Dairy Sci.*, 1971, 54, 1270-1274.

[2] Wang, Y; Gao, JY; Dubin, PL. *Biotechnol. Prog*, 1996, 12, 356-362.

[3] Dubin, PL; Gao, J; Mattison, K. *Sep. Purif. Methods*, 1994, *23*, 1-16.

[4] Strege, MA; Dubin, PL; West, JS; Flinta, CD. Protein separation via polyelectrolyte complexation In *Symposium on Protein Purification: From Molecular Mechanisms to Large-scale Processes*, 1990, Ladisch, M.; Willson, R.C.; Paint C.C.; Builder, S.E. eds, ACS Symposium Series 427, Chapter 5, p. 66. American Chemical Society: Washington.

[5] Antonov, YuA; Grinberg, VYa; Zhuravskaya, NA; Tolstoguzov, VB. *Carbohydrate polymers*, 1982, 2, 81-90.

[6] Antonov, YuA. Application of the Membraneless Osmosis Method for Protein Concentration from Molecular-dispersed and Colloidal dispersed Solutions. Review. *Applied Biochemistry and Microbiology* (Russia) *Engl.Transl.*, 2000, 36, 382-396.

[7] Kiknadze, EV; Antonov, YA. *Applied Biochemistry and Microbiology* (Russia) Engl.Transl., 1998, 34, 462-465.

[8] Xia, J; Dubin, PL. Protein-polyelectrolyte complexes In: *Macromolecular Complexes in Chemistry and Biology*, 1994, Dubin P.L.; Davis R.M.;.Schultz, D.; Thies C. (eds), Springer-Verlag, Berlin, Chapter 15.

[9] Ottenbrite, RM; Kaplan, AM. *Ann. N. Y. Acad. Sci.*, 1985, 446, 160–168.

[10] Magdassi, S; Vinetsky, Y. Microencapsulation: methods and industrial applications.In *Microencapsulation of oil-in-water emulsions by proteins*, 1997, Benita, S. (es) Marcel Dekker Inc.: New York, p. 21-33.

[11] Regelson, W. *Journal of Bioactive and Compatible Polymers,* 1991, 6, 178-216.

[12] Albertsson, P-Å; Johansson, G; Tjerneld, F. In *Separation Processes in Biotechnology*, 1990, Asenjo, J.A. (ed.), Marcel Dekker, New York, 287–327.

[13] Harding, S; Hill, SE; Mitchell, JR. *Biopolymer Mixtures*, 1995, University Press, Nottingham, 499 pp.

[14] Tolstoguzov, VB. Functional properties of protein-polysaccharide mixtures. In: *Functional properties of food macromolecules*, 1998, Hill, S.E.; Ledward, D.A.; Mitchel, J.R., Gaitherburg, Maryland: Aspen Publishers Inc.

[15] Piculell, L; Bergfeld, K; Nilsson, S. Factors determining phase behaviour of multi-component polymer systems. In *Biopolymer Mixtures*, 1995, Harding, S.E.; Hill, S.E.; Mitchell, J.R. (ed) Nottingham University Press, Nottingham, 13-36.
[16] Antonov, YA; Van Puyvelde, P; Moldenaers, P. *Biomacromolecules*, 2004, 5(2), 276-283.
[17] Antonov, YA; Van Puyvelde, P; Moldenaers, P. *IJBM*, 2004, 34, 29-35.
[18] Antonov, YA; Wolf, B. *Biomacromolecules*, 2006, 7(5), 1582-1567.
[19] Walter, H; Brooks, DE; Fisher, D. In *Partitioning in Aqueous Two-Phase Systems: Theory, Methods, Uses and Applications to Biotechnology*, 1985. Academic Press, London.
[20] Hellebust, S; Nilsson, S; Blokhus, AM. *Macromolecules*, 2003, 36, 5372-5382.
[21] Scott, R. *J. Chem. Phys*. 1949, 17, 3, 268-279.
[22] Tompa, H. *Polymer Solutions*, 1956, Butterworth: London.
[23] Flory, PJ. *Principles of Polymer Chemistry*, 1953. Cornell University Press: Ithaca, NY.
[24] Prigogine, I. *The Molecular Theory of Solution*, 1967, North- Holland: Amsterdam.
[25] Patterson, D. *Polym. Eng. Sci.*, 1982, 22 (2), 64-73.
[26] Gottschalk, M; Linse, P; Piculell, L. *Macromolecules*, 1998, 31, 8407-8416.
[27] Lindvig, T; Michelsen, ML; Kontogeorgis, GM. *Fluid Phase Equilib.*, 2002, 203, 247-260.
[28] Piculell, L; Nilsson, S; Falck, L; Tjerneld, F. *Polym. Commun.*, 1991, 32, 158-160.
[29] Bergfeldt, K; Piculell, L; Tjerneld, F. *Macromolecules*, 1995, 28, 3360-2270.
[30] Bergfeldt, K; Piculell, LJ. *Phys. Chem.*, 1996, 14, 5935-5840.
[31] Antonov, YA; Grinberg, VYa; Tolstoguzov, VB. *Starke*, 1975, 27, 424-431.
[32] Antonov, YA; Grinberg, VYa; Zhuravskaya, NA; Tolstoguzov, VB. *J. Texture Studies*, 1980, 11(3), 199-215.
[33] Rha, CK; Pradipasena, P. In *Functional Properties of Food Macromolecules*, 1986 Mitchell, J. R.; Ledward, D.A. (eds), Elsevier Applied Science: London, 79-119.
[34] Whistler, RL. *Industrial Gums*, *2nd edn.*, 1973 Academic Press, New York.
[35] Capron, I; Costeux, S; Djabourov, M. *Rheol. Acta*, 2001, 40, 441-456.
[36] Van Puyvelde, P; Antonov, YA; Moldenaers, P. *Food Hydrocolloids*, 2003, 17, 327–332.
[37] Koningsveld, R; Staverman, AJ. *Journal of Polymer Science A-2*, 1968, 305–323.
[38] Polyakov, VI; Grinberg, VYa; Tolstoguzov, VB. *Polymer Bulletin*, 1980, 2, 760–767.
[39] Dubois, M; Gilles, KA; Hamilton, JK; Revers, PP; Smith, T. *Anal.Chem.*, 1956, 18, 350-356.
[40] Zaslavsky, BY. *Aqueous two phase partitioning: Physical Chemistry and Bioanalytical Applications*, 1995, Marcel Dekker: New York.
[41] Grinberg, VYa; Tolstoguzov, VB. *Food Hydrocolloids*, 1997, 11, 145-158.
[42] Van Puyvelde, P; Antonov, YA; Moldenaers, P. *Food Hydrocolloids*, 2003, 17, 327–332.
[43] Overbeek, JTG; Voorn, MJJ. *Cell Comp. Physiol.*, 1957, 49, 7–39.
[44] Kaibara, K; Okazaki, T; Bohidar, HP; Dubin, P. *Biomacromolecules*, 2000, 1, 100-107.
[45] Antonov, YuA; Dmitrochenko, AP; Leontiev, AL. *Int. Journal Biol. Macromol.*, 2006, 38 (1), 18-24.
[46] Antonov, YA; Gonçalves, MP. *Food Hydrocolloids*, 1999, 13, 517-524.
[47] Michon,C; Konate, K; Cuvelier, G; Launay, B. *Food Hydrocolloids*, 2002, 16, 613-618.

[48] Galazka, VB; Ledward, DA; Sumner, IG; Dickinson, E. *Agric. Food Chem.*, 1997, 45, 3465-3471.

[49] Dickinson, E. *Trends in Food Science and Technology*, 1988, 9, 347-354.

[50] Snoeren, THM; Payens, TAJ; Jevnink, J; Both, P. *Milchwissenschaft*, 1975, 30, 393-395.

[51] Kabanov, VA; Evdakov, VP; Mustafaev, MI; Antipina, AD. *Mol. Biol.*, 1977, 11, 582-597.

[52] Noguchi, H. *Biochim. Biophys. Acta*, 1956, 22, 459-462.

[53] Kumihiko, G; Noguchi, H. *Agriculture and Food Science*, 1978, 26, 1409-1414.

[54] Sato, H; Nakajima, A. *Colloid and Polymer Science*, 1974, 252, 294-297.

[55] Sato, H; Maeda, M; Nakajima, A. *Journal of Applied Polymer Science*, 1979, 23, 1759-1767.

[56] Bungenberg de Jong, HG. Crystallisation- Coacervation- Flocculation. In *Colloid Science*; Kruyt H. R. ed.; Elsevier Publishing Company: Amsterdam, 1949, Vol. II, Chapter VIII, 232-258.

[57] Gurov, AN; Wajnerman, ES; Tolstoguzov, VB. *Stärke*, 1977, 29 (6), 186-190.

[58] Antonov, YA; Lefebvre, J; Doublier, JL. *Polym. Bull.* 2007, 58, 723-730.

[59] Antonov, YA; Lashko, NP; Glotova, YK; Malovikova, A; Markovich, O. *Food Hydrocolloids*, 1996, 10 (l), 1-9.

[60] Polyalov, VI; Grinberg, VYa; Antonov, YA; Tolstoguzov, VB. *Polymer Bulletin*, 1979, 1, 593-597.

[61] Polyalov, VI; Kireeva, OK; Grinberg, VYa; Tolstoguzov, VB. *Nahrung*, 1985, 29, 153-160.

[62] Swaisgood, HE. In *Advanced Dairy Chemistry-1:Proteins*, 1992, Fox, P. F. (Ed)., Elsevier Applied Science, London,

[63] Langendorff, V; Cuvelier, G; Launay, B; Michon, C; Parker, A; De Kruif, CG. *Food Hydrocolloids*, 1999, **13**, 211–218.

[64] Bowman, WA; Rubinstein, M; Tan, JS. *Macromolecules*, 1997, 30 (11), 3262-3270

[65] Junhwan, J; Dobrynin, A. *Macromolecules*, 2005, 38 (12), 5300–5312.

[66] Smid, J; Fish, D. In: *Encyclopedia of polymer science and engineering, 2nd ed.Polyelectrolyte complexes*, 1988, Mark, H.F.; Bikales. N.M.; Overberger, C.G.; Menges, G. (Eds)., vol. 11. New York: Wiley/Interscience;. p. 720.

[67] Striolo, A; Ward, J; Prausnitz, JM; Parak, WJ; Zanchet, D; Gerion, D; Milliron, D; Alivisatos, AP. *J.Phys. Chem.*, 2002. B, 106, (21), 5500-5505.

[68] Antonov, YA; Van Puyvelde P; Moldenaers, P. *Biomacromolecules*, 2004, 5(2), 276-283.

[69] Antonov, YA; Van Puyvelde, P; Moldenaers, P. *Food Hydrocolloids* 2009, 23, 262– 270.

In: News in Chemistry, Biochemistry and Biotechnology ISBN: 978-1-63117-273-1
Editors: G. E. Zaikov, G. Nyszko, L. P. Krylova et al.

Chapter 17

PHASE BEHAVIOUR AND STRUCTURE FORMATION IN AQUEOUS SOLUTIONS OF BOVINE SERUM ALBUMIN

***Y. A. Antonov*[1] *and Bernhard A. Wolf*[2]**

[1] N.M. Emanuel Institute of Biochemical Physics, Russian Academy of Sciences, Moscow, Russia

[2] Institut für Physikalische Chemie der Johannes Gutenberg-Universität Mainz, Mainz, Germany

ABSTRACT

The thermodynamic behavior of the system H_2O/BSA was studied at 25 °C within the entire composition range: Vapor pressure measurements via head space sampling gas chromatography demonstrate that the attainment of equilibria takes more than one week. A miscibility gap was detected via turbidity and the coexisting phases were analyzed. At 6 °C the two phase region extends from ca. 34 to 40 wt% BSA; it shrinks upon heating. The polymer rich phase is locally ordered, as can be seen under the optical microscope using crossed polarizers. The Flory-Huggins theory turns out to be inappropriate for the modeling of experimental results. A phenomenological expression is employed which uses three adjustable parameters and describes the vapor pressures quantitatively; it also forecasts the existence of a miscibility gap.

INTRODUCTION

Despite the long-standing engagement in the thermodynamics of polymer containing liquids our knowledge in some interesting and biologically important areas is still rudimentary. One such example concerns the phase separation behavior of joint solutions of different types of polymers. This statement does not imply the absence of research on such systems; numerous studies have been performed on biological systems as described in several overviews. [1-3] The difficulty with the reported knowledge lies in the fact that it refers to biological systems, which are by nature very complex and contain a multitude of different components.

For a more comprehensive understanding of the above-mentioned multinary systems it appears mandatory to dispose of reliable information concerning the thermodynamic behavior of the corresponding binary subsystems. Only under this condition it is for instance possible to assess whether the interactions between the different components observed in binary mixtures remain unchanged in the case of a higher number of components or whether special interactions between two or more solutes change the behavior fundamentally. This situation has become very clear to us when we wanted to interpret ongoing experiments with the ternary system water/dextran/bovine serum albumin (H_2O/DEX/BSA). For that reason we searched for information concerning aqueous solutions of the two types of polymers. In the case of H_2O/DEX the required data have already been published [4], in contrast to the subsystem H_2O/BSA, for which we could not find the required information. This is the reason why we have conducted the study reported here.

1. Experimental Section

1.1. Materials

BSA, Fraction V, pH 5 (Lot A018080301), was obtained from Across Organics Chemical Co. (protein content = 98-99%; trace analysis, Na < 5000 ppm, Cl < 3000 ppm, no fatty acids detectible). According to literature [5] the molar mass of BSA is 66.4 kDa. The isoelectric point of the protein amounts to 4.8-5.0 and the radius of gyration [6] at pH 5.3 is 30.6 Å. Millipore-quality water was used throughout the experiments. All measurements were performed at pH 5.4, because serum albumin undergoes conformational isomerization and changes in the conformation state and secondary structure with changes in pH from pH 5-5.5 to acid and alkaline region. [6] The extinction of 1% BSA solution at 279 nm was $A^{1cm}{}_{279}$= 6.70; this value is very close to the tabulated value [7] of 6.67.

1.2. Solution Preparation

To prepare BSA stock solutions, the biopolymer was gradually added to the deionized water and stirred at 298 K for 2 h. The solutions were then centrifuged at 13 000*g* for 1 h at 298 K to remove insoluble particles. Subsequently, the concentration of the biopolymer was determined by measuring the dry weight residue. The content of protein nitrogen in the dry BSA samples was always taken into account calculating the concentration of protein in solution. In some cases, the final protein concentration was determined also by spectrophotometric measurements.

1.3. Vapor Pressures

Vapor pressure measurements were carried out as described in the literature [8] for volume fractions φ of the polymer up to 0.968 by means of an apparatus consisting of the headspace-sampler *Dani HSS 3950, Milano (Italy)* and a gas chromatograph *Shimadzu GC*

14B, Kyoto (Japan). In these measurements a constant volume of the equilibrium vapor phase is taken out of sealed vials by means of a syringe through a septum and analyzed in a gas chromatograph. In this manner it is possible to calculate the partial vapor pressures of the volatiles. No corrections for imperfections of the gas were necessary with the present systems to obtain fugacities. In all cases we made allowance for the amount of gas that is contained in the vapor phase, when calculating the composition of the liquid mixture. In order to promote the attainment of equilibria at high polymer concentrations we have prepared thin films (5 to 20 μm thick) on glass beads of 4 mm diameter. To this end the voids between the beads were filled with a sufficiently viscous aqueous solution of BSA (ca. 20 wt%) and the desired final composition of the solutions was established stripping off the excess solvent at room temperature by applying vacuum. For the highest concentrations the solvent was totally removed and then the dry films were either loaded with water via the gas phase or by adding the required amount of liquid water. The establishment of equilibria was checked by measuring the vapor pressures as a function of time up to three weeks. For organic solvents the error of the vapor pressures is typically on the order of 1% - 2%; with the present aqueous solutions the errors are markedly larger, particularly in the range of low φ values as indicated in Figure 1.

1.4. Phase Behavior

Aqueous solutions containing approximately 25wt% BSA were prepared by gradually adding the appropriate amount of the biopolymer to the deionized water under stirring at the temperature of interest. After three hours the solutions were centrifuged at 13.000 g for 1 hour at the same temperature to remove insoluble particles. The BSA concentration in the thus obtained solutions was determined by measuring the dry weight residue.

Cloud point concentrations were determined by removing water from 24.5 wt% solutions of BSA at constant temperature. To this end the solutions were kept in open vials in a thermostat and stirred until the liquids became turbid. Depending on temperature this procedure took 18-36 hours. Stripping off more solvent leads to the segregation of a polymer rich phase; its separation by centrifugation and the analysis of the coexisting phases yields the tie lines of the system at the given temperature. The BSA rich phase appears slightly opaque and was therefore inspected under the Olympus CX31-P Polarized Light Microscope.

2. Results

2.1. Vapor Pressures

Figure 1 shows how the reduced vapor pressure of water (p/p_o, where p_o is the vapor pressure of pure water) decreases as the volume fraction, φ, of BSA becomes larger. It also demonstrates a pronounced influence of time: Within some medium range of φ values the vapor pressures decline upon standing; in order to reach equilibria one needs to wait at least two weeks. Both curves, the one for 1 week and the one for 3 weeks, exhibit extrema; this observation clearly indicates the existence of a miscibility gap between water and BSA: The vapor pressure may become identical for different polymer concentrations only if two liquid phases coexist.

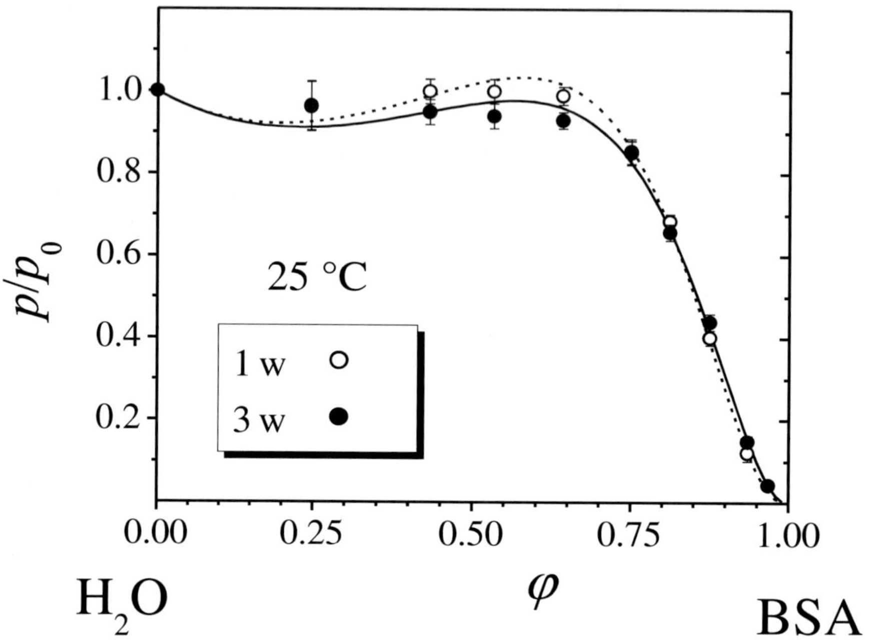

Figure 1. Reduced vapor pressures as a function of polymer concentration (volume fractions) for different equilibration times. The curves are calculated by means of eq (4) plus the parameters specified in the legend of Figure 5.

The observed time dependence of the vapor pressures, most pronounced in the middle range of composition, is untypical for solutions of uncharged chain molecules. Slow equilibration of the present order was, however, observed when mixing dilute solutions of oppositely charged polyelectrolytes. [9-11] The molecular explanation is the following: After a first rapid step of equilibration - consisting in contact formation between the two types of solutes - slow rearrangements are required to attain the minimum Gibbs energy of the system. It does not appear unreasonable to assume that similar processes are necessary to transfer parts of the BSA molecule that interact most favorably with water to the outer regions. This reasoning is backed by the observation that the vapor pressures decrease with time, indicating an improvement in the thermodynamic quality of water for BSA. The explanation why this effect dies out at very dilute and very concentration mixtures is trivial. On one end of the composition scale the pure solvent becomes dominant for the measured vapor pressure and on the other end water will interact almost exclusively with the most favorable site of BSA, which may always be readily accessible.

2.2. Demixing Behavior

Indirect information concerning limitations in the miscibility of BSA and water obtained from the vapor pressure date were checked by directed experiments. The coexisting phases, segregated by centrifugation, look very different. The upper phase of lower BSA content flows readily, whereas the lower phase is highly viscous and has a gel like appearance; it is slightly opaque and exhibits a complex rheological behavior.

Figure 2 shows how the compositions of the coexisting phases vary with temperature. The fact that the miscibility gap narrows as T rises indicates endothermal mixing of the components. This graph also displays the measured cloud points and demonstrates that they do not fall on the coexistence curve. In view of the experimental procedure it is highly probable that they represent spinodal points. Due to the extensive purification of the starting BSA solutions the concentration of nuclei, promoting the segregation of a second phase, is expected to be very low. This means that the homogenous mixtures do not demix when crossing of stability limits upon the stripping of water and increasing the BSA concentrations. Instead they remain one phase within the metastable regime. It is only at the boarder between metastability and instability that phase becomes inevitable and the solutions get turbid. In other words: The temperature/composition range located in the phase diagram between the left hand branch of the coexistence curve and the dotted line represents the metastable region.

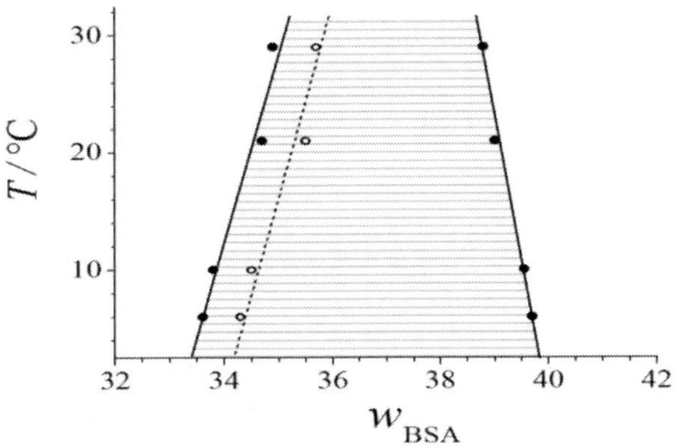

Figure 2. Phase diagram of the system H_2O/BSA. Full circles: composition of the coexisting phases, open circles: spinodal points (cf. text).

Figure 3. Micrograph of the BSA rich coexisting phase; the bar indicates 200 μm. The bright parts indicate the ordered regions of the solution.

A closer inspection of the BSA rich coexisting phase under crossed polarizers reveals the existence of ordered regions; Figure 3 shows a typical example of these images.

3. DISCUSSION

The Flory-Huggins theory [12] acquires the parameter χ from experimental data on the reduced vapor pressures of the solvent according to the following relation

$$\ln \frac{p}{p_o} = \chi \varphi^2 + \ln(1-\varphi) + \left(1 - \frac{1}{N}\right)\varphi \tag{1}$$

φ stands for the volume fraction of the polymer, N for the number of polymer segments (normally the ratio of the polymer molar volume divided by the molar volume of the solvent), p is the vapor pressure at a given value of φ and p_o that of the pure solvent. In the general case χ depends markedly on φ, which means that the integral Flory-Huggins interaction parameter g (required for instance for the calculation of phase diagrams) is not identical with χ. Knowing $\chi(\varphi)$, the integral parameter is accessible via the expression

$$g = -\frac{1}{1-\varphi}\int_1^{\varphi} \chi(\varphi)\, d\varphi \tag{2}$$

Figure 4 shows the Flory-Huggins interaction parameter χ as a function of composition, calculated according to eq (1) from the vapor pressure measurement after three weeks of equilibration. The curve shown in this graph is modeled by means of the following series expansion of χ

$$\chi = \sum_{i=0}^{n} \chi_i \varphi^i \tag{3}$$

For most polymer solutions it suffices to account for three terms (i = 2), however in the present case we had to use one more.

The main problem with the evaluation along the traditional routes outlined above lies in the fact that the present solute is a globular macromolecule and that approaches developed for chain molecules are inadequate by nature. The modeling remains rather inaccurate even if higher members of the series expansion are included and it does not predict the experimentally observed phase equilibria. Considerations similar to the ones described above also hold true for the application of a newer approach [13] eliminating some deficiencies of the original Flory-Huggins theory. This can for instance be seen from the fact that the normally concentration independent parameter ν of this approach (accounting for the different surfaces of solvent molecules and polymer segments) must be treated as concentration dependent in order to model the present results with similar quality as the Flory-Huggins theory. Flory's theory for stiff rods [14] as well as the theory of Semenov and Rubinstein [15,16] fail too.

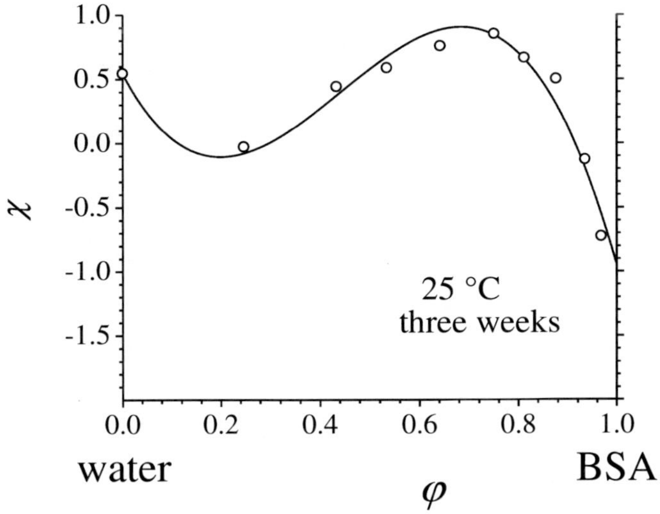

Figure 4. Composition dependence of interaction parameter water/BSA obtained from the measured vapor pressures according to eq (1), the data point for vanishing polymer concentration stems from light scattering measurements. The curve is modeled by means of a series expansion of χ (cf. eq (3)) up to i=3 ($\chi_o = 0.545$, $\chi_1 = -7.242$, $\chi_2 = 23.50$, $\chi_3 = -17.75$).

For the reasons described above we have worked out a purely phenomenological approach, [17] which is also capable of modeling solutions of globular or charged macromolecules. The relation for the composition dependence of the vapor pressures corresponding to eq (1) reads.

$$\ln\frac{p}{p_o} = (z-k)\varphi + b\varphi^2 + c\varphi^2(1-2\varphi) + z\ln(1-\varphi) \tag{4}$$

where the parameter k is calculated from the molar mass M of the polymer and its density ρ plus the molar volume of the solvent according to

$$k = \frac{\overline{V}_1\rho}{M} \tag{5}$$

The parameters b and c originate from a series expansion of the Gibbs energy of mixing; b quantifies binary interactions solvent/polymer and c ternary interactions of the type solvent/polymer/polymer. Formally the inverse of the parameter z corresponds to an effective number of solvent segments, which is in the Flory-Huggins theory by definition set equal to unity.

The curves connecting the measured vapor pressure data in Figure 5 are modeled by means of eq (4) adjusting the parameters b, c and z. This graph demonstrates that despite the use of three parameters only (instead of four in the case of the series expansion) the functions deviate less from the data points.

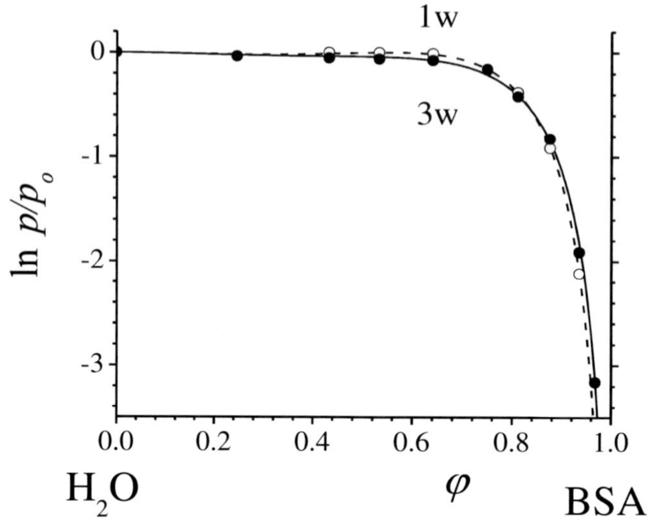

Figure 5. Evaluation of the composition dependent reduced vapor pressures according to eq (4). For 1 week the parameters are: b=2.26, c= - 1.57, z=2.94 and for 3 weeks they are b=1.68, c= - 1.25, z=2.36 (k=2.7 10^{-4} in both cases).

In the following we check to which extent the thermodynamic information acquired from vapor pressures can model the observed phase behavior. To that end it is not only necessary to know the Gibbs energy of *dilution* (eq (4)) but also the corresponding Gibbs energy of *mixing*, $\Delta\overline{G}$. For the present approach this relation reads [17]

$$\Delta G_1 = \Delta G - \varphi \frac{\partial \Delta G}{\partial \varphi} \tag{6}$$

$$\frac{\Delta\overline{G}}{RT} = z(1-\varphi)\ln(1-\varphi) + k\varphi\ln\varphi + b\varphi(1-\varphi) - c(1-\varphi)\varphi^2 \tag{7}$$

The easiest way to recognize the existence or absence of miscibility gaps is provided by the calculation of the spinodal conditions (e.g., the limits between metastability and instability of the mixture). Under these circumstances the second derivative of $\Delta\overline{G}$ with respect to φ becomes zero. From eq. (7) one obtains the following expression

$$\frac{\partial^2\left(\Delta\overline{G}/RT\right)}{\partial\varphi^2} = -2(b+c) + \frac{k}{\varphi} + \frac{z}{1-\varphi} + 6c\varphi \tag{6}$$

Figure 6 displays the second derivatives calculated by means of the parameters following from the vapor pressure data (cf. legend of Figure 4). According to the broken curve (equilibration of one week) the system H_2O/BSA should be unstable within the composition range from $\varphi \approx 0.3$ to 0.56; this prediction exaggerates the extension of the miscibility gap considerably. It does, however, not yet refer to equilibrium conditions, which are prevailing after three weeks. Figure 6 shows these data by the full line, which comes very close to zero but does not fall below this value. Keeping the experimental uncertainties in mind and allowing for the necessity to integrate and to differentiate twice in order to obtain eq (6), the prediction is noteworthy close to reality. The shift in the second derivatives with time indicates that the miscibility gap narrows during the equilibration period. This observation is consistent with the lower vapor pressures (more favorable interaction of the solvent with the polymer) after three weeks as compared with that for one week (Figure 1).

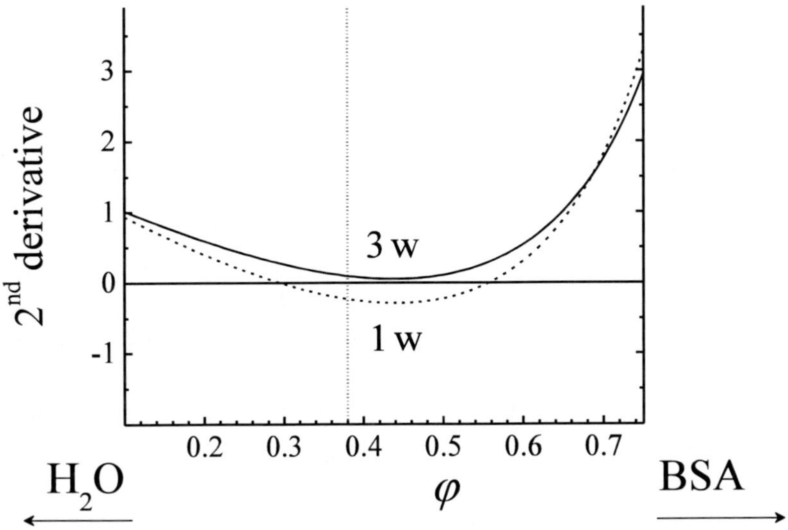

Figure 6. Second derivative of the Gibbs energy of mixing as a function of composition calculated according to eq (6) with the parameters of Figure 5; the dotted vertical line displays the center of the experimentally observed miscibility gap.

The phase behavior of the present system differs fundamentally from that observed with solutions of uncharged chain molecules by the fact that the polymer rich coexisting phase exhibits microscopic inhomogeneities as documented by the micrograph of Figure 3. Such pictures are typical for solutions of colloids or ionic solutes and have already been discussed extensively in the literature. [18] For the present case we assume that locally ordered volume elements lead to birefringence.

According the above cited investigations free particles are roughly speaking moving in a Brownian manner, whereas particles located within the ordered regions oscillate around their lattice point, the structure as a whole moving rather slowly. This situation implies that at least two diffusion processes exist: a fast one for free particles and a slow one for particles caught in ordered regions. It is the latter mode that could be responsible for the slow equilibration process observed in the context of the vapor pressure measurements.

For the discussion of the temperature influences on the extension of the miscibility gap shown in Figure 2 it would appears interesting to have a look on the heat effects associated with the formation of the ordered structures. Unfortunately such information is presently not available; according to theoretical considerations [19] the potential is not very deep, which means that thermal measurements need to be very accurate to yield reliable data. Conclusions based on the observed phase diagram (phase separation upon cooling) indicate moderate to small endothermal heats of mixing. This seems to contradict structure formation, requiring that the loss of entropy is overcompensated by favorable exothermal heats of mixing. However, in view of the complexity of the present system, i.e., possible restructuring of the biopolymer and changes in the structure of water by the solute the above qualitative reasoning is not conclusive.

The observed microscopic inhomogeneities (birefringent parts of Figure 3) are according to the above discussion attributed to weak favorable interactions between the solute. [18] This line of reasoning implies that the ordered structures can be readily broken by shear. This inference is in good agreement with the pronounced shear thinning behavior observed with the present system. [20-22]

CONCLUSION

One central finding of the present vapor pressure measurements concerns their pronounced time dependence, which – to our knowledge – has not yet been reported for protein solutions. Such effects have, however, been observed when studying the formation of polyelectrolyte complexes [23], where this phenomenon was attributed to slow rearrangements of the segments of the oppositely charged macromolecules. Analogous changes in the location of hydrophilic and hydrophobic parts of BSA are probably also required to attain the minima of the Gibbs energy.

The measured equilibrium vapor pressures yield the Gibbs energy of dilution as a function of composition. For solutions of chain molecules it is normally possible to describe them by means of the Flory-Huggins theory, for BSA this approach fails. Modeling attempts show that even a series expansion of four terms does not suffice. Moreover, the composition dependence of the Gibbs energy does not forecast the experimentally observed miscibility gap between water and BSA. These observations are, however, not surprising in view of the fact that BSA is a globular macromolecule, i.e., that the laws established for chain molecules become obsolete.

In view of the above described impracticality of the Flory-Huggins theory we have designed a phenomenological approach, which is capable to reproduce the composition dependence of the measured chemical potential of the solvent by means of three adjustable parameters. This relation complies with all laws of thermodynamics and predicts the existence of a miscibility gap between the components. According to the so far available experimental information on other protein solutions and on polyelectrolyte solutions, the present approach is generally applicable.

ACKNOWLEDGMENTS

The authors are grateful to the DAAD and to the DFG for their support. Furthermore they would like to thank the coworkers of Prof. R. Zentel (department of organic chemistry) for helping us with their microscopy experience.

REFERENCES

[1] Albertsson, P.-A. Partition of Cell Particles and Macromolecules, 1986, Wiley-Interscience, New York, NY.
[2] Tolstoguzov, V.B. Food Hydrocolloids 1991, 4, 429-468.
[3] Antonov,Y.A.; Grinberg, V.Ya.; Tolstoguzov, V.B. *Vysokomolekularnie Soedineniya (Macromolecules USSR)*, 1976, B 18, 566-569.
[4] Eckelt, J.; Sugaya, R.; Wolf, B. A. *Biomacromolecules* 2008, 9, 1691-1697.
[5] Hiryama, K. *BBRC* 1990, 173, (2), 639.
[6] Foster, J. F., In *Albumin Structure, Function and Uses*, Pergamon: Oxford, 1977; pp 53-84.
[7] Kirschenbaum, D. M. *Analytical Biochemistry* 1977, 81, (1), 220-246.
[8] Petri, H.-M.; Wolf, B. A. *Macromolecules* 1994, 27, 2714-2718.
[9] Zintchenko, A.; Rother, G.; Dautzenberg, H. *Langmuir* 2003, 19, (6), 2507-2513.
[10] Bakeev, K.; Izumrudov, V.; Kuchanov, S.; Zezin, A.; Kabanov, V. *Macromolecules* 1992, 25, (17), 4249-4254.
[11] Bercea, M.; Nichifor, M.; Eckelt, J.; Wolf, B. A. *Macromol Chem Phys* 2011, 212, (17), 1932-1940.
[12] Koningsveld, R.; Stockmayer, W. H.; Nies, E., *Polymer phase diagrams a textbook.* Oxford University Press: Oxford ; New York, 2001; p xvii, 341 p.
[13] Wolf, B. A. *Adv Polym Sci* 2011, 238, 1-66.
[14] Flory, P. J.; Ronca, G. *Mol Cryst Liq Cryst* 1979, 54, (3-4), 311-330.
[15] Rubinstein, M.; Semenov, A. N. *Macromolecules* 1998, 31, (4), 1386-1397.
[16] Semenov, A. N.; Rubinstein, M. *Macromolecules* 1998, 31, (4), 1373-1385.
[17] Wolf, B. A. *Macromolecules* 2012 Dec 15, submitted
[18] Ise, N. *Proceedings of the Japan Academy Series B-Physical and Biological Sciences* 2002, 78, (6), 129-137.
[19] Sogami, I.; Ise, N. *J Chem Phys* 1984, 81, (12), 6320-6332.
[20] Lefebvre, J.; Riot, A.-S. *Conference proceedings 1st International Symposium on Food Rheology and Structure, Swiss Federal Institute of Technology, Zurich, Switzerland* 1997, 175-179.
[21] Ikeda, S.; Nishinari, K. *Biomacromolecules* 2000, 1, (4), 757-763.
[22] Sharma, V.; Jaishankar, A.; Wang, Y.-C.; Mckinley, G. H. *Soft Matter* 2011, 7, (11), 5150-5160.
[23] Bercea, M.; Nita, L.-E.; Eckelt, J.; Wolf, B. A. *Macromol Chem Phys* 2012, 213, (23), 2504–2513.

In: News in Chemistry, Biochemistry and Biotechnology ISBN: 978-1-63117-273-1
Editors: G. E. Zaikov, G. Nyszko, L. P. Krylova et al.

Chapter 18

PHASE TRANSITIONS IN WATER-IN-WATER BSA/DEXTRAN EMULSION IN THE PRESENCE OF STRONG POLYELECTROLYTES

***Y. A. Antonov*[*,1] *and P. Moldenaers*[2]**

[1]N.M. Emanuel Institute of Biochemical Physics, Russian Academy of Sciences, Moscow, Russia

[2]K.U. Leuven, Department of Chemical Engineering, Leuven, Belgium

ABSTRACT

We examine whether a small amount of strong polyelectrolyte (dextran sulfate sodium salt /DSS/) can induce mixing of BSA and dextran in aqueous water-in-water BSA/dextran emulsion and how intermacromolecular interactions affect its rheological properties.

Addition of DSS to water-in-water emulsion at pH 5.4 leads to its mixing, a noticeable increase in viscosity and module (G′). Mixing is observed at the DSS/BSA weight ratio, $q_{BSA} \geq 0.07$. Increasing the ionic strength in the resulting single-phase system induces phase separation. Our results show that the increase in viscoelasticity results from the interaction of DSS with both macromolecular components. The interaction of DSS with BSA leads to the screening of BSA tryptophanyls from the aqueous environment.

Such interaction is not accompanied by the polarization of the protein, whereas the affinity of DSS to dextran results in an increase of viscoelasticity and in an appreciable change in the microstructure of the DSS/dextran mixture.

It was assumed that similar to compatibilization of polymer blends by diblock copolymers, the driving force for inducing mixing of water/BSA/dextran emulsions by DSS results from the affinity of strong polyelectrolytes to both macromolecular components of the mixture.

---

* Corresponding author: Yurij A. Antonov e-mail:chehonter@yandex.ru.

I. INTRODUCTION

Biopolymer incompatibility and complex formation are fundamental physicochemical phenomena that determine the structure and physical properties of biopolymers mixtures [1-3] playing an important role in protein processing [6-8]. Polymer incompatibility is a consequence of the normally unfavorable interactions between polymer species. Even small positive values of the Flory-Huggins interaction parameter between different polymer species can result in phase separation, due to the small entropy gain upon mixing these macromolecules [9-11]. Therefore, even minute differences in the structure of macromolecules may result in phase separation. The introduction of charges on a water-insoluble polymer generally increases its solubility in water. Similarly, the introduction of charges on one of the two polymers in an aqueous polymer mixture increases the miscibility of the two polymers. [12, 13]. Added salt reduces the effects of the charges. For ternary systems solvent/polymer A/polymer B the introduction of charges of the same sign on both polymers reduces the favorable miscibility effect seen when charging only one of the polymers [14-16]. These observations indicate that the mixing entropy of the small counterions of the polyelectrolytes plays an important role for the solubility and for the miscibility in polyelectrolyte systems, and they have inspired theoretical investigations on the Flory-Huggins level of the solubility and miscibility of polyelectrolyte solutions. The behavior of blends of neutral and polyelectrolyte chains in a common solvent has been investigated theoretically by Khokhlov and coworkers [12, 17]. They studied the dependence of free energy on concentration fluctuations, systematically accounting for the translational entropy of polymers and counterions, the interfacial tension, and the electrostatic interactions between the charged species. Their study showes that increasing the charge of the polyelectrolyte has the following effects:

1. It stabilizes the mixed phase, moving the spinodal point to lower temperatures, and increases the extension of the homogeneous region
2. It changes the character of the transition from macro- to micro phase separation, the characteristic length of the micro phase separation being dependent on the polyion charge and on the amount of added salt.

The former effect is due to counterion entropy, whereas the latter results from the new length scale introduced by the electrostatic interactions, the Debye screening length.

Independently, Nilsson [14, 15] and later Johansson et al. [18] numerically solved models based on an extended version of the Flory-Huggins theory, where the counterions were explicitly included as a separate component on the same level as the solvent and the chain molecules. In the spirit of the random-mixing approximation, the charges of each component in a phase are uniformly distributed; leading to zero electrostatic energy and the effect of the electrostatic interactions among the charges enters the model only by imposing that each phase should be electroneutral. Both approaches predict results in qualitative agreement with experimental data [14-17].

The thermodynamic behavior of aqueous systems containing one or two types of charged biopolymers is less clearly analyzable, because the structure of theses macromolecules is more complicated than that of synthetic polymers. In contrast to ternary solvent/polymer/

polymer systems, an introduction of charges of the same sign on both biopolymers frequently does not reduce the miscibility. For example, similarly charged ternary water/linear anionic polysaccharide-1/linear anionic polysaccharide-2 systems are single phase one in all the concentration range, pH and ionic strength [19]. Moreover, data obtained for water-protein-uncharged polysaccharide systems show a variety of different phase behaviours. Some show single phase behaviour at a low ionic strength. However, unsimilar to ternary polymer systems the dominant mechanism responsible for single phase state of such systems at low ionic strength (below 0.01) involves the formation of water-soluble weak interpolymer complexes, which may be destroyed by increasing ionic strength [20]. The other water/protein/uncharged polysaccharide systems are two phase ones at a low ionic strength at relatively high concentrations of biopolymers. Typical examples of such systems are BSA/dextran system. This system is two phase one at pH 5.4 and at relatively high enough concentrations [5]. At low concentrations dextran molecules are able to form interpolymeric complexes with BSA in water if the polysaccharide is in excess and if the protein exists in its associated state [21]. The other example is water/gelatin/dextran system. This system is two phase in water and single phase in the acidic or alkaline range. The possibility of complex formation in this system in the acidic range has been discussed by Woodside and colleagues [22, 23] , and by Grinberg and Tolstoguzov [24]. These authors analyzed the thermodynamic behavior of gelatin/D-glucan mixtures. The existence of gelatin- D-glucan complexes was inferred from the considerable solubility of D-glucans in acidic ethanol in the presence of gelatin [22, 23] and from nephelometric and viscosimetric data [24]. Thus, phase behaviour in the system water-charged biopolymer 1/uncharged biopolymer-2 depends strongly on weak intermacromolecular interactions, the origin of which is not quite clear till now.

Although phase separation in biopolymer systems frequently allows control of the morphology and, hence, the rheology/texture of biopolymer systems [1, 25-27], in many cases it leads also to a spontaneous separation into two layers, which is not desired in many food and biotechnology processes, for example, blood substitutes production.

In polymer processing compatibilizers are frequently used [28] to stabilize a fine microstructure of synthetic polymer blends and to increase the interfacial adhesion between their phases. Normally these compatibilisers are copolymers, containing two blocks each compatible with one of the polymers, or so called ionomers containing both nonionic repeat units, and a small amount of ion-containing repeat units. It is less clear if it is possible to compatibilise two phase biopolymer mixtures, because experimental observations in this field are lacking.

Our recent studies show [29, 30] that intermacromolecular interactions caused by the presence of a complexing agent (dextran sulfate sodium salt /DSS/) in semi-dilute single-phase protein-anionic polysaccharide mixtures can induce phase separation and structure formation at pH values above the isoelectric point of the protein, due to water soluble protein/DSS associations resulting in network formation.

The aim of this work is to examine whether a strong polyelectrolyte can lead to the opposite phenomenon, i. e. mixing in concentrated globular protein-polysaccharide mixtures, if the capacity of the protein to form large interpolymer associates is weak and if the polysaccharide does not contain charged functional groups. Assuming that the question of weak interactions between different macromolecular species is important for understanding the phenomena of the incompatibility of biopolymers, and taking into account the lack of experimental data in this area, the present study deals with the relationship between the phase state of protein–neutral polysaccharide mixtures in the presence of a strong polyelectrolyte, and possible interactions between these biomacromolecules.

We focused on aqueous two-phase systems formed by bovine serum albumin (BSA) and high molecular weight dextran (a highly branched nonionic polysaccharide). The water/BSA/dextran system is chosen because of the low compatibility of these macromolecules under the selected conditions. Moreover, these biopolymers are widely used for biomedical and bioseparation purposes [33], and the thermodynamic behavior and phase diagram of the ternary water/BSA/dextran system has been described in past [5].

The effect of a sulfate polysaccharide (DSS) on phase equilibrium, macromolecular interactions and structure of the two-phase water-gelatin-dextran system was studied using a rheo-small angle light scattering (SALS) device, optical microscopy (OM), phase analysis, static light scattering (SLS), rheology, and isoelectric focusing.

II. Materials and Methods

Materials

The BSA Fraction V, pH 5 (Lot A018080301), was obtained from Across Organics Chemical Co. (protein content = 98-99%; trace analysis, Na < 5000 ppm, CI < 3000 ppm, no fat acids were detected). The isoelectric point of the protein is about 4.8-5.0 [34], and the radius of gyration at pH 5.4 is equal to 30.6 Å [34]. The water used for solution preparation was distilled three times. Measurements were performed at pH 5.4, because serum albumin undergoes conformational isomerization and changes in the conformation state and secondary structure with changes in pH from pH 5-5.5 to acid and alkaline region [35]. The extinction of 1% BSA solution at 279 nm was $A^{1cm}{}_{279}$= 6.70, and that value is very close to the tabulated value of 6.67 [35]. The high molecular weight dextran T-2000 sample was purchased from Amersham Pharmacia Biotech AB. Its radius of gyration, Stokes radius, intrinsic viscosity in water at 20°C and weight average molecular mass, reported by the manufacturer, are 380 Å, 270 Å, 0.9 dL/g and $2 \cdot 10^6$ Da respectively. Dextran 2000 behaves in solution as a highly branched expanded coil. Dextran sulfate (DSS) (M_W = 500 kDa, M_n = 166 kDa, intrinsic viscosity η (in 0.01 M NaCl) = 0.5dL/g, 17% sulfate content, free SO_4 less than 0.5%) was produced by Fluka, Sweden (Reg. No. 61708061 A, Lot No. 438892/1).

Preparation of the protein/polysaccharide mixtures. To prepare BSA and dextran stock solutions, the biopolymer was gradually added to the three times distilled water and stirred at 298 K for 2 hour. The solutions were centrifuged at 13.000 g for 1 hour at 298 K to remove insoluble particles. Subsequently, the concentration of the biopolymer was determined by measuring the dry weight residue. In the case of BSA the content of the protein nitrogen in dry BSA sample was always taken into account in order to calculate the concentration of protein in solution. In some cases, the final protein concentration was determined also by spectrophotometric measurements.

All the measurements were performed after equilibrating the biopolymer solutions and their mixtures for12- 15 h. Both, the BSA and the dextran solutions show Newtonian behavior (at temperatures of 298 K and at shear rates up to 30 s^{-1}) in the concentration range from 25-30 wt% within which the hydrodynamic volumes of the different macromolecules already overlap [36].

Methods

Turbidimetry at rest. Cloud points of aqueous solutions of BSA plus DEX, and BSA plus DEX plus DSS were determined by measuring their light transmittance I either as a function of temperature or of composition. A laser beam (5mW, He-Ne laser, 632.8 nm) of intensity I_o passes the solution and the intensity, I of the transmitted light is determined.

The apparatuses and the procedures for the variation of temperature have already been described earlier in detail [37]. The demixing temperatures, T_1, are defined as the intercept of the tangent to the transmittance curves (I/I_o as a function of T) in the point of inflection, with $I/I_o = 1$. Isothermal measurements were performed by titrating solutions of BSA with solutions of DEX, and (BSA and DSS)+(DEX+DSS). In this case one determines the phase separation conditions from plots of the transmittance as a function of the amount the DEX solution added. Characteristic demixing compositions were determined as the intercept of the tangent to the transmittance curves (I/I_o as a function of the concentration of the DEX solution added) in the point of inflection, with $I/I_o = 1$.

Rheo-optical study A rheo-optical methodology based on small angle light scattering (SALS) during flow, is applied to study in-situ and on a time-resolved basis the structure evolution. Light scattering experiments were conducted using a Linkam CSS450 flow cell with parallel-plate geometry. The CSS450 uses two highly polished quartz plates that are parallel to within 2um. Each plate is in thermal contact with an independently controlled pure silver heater utilizing platinum resistors sensitive to 0.1°C. A 5 mW He-Ne laser (wavelength of 633 nm) was used as light source. The 2D scattering patterns were collected on a screen by semi transparent paper with a beam stop and recorded with a 10-bit progressive scan digital camera (Pulnix TM-1300). Images were stored on a computer with the help of a digital frame grabber (Coreco Tci-Digital SE). The optical acquisition set-up has been validated for scattering angles up to 18°. The gap between the plates has been set at 1 mm and the temperature was kept 20°C by means of a temperature-controlled water bath. In house developed software was used to obtain intensity profiles and contour plots of the images (New SALS SOFT-WARE-K.U.L.). Turbidity measurements have been performed by means of a photo diode.

Bright light microscopy is used to visualize particles distributed in the coexisting phases, using an Olympus BX51W1 fixed stage microscope equipped with a high resolution CCD-camera, (1000x1000 pixels, C-8800-21, Hamamatsu).

DLS. Determination of Intensity- size distribution, and volume-size distribution functions, as well as zeta potentials of BSA, DSS, and BSA+DSS particles were performed, by the Malvern Zetasizer Nano instrument (England), using a rectangular quartz capillary cell. The concentration of BSA in the water- BSA/DSS mixtures was kept 0.20% (w/w). For each sample the measurement was repeated 3 times. The samples were filtered before measurement through DISMIC-25cs (cellulose acetate) filters (sizes hole of 0.22 μm for the binary water-protein solutions. Subsequently the samples were centrifuged for 30 seconds at

4000 g to remove air bubbles, and placed in the cuvette housing. The detected scattering light intensity was processed by Malvern Zetasizer Nano software.

Fluorescence emission spectra between 280 and 420 nm were recorded on a RF 5301 PC Spectro-fluorimeter (Shimadzu, Japan) at 25 °C with the excitation wavelength set to 270 nm, slit widths of 3 nm for both excitation and emission, and an integration time of 0.5 s. The experimental error was approximately 2%.

Rheological measurements were performed using a Physica Rheometer, type MCR 501, with a cone-plate geometry CP50-1/Ti (cone diameter 5 cm, angle 0,993°), Anton Paar. The temperature was controlled at 18 °C by using a Peltier element. For each sample, flow curves were recorded in the shear rate range 0.1 to 150 s^{-1}. The ramp mode was logarithmic. The mechanical spectra were recorded over the 0.1–200 rad/s frequency range; the dynamic moduli G' (storage) and G'' (loss) were measured at a constant strain amplitude of 3%, which was checked as being within the linear regime. During the rheological measurements, all samples were covered with paraffin oil to avoid drying.

Environment scanning electron microscopy (ESEM). Micro structural investigation was performed with the environment scanning electron microscope Philips XL30 ESEM FEG. The samples were freeze-fractured in freon and immediately placed in the ESEM. Relative humidity in the ESEM chamber (100%) was maintained using a Peltier stage. Such conditions were applied to minimize solvent loss and condensation, and control etching of the sample. Images were obtained within less than 5 minutes of the sample reaching the chamber. The ESEM images were recorded multiple times and on multiple samples in order to test reproducibility.

Iso-electric focusing (IEF) was performed with the IPGphor isoelectric focusing system (Amersham Biosciences) at 45°C using 2 µl IEF Standard (Bio-rad #161-0310) range pI 4.45-9.6. Run conditions: 51V for 1 hour, 200 V for 1.5 hour, 200 V for 1 hour, 500 V for 30 minutes. Protein was fixed in a bath containing 40% methanol, 10 % trichloro acetic acid. Gels were stained overnight in a Sypro Ruby fluorescent protein stain (Invitrogen) bath and then scanned with Typhoon Imager Analyzer (GE Healthcare).

III. Results and Discussion

DSS-Induced Mixing

The experimental results shown in this section have been obtained on water(76.90 wt%) /BSA(7.70 wt%)/dextran(15.40 wt% 2000 systems located in the two-phase region far from the binodal line [5].

The volume fraction of the BSA-enriched disperse phase was 0.11 [38]. To study the effect DSS on the phase behavior, the mixtures water (76.90% - X wt%)/BSA (7.7 wt%)/dextran (15.4 wt %) /DSS (X wt%) were prepared. Here X corresponds to a variable amount of DSS. The system was hand mixed prior to loading into the rheo-optical device. First, a preshear of 0.5 s^{-1} was applied for 1000 s (500 strain units). Subsequently, this preshear was stopped, and the sample was allowed to relax for 30 s, leaving enough time for full relaxation of deformed droplets. Then SALS patterns are monitored. The SALS patterns and the scattering intensity upon adding different amount of DSS are shown in Figure 1, starting from of DSS concentrations as low as 0.01 wt%. In each experiment, a freshly loaded sample has been used.

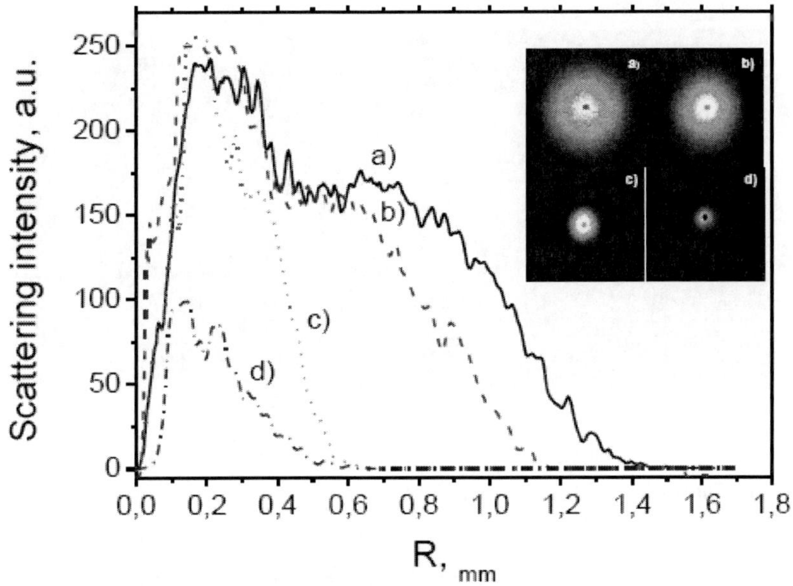

Figure 1. SALS patterns and scattering intensity of water (76.9 %)/BSA (7.7 wt%)/dextran (15,4 wt %) emulsion. upon adding different amount of DSS. 20°C Concentration of DSS (wt%) : a) 0.0, b) 0.03, c) 0.05, d) 0.08. The SALS patterns were obtained after shearing at 0.5 s^{-1} for 1000 s and subsequently left to rest for 30 s before measurements.

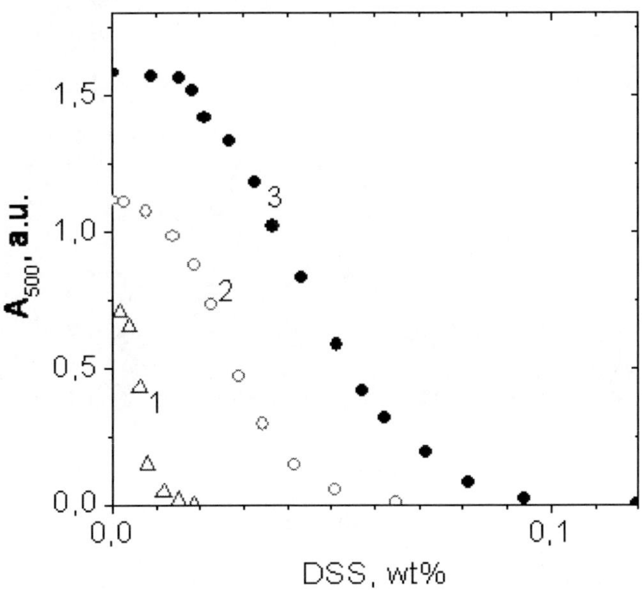

Figure 2. Dependences of light absorbance at 500 nm on concentration of DSS in the emulsion for BSA(variable)/dextran(15.4 wt%) emulsions. Concentration of BSA in the emulsion: 1)-10 wt%; 2)-5.0 wt%; 3)-2.5 wt%. The absorbance of pure BSA solution at the same concentrations was subtracted. 20°C.

The presence of 0.03 wt% of DSS in the two phase system led to appreciable decrease in the SALS signal (Figure 1, curves a,b). The higher the concentration of DSS, the weaker the scattering intensity becomes (Figure 1, curves b,c,d).

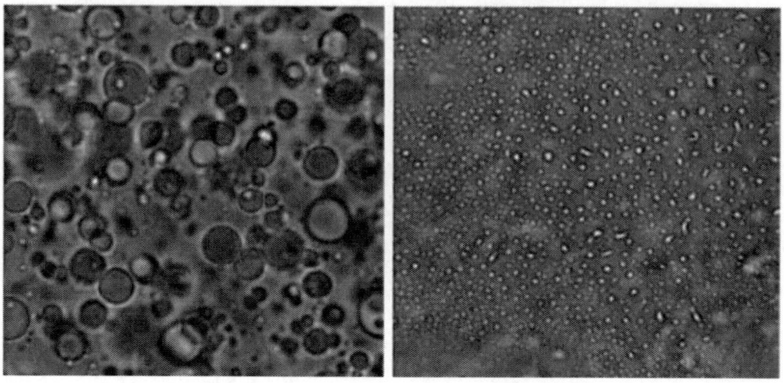

Figure 3. Photographs of water(76.9 wt%)/BSA(7.7 wt%)/Dextran(15.4 wt%) emulsion without DSS (Left); with 0.07 wt% of DSS (Right). The full length of each image corresponds to 0.2 mm. Images obtained after stop of preshear at 0.5 s^{-1} for 1000 s and subsequent 30 s rest period. The full length of each image corresponds to 0.2 mm.

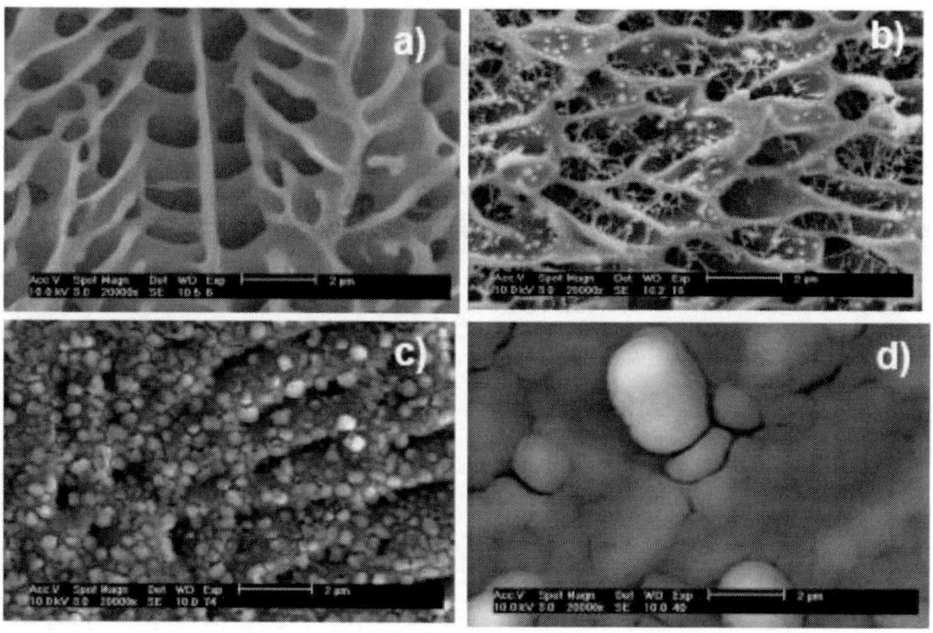

Figure 4. ESEM images of the water(76.9 wt%)/dextran (15.4 wt%) system, (a); water(98 wt%) / DSS(2 wt%) system, (b); water(92.3 wt%)/BSA(7.7 wt%)/DSS(0.03 wt%) system, (c); water/BSA(7.7 wt%)/dextran(15.4 wt%)/DSS(0.14 wt%) system, (d). pH 5.4.

This indicates that the sample is less and less heterogeneous on the length scales probed by light scattering (in the order of 0.5 micron). It can be argued that the scattering power of small structures decreases significantly and therefore the sensitivity of the camera becomes insufficient to pick up the scattering patterns. Therefore, in addition to SALS experiments, the effect of the presence of DSS on the absorption values (A_{500}) of water/BSA(variable)/dextran

(15.4 wt%), containing constant concentration of dextran and different concentrations of BSA, was examined (Figure 2).

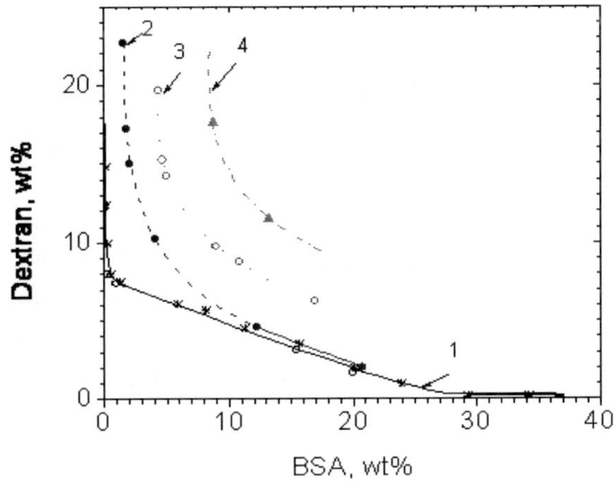

Figure 5. Cloud point curves of the water/BSA/dextran/DSS systen at 25°C. pH=5.4. Concentration of DSS in the water/BSA/DSS and water/dextran/DSS solutions (wt%) before mixing are 0.0, (1); 0.015,-(2); 0.03, (3); and 0.08 (4) respectively.

The concentration of DSS at which the absorption value reaches the same value as the one measured for the external / continuous/ phase is in agreement with the concentration of DSS at which the SALS pattern disappears. The presence of DSS in the water-in-water BSA/dextran emulsions results in dramatic changes in its morphology. The corresponding microscopy images for the same concentrations DSS and the same flow conditions are shown in Figure 3. After the addition of 0.07 wt% DSS to the emulsion, the droplets became much smaller and their volume fraction decreased, meaning a strong reduction of the interfacial tension and an increasing the thermodynamic compatibility of the biopolymer pair. The ESEM images of water/dextran (15.4 wt %) system, water/DSS(2.0 wt%) system, water/BSA(7.7 wt%)/dextran(15.4 wt%)-DSS(0.03 wt%) system, and water/BSA(7.7 wt%)/dextran(15.4 wt%)/DSS(0.14 wt%) system are shown in Figure 4. Dextran develops a skeleton-like structure (Figure 4a) whereas DSS shows the the skeleton-like structure with the cobweb of the sulphate functional groups (Figure 4 b). In the joint solution of BSA and dextran containing small amount of DSS (0.03 wt%) skeleton-like structure of polysaccharides and close to spherical structure of BSA were registered, whereas at a higher concentrations of DSS the system develops (Figure 4 d) an amorphous structure similar to that observed for compatibilized polymer blends [38].

Centrifugation of the water-BSA(7.7 wt%)/dextran(15.4 wt%)/DSS(0.14 wt%) system (120 min. 60.000 g, 45°C) prepared in the same way did not result in macroscopic phase separation. This indicates that the presence of DSS moved the system on the phase diagram outside the two-phase range. In order to quantify the effect of DSS on phase equilibrium in the system BSA/dextran, the cloud point curves of the water/BSA/dextran system were determined at different concentrations of DSS at pH 5.4 and 25°C. (Figure 5). It is important to note that our preliminary experiments showed that DSS is contained almost quantitatively in the BSA-rich phase. The total concentration of the biopolymers at the threshold point increases from 8.76 wt % in the absence of DSS to 24.4 wt % as the DSS concentration reaches 0.045 wt%. At higher DSS concentrations, phase separation was not observed, whatever the total biopolymers concentration in the range studied (6.0 to 30.0 wt %).

EFFECT OF DSS ON RHEOLOGICAL PROPERTIES OF BSA/DEXTRAN WATER IN WATER EMULSIONS

We also characterized the phase state of the systems by means of their viscoelastic behaviour. It has been shown [25] that at moderately low shear rates, the biopolymer emulsions can be regarded as conventional emulsions and various structural models are available in the literature to prediction the morphology. The evolution of the mechanical spectrum was investigated as a function of DSS concentration. The rheology of the water/BSA(8.46 wt%)/Dextran (11.8 wt%) system containing 15 vol % BSA enriched phase was examined before and after addition of DSS at concentrations from 0.2 to 2.13 wt%? (q_{BSA} values from 0.02 to 0.25).

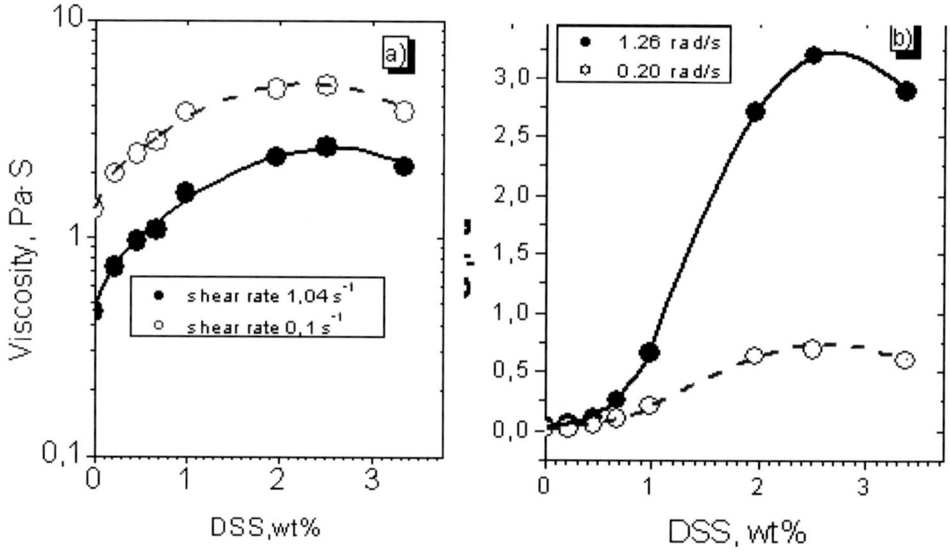

Figure 6. Dynamic viscosity of the water (79.74 wt%)/BSA(8.46 wt%)/dextran(11.8 wt%)/DSS (variable) system as a function of the concentration of DSS at a shear rate 0.1 s^{-1} and 1.04 s^{-1}, (a); the values of G' of the same system as a function of the concentration of DSS at a frequency 0.2 rad/s and 1.26 rad/s. (b). 18°C, pH 5.4.

The experimental flow protocol was the same as the one used for the rheo-SALS measurements. The viscosity of the system as a function of the concentration of DSS at a shear rate 0.1 s^{-1} and 1.04 s^{-1}, and the values of G' of the system as a function of the concentration of DSS at a frequency 0.2 rad/s and 1.26 rad/s are shown in Figure 6 a,b. In the presence of DSS, the system undergoes mixing, and this transition leads to an appreciable increase of the moduli.

The viscoelastcity of the water/BSA/dextran system, which is very low in absence of DSS, increases markedly by the presence of DSS. The growth of viscoelasticity peaks for 2.5 wt% DSS (q_{BSA}=0.295), concentration which 4 times higher then, according to rheo-SALS results (Figure 1), to the transition of the system from two-phase state to single-phase state. Thus, at 1 rad/s, G´ value is more than 300 times higher than that of the emulsion without DSS (Figure 6 a). In the presence of high ionic strength (0.25/ NaCl/), i. e. when electrostatic interactions are suppressed, the mechanical spectrum of the system becomes insensitive to the presence of DSS (data are not presented). Similar changes were observed for the viscosity. At q_{BSA}=0.04 and a shear rate of 1.0 s^{-1} the viscosity is 5,8 times higher compared with that of the single-phase system with almost the same composition (Figure 6 b). It is important to note that in the shear rate range from 0.1 to 150 s^{-1} flow curves obtained with ascending and descending ramps superimposed (data not presented).

Intermacromolecular Interactions As a Driving Force of Mixing

To understand the reasons for such dramatic effects, the dynamic modules and the viscosity of the water-BSA-DSS and water/dextran/DSS systems were measured as a function of the DSS/BSA, and DSS/dextran weight ratio (q_{BSA} and q_{DEX} respectively)-(Figures 7, and 8).

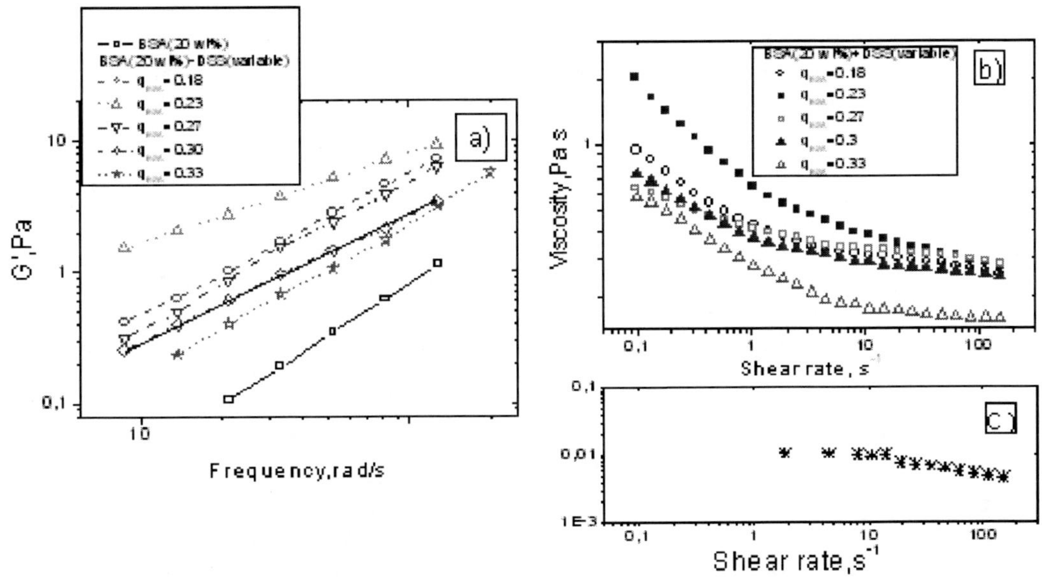

Figure 7. (a)The dynamic module G′ as a function of frequency for the water/BSA/DSS system after preshearing at 0.5 s^{-1} for 1000 s and subsequent 30 s rest period, and (b) the flow curves of the water/BSA(20 wt%)/DSS(variable) systems at different DSS/BSA weight ratio (q_{BSA}), and (c) flow curve of the water/DSS(18wt%) system. pH 5.4 18°C.

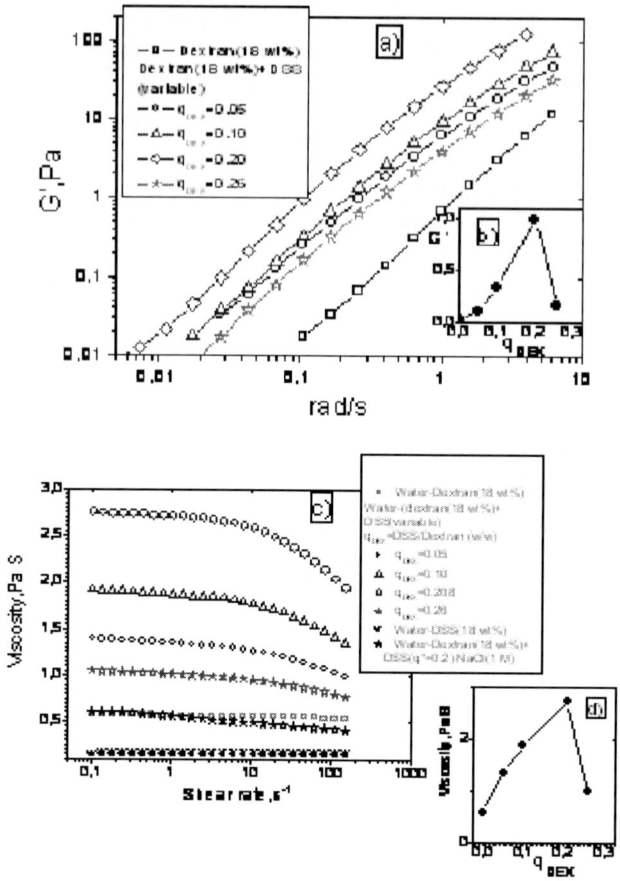

Figure 8. (a) The G′ values as a function of frequency for the water/dextran (18 wt%)/DSS (variable) system at different q_{DEX} values. (b) G′ values of the same system at a frequence 0.1 rad/s as a function of q_{DEX} . After preshearing at 0.5 s^{-1} for 1000 s and subsequent 30 s rest period. (c) flow curves of the same systems at different q_{DEX} values , (d) viscosity of the water/dextran (18 wt%)/DSS (variable) system as a function of q_{DEX} at a shear rate 1 s^{-1}. pH 5.4 18°C.

The dynamic module G′, and the viscosity of the water-BSA-DSS systems were measured as a function of the DSS/BSA weight ratio, q_{BSA} (Figures 7). As can be seen, (Figure 7 a,b,c) the G′, and the viscosity of BSA in presence of DSS depends pronouncedly on q_{BSA}, and shows a maximum at $q_{BSA} \cong 0.25$. At q_{BSA}=0.25. The viscosity at shear rate 1 s^{-1}, and G′ values at 0.1 rad/s are more than 60 and 27 times higher, respectively, than those of the pure BSA solution. Unexpectedly, similar dependences of the G′, and viscosity of the water/dextran/DSS system as a function of q_{DEX} were detected (Figure 8). These dependences show a maximum at q_{DEX} =0.2. At q_{DEX}=0.2, the viscosity at shear rate 1 s^{-1}, and G′ values at 0.1 rad/s are more than 17 and 57 times, respectively, those of the pure dextran solution. From theory [39], we know that the dependences of G′ and of η for aqueous polymer system as a function of the concentration of the other polymer in the same solution is typical for the formation of inter-polymer complexes Therefore, it can be assumed that the dramatic changes in rheological behaviour of the water-BSA and water-dextran systems (Figures 7, 8) in the presence of DSS are due to interactions of DSS with BSA and with dextran.

Since intermacromolecular interactions takes place in the water-BSA-dextran-DSS system it will be useful to understand how such interactions affect the size particles, structure development and possible polarization of BSA in the presence of DSS. The intensity size distributions functions for BSA (0.20 wt %), DSS (0.20 wt %), and their mixtures at different q_{BSA} values. pH 5.4. 25°C. are presented in Figure 9. It can be seen that at q_{BSA}=0.07 and 0.14, when according to SALS data the water-BSA-dextran system is homogenized (Figure 1), the average size of the complex particles is only slightly higher than that of the DSS particles. It means that the size of the complex particles is determined mainly by the size of the DSS molecules.

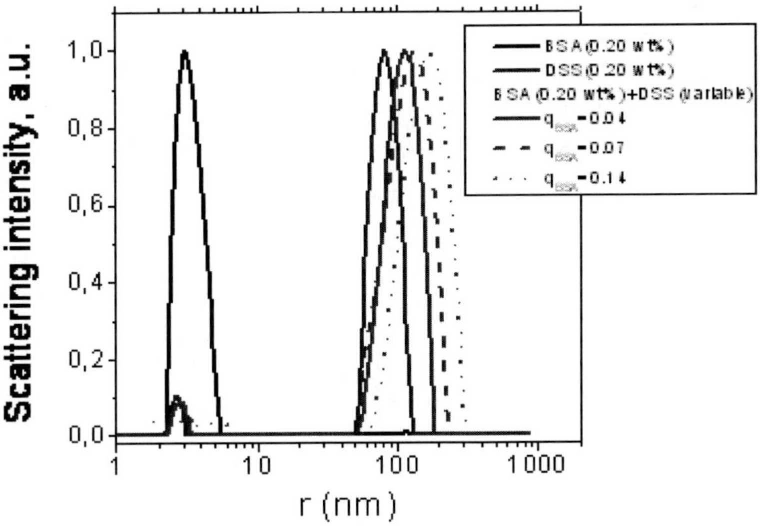

Figure 9. The intensity size distributions function for BSA (0.20 wt %), DSS (0.20 wt %), and their mixtures at different q_{BSA} values. pH 5.4. 20°C.

Such intermacromolecular interactions can affect the structure of the BSA macroions and their isoelectric point due to polarization, and as consequence, change its compatibility with dextran. Let us to analyze this assumption in detail.

Measurements of the fluorescence intensity are frequently used to study possible structural changes of proteins in processes of their complex formation with other polymers. The changes of protein fluorescence may be characterized by the wavelength at the maximum emission (λmax) and the maximum fluorescence intensity (Imax). The fluorescence intensity of proteins can be decreased by a variety of molecular interactions, including excited-state reactions, molecular rearrangements, energy transfer, ground state complex formation and collision quenching [40].

Fluorescence spectra of pure BSA, and BSA in the presence of DSS are given in Figure 10. It is well-known that tryptophan (Trp) fluorescence of proteins varies with conformational changes of these biopolymers resulting in changing of fluorescence parameters, such as emission maximum (λmax), quantum yield, lifetime, and others [40, 41]. The wavelength of maximum emission (λmax) of pure BSA was found to be 339-340 nm. The emission maximum is usually shifted from shorter wavelengths to about 350 nm upon protein unfolding, which corresponds to the fluorescence maximum of pure tryptophan in aqueous solution. As shown in Figure 10 the fluorescence intensity (I_{max}) of BSA decreases (quenching) in the presence of DSS and λ_{max} shows a clear shift as the weight ratio of DSS/BSA increases in mixture. At the maximal q_{BSA} values, blue spectral shift of BSA fluorescence reaches λmax=324 nm, which suggests the screening of BSA tryptophanyls from water environment [41]. BSA tryptophanyls become less accessible in water solution, which may be the result of increasingly tight binding of DSS with protein. We interprete these blue shifts as the result of shielding of tryptophan residues from aqueous media by the complexation of protein globules with DSS chains.

The most intriguing type of affinity is that of dextran and DSS shown in Figure 8. We did not find any literature reports on the existence of such type of complexes. However, there are indications for some interaction of dextran with polyampholytes. The possibilities of complex formation in the systems containing the charge and neutral polysaccharide have been discussed by Woodside and colleagues [22, 23], Grinberg and Tolstoguzov [24], and Antonov [42] on the basis of the analysis of the phase behavior of dextran-gelatin, and dextran-caseinate mixtures. Unfortunately most data in this field stem from studies performed a long time ago, when structural methods where less available [22-24, 44, 6].

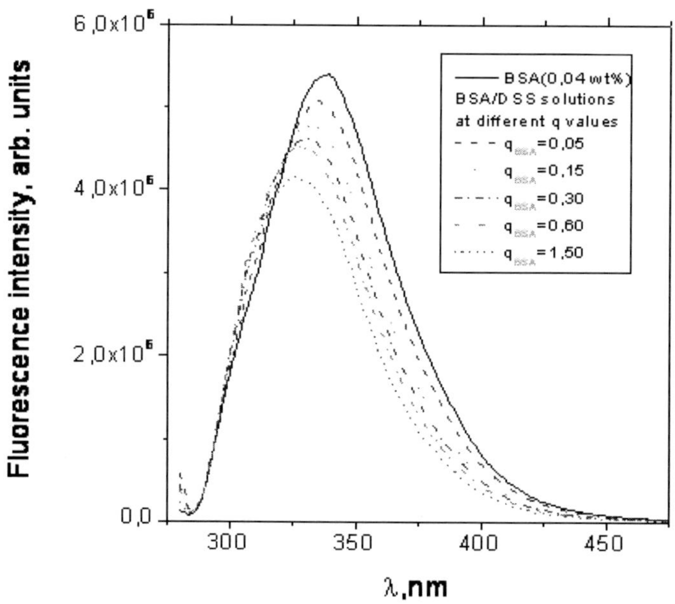

Figure 10. The effect of the presence of DSS on the fluorescence intensity of BSA at the excitation wavelength 270 nm. 20°C.Concentration of BSA = 0.04 wt%.

Note, that it has been shown long ago [6] that at a low ionic strength most proteins accumulate in the dextran rich phase of water-dextran-PEG system, whereas at a high ionic strength situation is reverse; proteins concentrate in the PEG enriched phase. In order to understand the origin of the interaction in the solutions containing dextran and DSS, the viscosity of the water/dextran/DSS systems was measured as a function of shear rate at different concentrations of NaCI (Figures 11).

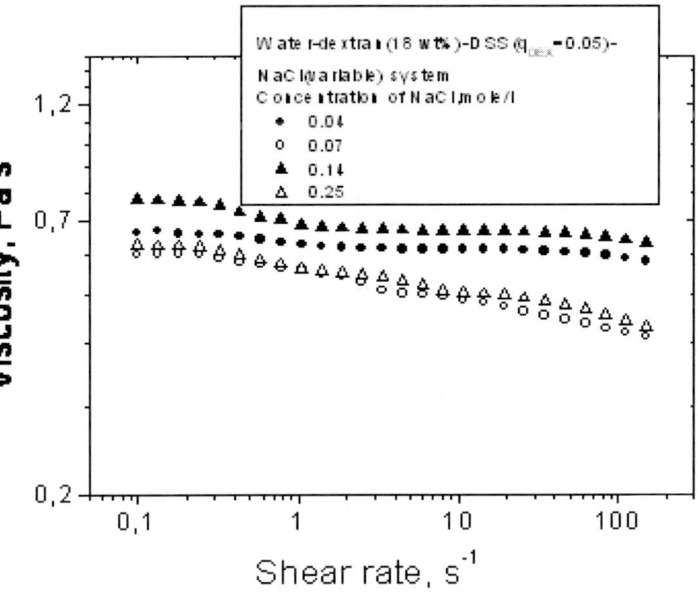

Figure 11. Flow curves of the water/dextran(18wt%)/DSS(q_{DEX}=0.05)/NaCl(variable) systems. pH 5.4 18°C.

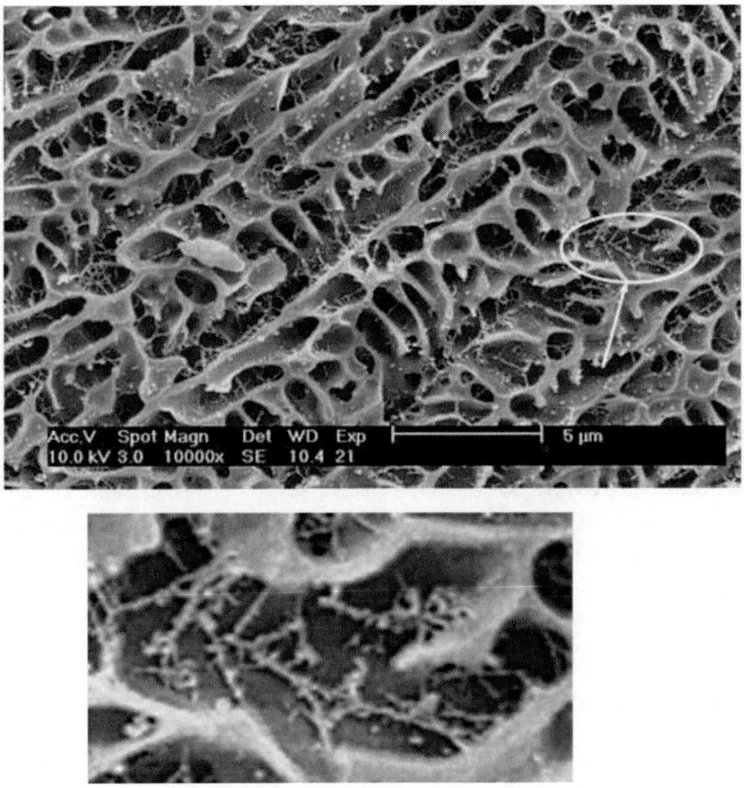

Figure 12. The ESEM images of water/dextran (15.4 wt%)/DSS(3.08 wt%) system (q_{DEX}=0.2).

At ionic strengths equal and higher 0.14, the water/dextran/DSS system becomes insensitive to the presence of DSS. It means that affinity of DSS to dextran has electrostatic origin. In order to see how microstracture of dextran changes in the presence of DSS the ESEM images of water/dextran (15.4 wt%)/DSS(3.08 wt%) system (q_{DEX}=0.2) were obtained (Figure 12).

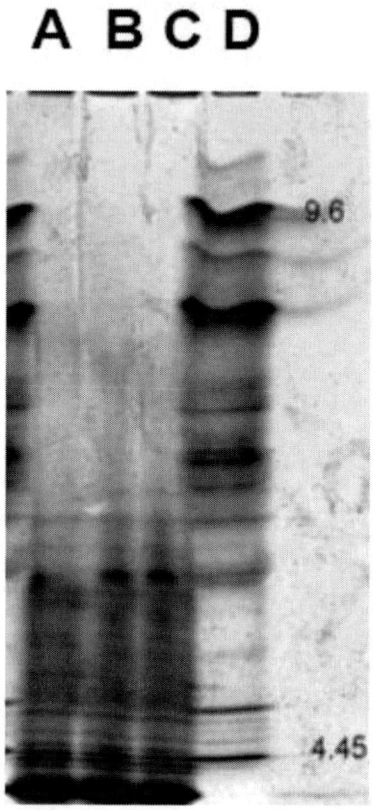

Figure 13. Iso-electric focusing patterns of BSA solutions before and after addition of different amounts of DSS. Concentration of BSA 0.5 wt%. (A)- pure BSA, (B,C) –BSA/DSS systems with q_{BSA} values 0.07, and 0.14, respectively, D- Standard protein kit.

It can be seen that the system develops both the skeleton-like structure of dextran and the skeleton-like structure shown by the cobweb of the sulphate functional groups. On the enlarge photography we observed formation of some kind of weak network on the basis of negatively charged DSS and dextran. At present time it is difficult to say something definite about the origin of such interaction. Nevertheless, one can imagine that adding small amounts of DSS could modify the state of dextran in highly concentrated solutions (Figures 8, 12).

What is the diving force for DSS-induced mixing in water-BSA –dextran systems? How protein-polyelectrolyte interaction affects the thermodynamic compatibility of BSA with dextran? Bowman et al. [43], characterizing complex formation between a negatively charged polyelectrolyte (sodium polystyrene sulfonate) and a negatively charged gelatin, suggested that the protein is polarized in the presence of strong polyelectrolyte. Therefore, if polarization of BSA in the presence DSS takes place, i.e., if surface and total charges of the BSA increase in the presence of DSS, then increased compatibility in BSA-dextran systems could be explained by the theory developed by Khoklov and coworkers [12, 17]. Let us to consider this possibility.

Figure 13 shows the iso-electric focusing patterns BSA before and after adding different amounts of DSS. The BSA bands remains stable and don't move appreciably as the DSS/BSA ratio increases from 0.07 to 0.14. The results demonstrate that the presence of DSS doesn't leads to significant polarization of BSA in solution probably due to compact structure of BSA and screening of the BSA charged functional groups in the presence of DSS. Moreover, as it can be seen from Figure 10, interaction of DSS with BSA leads to screening of BSA tryptophanyls from water environment. Therefore polarization of BSA in the presence of DSS is unprobable.

Based on our results, we assume that the phenomenon of the DSS induced homogenization in water BSA-dextran emulsion is the result of the affinity of DSS molecules to both, BSA and dextran molecules. It seems, there is a clear analogy in the mechanisn of compatibilization for polymer blends by diblock copolymers and homogenization of aqueous BSA/dextran emulsion in the presence of sulphate polysaccharide.

Conclusion

We established experimental evidence for mixing of aqueous concentrated BSA/dextran system at pH 5.4 (slightly above the isoelectric point of BSA) in the presence of a strong polyelectrolyte, DSS, for the DSS/BSA weight ratio $q_{BSA} \geq 0.07$. Homogenization leads to a noticeable increase in viscosity and module (G′). The effect of mixing is reversible: increasing the ionic strength leads to phase separation of the water/BSA/dextran/DSS system. Increase in viscoelasticity is the result of interaction of DSS with the both macromolecular componets of emulsion. Interaction of DSS with BSA leads also to the screening of BSA tryptophanyls from water environment, and is not accompanied by the polarization of the protein, whereas the affinity of DSS to dextran results in increase of viscoelasticity of dextran+DSS mixtures and an appreciable change in microstructure of DSS/dextran mixture. The driving force for mixing in water-BSA-dextran system in the presence of DSS is the affinity of the strong polyelectrolyte to both macromolecular componets of the emulsion. Thus, we have obtained a new compatible biopolymer mixture that exhibits favorable rheological performance.

References

[1] V. B. Tolstoguzov, Functional properties of protein-polysaccharide mixtures. In: *Functional properties of food macromolecules* 1998, Hill, S. E., Ledward, D. A., Mitchel, J. R. (eds), Aspen Publishers Inc., Gaitherburg, Maryland.

[2] J. Xia, P.L. Dubin. Protein-polyelectrolyte complexes. In: *Macromolecular Complexes in Chemistry and Biology* 1994, Dubin P.L.; Davis R.M., Schultz, D.; Thies C. (eds), Springer-Verlag, Berlin, Chapter 15.

[3] Yu. A. Antonov, *Applied Biochemistry and Microbiology* (Russia) Engl.Transl., 2000, 36, 382-396.

[4] Y. A. Antonov, P. Van Puyvelde, P. Moldenaers, *Biomacromolecules,* 2004, 5, 276-283.
[5] Y. A. Antonov and B. A. Wolf, *Biomacromolecules*, 2006, 7, N5, 1582-1567.
[6] P. A. Albertsson, *Partition of Cell, Particles and Macromolecules*, 3rd edn., John Wiley-Interscience, NY, 1986.
[7] Y.A., Antonov, V. Y. Grinberg, N. A., Zhuravskaya, V. B. Tolstoguzov, *Carbohydr. Polym.* 1982, 2, 81-90.
[8] Y. A. Antonov, V. Y. Grinberg, , V. B. Tolstoguzov, *J. of Texture studies*, 1980, 11, 199-215.
[9] P. J. Flory, *Principles of Polymer Chemistry*; Cornell University, Press, Ithaca, NY, 1953.
[10] R. L. Scott, *J. Chem. Phys.* 1949, 17, 279-284.
[11] H. Tompa, *Trans. Faraday Soc.* 1949, 45, 1142-1152.
[12] V. V. Vasilevskaya, S. G. Starodoubtsev,. A. R. Khokhlov, *Vysokomolec. Soed.* 1987, *29B,* 390-397.
[13] I. Iliopoulos, D. Frugier, R. Audebert, *Polym. Prepr. (Am. Chem. Soc, Div. Polym. Chem.)* 1989**,** *30,* 371-273.
[14] L. Piculell, S. Nilsson, L. Falck, F. Tjemeld, *Polym. Commun.* 1991**,** *32,* 158-160.
[15] L. Piculell, I. Ilioupolos, P. Linse, S. Nilsson, T. Turquois, C. Viebke, W. Zhang, *Gums and Stabililisers for the Food Industry;* Oxford University Press: Oxford, 1994; 7, p 309.
[16] K. Bergfeldt, L. Piculell, *J. Phys. Chem.* 1996**,** *100,* 5935-5940.
[17] R. Khokhlov, I. A. Nyrkova, *Macromolecules* 1992**,** *25,* 1493-1502.
[18] H.-O. Johansson, G. Karlstrom, B. Mattiasson, F. Tjerneld, *Bioseparation* 1995**,** *5,* 269-279.
[19] Y. A. Antonov, M. G., Pletenko, V. B. Tolstoguzov, *Vysokomolec. Soed.* 1987, 29 A, N12, 2477-2481.
[20] Y. A Antonov,., J. Lefebvre., J.L Doublier,. *J. Appl. Polym. Sci.*, 1999, 71, 471–482
[21] Y. A. Antonov and B. A. Wolf, *Biomacromolecules* 2005, 6, 2980-2989.
[22] E. E. Woodside, G. F. Trott, R. J. Doyle, C. W. Fishel, *Arch. Biochem. Biophys.* 1966, *117*, 125-133.
[23] E. E. Woodside,. G. F. Trott,. Doyle, C. W. Fishel, *Carbohydr. Res.*, 1968, 6, 449-457.
[24] V. Y. Grinberg, V. B. Tolstoguzov, *Carbohydr. Res.* 1972, 25, 313-321.
[25] P. Van Puyvelde, Y.A. Antonov, P. Moldenaers. *Food Hydrocolloids* 2003, 17 327–332.
[26] J. R. Stokes, B. Wolf,., W.J. Frith, *J. Rheol.*, 2001, 45, 1173–1191.
[27] B. Wolf, & W.J. Frith, *J. Rheol.*, 2003, 47, 1151-1170.
[28] L. A. Utracki, *Polymer Blends Handbook*. 2-Volume Set. 2003.
[29] Y.A. Antonov, P. Moldenaers, Biomacromolecules, 2009,*10* (12),3235–3245.
[30] Y.A. Antonov and P. Moldenaers, *Food Hydrocolloids,* 2011, 25(3), 350-360.
[31] N. L. Abbott, D. Blankschtein T. A, Hatton,. *Macromolecules*, 1991, 24, 4334-4348.
[32] N. L. Abbott, D. Blankschtein, T. A. Hatton, *Macromolecules*, 1992, 25, 3917-3931.
[33] N. L. Abbott, D. Blankschtein, T. A Hatton,. *Macromolecules*, 1993, 26, 825-828.
[34] J. F Foster,. In *Albumin Structure, Function and Uses*; Roesnoer, V. M., Oratz, M., Rothschild, M. A., Eds.; Pergamon, Oxford, 1977. 53-84.
[35] D. M. Kirschenbaum, *Anal. Biochem.* 81 (1977) 220-246.

[36] J. Lefebvre, *Rheol. Acta* 1982, 21, 620-625.
[37] E. Djokpe, W. Vogt, *Macromol. Chem. Phys.* 2001, 202, 750-757.
[38] C. Yeung and K. Herrmann, *Macromolecules* 2003, 36, 229.
[39] J. T. G. Overbeek, M. J. Voorn, *J. Cell. Comp. Physiol.* 1957, 49, 7–39.
[40] J. R. Lakowicz, *Principles of Fluorescence Spectroscopy*, Springer Science/Business Media, NY, 2006.
[41] J. R. Lakowicz *Principles of fluorescence specytroscopy*. 1986, Plenum, New York and ondon, p 496.
[42] Y. A. Antonov, Thermodynamic compatibility of casein and soya bean globulins with acidic and neutral polysaccharides in aqueous medium. *PhD thesis*. 1977, Institute of Organo-element compounds, USSR Academy of sciences. Moscow, USSR.
[43] W. A. Bowman, M., Rubinstein, & J. S Tan, *Macromolecules,* 1997 30, 3262-3270.

In: News in Chemistry, Biochemistry and Biotechnology ISBN: 978-1-63117-273-1
Editors: G. E. Zaikov, G. Nyszko, L. P. Krylova et al.

Chapter 19

CRUCIAL ROLE FOR MILK XANTHINE OXIDOREDUCTASE IN CONVERSION OF TOXIC NITRATE AND NITRITE TO PHYSIOLOGICALLY IMPORTANT NITRIC OXIDE

A. Samarkanova, S. Altayuly* and Z. Alikulov
The L.N. Gumilov Eurasian National University, Astana, Kazakhstan

ABSTRACT

Molybdenum containing enzyme xanthine oxidase (XO) presents in milk fat globule membrane (MFGM) in the inactive form. Moreover, molybdenum content in milk XO tenth times lower than that in liver enzyme. Inactive xanthine oxidase may be released from milk fat globule membrane by the excess of phospholipids under high temperature. Presence of moderate concentration of molybdenum and natural reductants (ascorbic acid or glutathione) under such treatment restores the activity of the enzyme. XO activated after such treatment converts nitrate and nitrite in contaminated milk to physiological important NO.

Keywords: Xanthine oxidase, fat globule membrane, phospholipids, molybdenum, nitrate, nitrite, nitric oxide, glutathione, ascorbic acid

INTRODUCTION

Nitrogen is essential for all living things as it is a component of protein. Nitrogen exists in the environment in many forms and changes forms as it moves through the nitrogen cycle. Nitrate is a natural material in soils. It is primary source of nitrogen for plants and microorganisms. Probably more than 90 percent of the nitrogen absorbed by plants is in the nitrate form. Therefore, adequate supply of nitrate is necessary for good plant growth. Sources

* sagimbek@mail.ru.

of excess nitrate in water include fertilizers, septic systems, wastewater treatment effluent, animal wastes, industrial wastes and food processing wastes. Chemical fertilizers may be composed of ammonium nitrate, ammonium phosphates, ammonium sulfate, various nitrate salts, urea and other organic forms of nitrogen. Animal manure is an excellent source of nitrogen and can contribute significantly to soil improvement. Decomposition of plant residues and animal waste by soil microorganisms results in the formation of the ammonium form (NH_4^+). Specific soil microorganisms oxidize the ammonium form to nitrate [1, 2]. Aeration of soil by cultivation can speed up the formation of nitrates. Nitrogen in the ammonium is strongly held by the negative charges of clay and soil organic matter colloids until converted to the nitrate form by bacteria. This is desirable as the majority of the nitrogen used by plants is absorbed in the nitrate form. Thus, the formation of nitrates is an integral part of the nitrogen cycle in our environment. Nitrate-nitrogen is soluble in water and moves with soil moisture. Nitrate levels can be high in streams and rivers due to runoff of nitrogen fertilizer from agricultural fields and urban lawns. Groundwater is susceptible to contamination from many different chemicals, including nitrate fertilizers, especially where the water table is shallow and there are no confining units to reduce migration downward. Most of these contaminated groundwaters flow into streams and rivers, causing elevated nitrate levels in those water bodies downstream. By applying fertilizers and burning fossil fuels human have doubled the rate of nitrogen deposition onto land over the past 50 years.

NITRATE AND NITRITE IN ANIMALS

Nitrate is of special concern in animal production and in human foods because of its potential toxicity when excessive amounts are ingested. Nitrate levels of up to 3 parts-per-million (ppm) in well water may be naturally-occuring or possibly indicates some low level of contamination, but are considered to be safe for consumption. The Environment Protection Agency (EPA) has set a maximum contamination level (MGL) of 10 ppm for nitrate for drinking water [3]. Nitrate levels above 10 ppm may present a serious health concern for infants and pregnant or nursing women [5]. Adults receive more nitrate exposure from food than from water. Infants, however, receive the greatest exposure from drinking water because most of their food is liquid form. This is especially true for bottle-fed infants whose formula is reconstituted with drinking water with high nitrate concentrations. Pregnant women may be less able to tolerate nitrate, and nitrate in the milk of nursing mothers may affect infant directly. These persons should not consume water containing more than 10 ppm nitrate directly, added to food products, or beverages (especially in baby formula). Thus, infants, pregnant women, nursing mothers, or elderly people are the most susceptible to nitrate or nitrite contamination [4].

A potential cancer risk from nitrate (and nitrite) in water and food has been reported. Recent human epidemiology studies have shown that nitrate ingestion may be linked to gastric or bladder cancer. The most likely mechanism for human cancer related to nitrate is the body's formation of N-nitrosamines [6]. Carcinogenic nitrosamines are formed when amines that occur naturally in food react with nitrite: R_2NH (amines) + $NaNO_2$ (nitrite) $\rightarrow$ $R_2N\text{-}N{=}O$ (nitrosamine). Nitrite reacts in the acidic stomach to form nitrosating agents that then react with certain compounds from protein or other sources such as medications to form nitrosamines. Nitrosamines have been shown to cause tumors at multiple organ sites in every animal species tested [7, 8]. Certain nitrosamines such as N-nitrosodimethylamine and N-nitrosopyrrolidine form carbocations that react with biological nucleophiles (such as DNA or an enzyme) in the cell. If this nucleophilic substitution reaction occurs at a crucial site in a biomolecule, it can disrupt normal cell functioning leading to cancer or cell death.

The primary health hazard from drinking water with nitrate occurs when nitrate is transformed to nitrite in the digestive system. The nitrite oxidizes iron in the hemoglobin of the red blood cells to form methemoglobin, which lacks the oxygen-carrying ability of hemoglobin. This creates the condition known as methemoglobinemia (sometimes referred to as "blue baby syndrome"), in which blood lacks the ability to carry sufficient oxygen to the individual body cells causing the veins and skin to appear blue. Most humans over one year of age have the ability to rapidly convert methemoglobin back to oxyhemoglobin; hence, the total amount of methemoglobin within red blood cells remains low in spite of relatively high levels of nitrate/nitrite uptake. However in infants under six months of age, the enzyme (NADH-methemoglobin reductase) systems for reducing methemoglobin to oxyhemoglobin are incompletely developed and methemoglobinemia can occur [9]. This also may happen in older individuals who have genetically impaired enzyme systems for metabolizing methemoglobin. Definitive guidelines for determining susceptibility to methemoglobinemia have not been developed.

Nitrate contamination in groundwater from fertilizer and animal manure is severe and getting worse for hundreds of thousands of residents in Kazakhstan. Nitrate-contaminated water is a well-documented fact in many of Kazakh farming communities. For example, Oskemen is recognized as the most polluted town in the Republic. It has the highest rate of oncology-related and respiratory diseases. Percentage of nitrates exceeds MGL 27 times. In moderate amounts, nitrate is a harmless constituent of food and water. Nitrate poisoning is a problem that all horse and livestock owners should be aware of. Forages can be tested for nitrate content at no cost and it is recommended that all forages be tested before being fed to horses and livestock.

High nitrates in forages can cause reduced feed consumption and growth rates, lowered milk production, and abortions. Ruminant animals (cattle, sheep) are susceptible to nitrate poisoning because bacteria present in the rumen convert nitrate to nitrite. Nonruminant animals (swine, chickens) rapidly eliminate nitrate in their urine. Horses are monogastric, but their large cecum acts much like a rumen [5]. This makes them more susceptible to nitrate poisoning than other monogastric animals. If nitrates reach dangerously high levels, it can cause death.

Nitrate-poisoned animals show symptoms of suffocation, including labored breathing, lack of coordination, and blue mucous membranes. Pregnant animals may abort within a few days. The most reliable symptom of nitrate toxicity is a chocolate brown coloration of the blood. Other signs include: diarrhea, frequent urination, muscular weakness or poor coordination and frothing at the mouth. Young animals are affected by nitrates the same way as human babies. A few hundred milligrams of nitrate may cause poisoning if consumed in a few hours [1, 3].

Xanthine Oxidoreductase (XOR) Converts Nitrate and Nitrite to Nitric Oxide

A long time ago, milk xanthine oxidase has been shown to catalyze the disappearance of nitrates and nitrites in the reaction mixture [10]. More later, it was found that both purified and tissue containing XO catalyze the reduction of nitrate and nitrite to NO [11, 12] has also been shown that XOR is able to convert nitrates and nitrites to NO, an important signaling molecule in its own right and the source of other, potentially destructive reactive nitrogen species (RNS), such as peroxynitrite [1, 2].

Nitric oxide (NO) synthesis is now well-known to result from the oxidation of L-arginine by an enzyme family of NO-synthases (NOS) in the presence of oxygen [2]. Therefore, in case of low oxygen such as restricted blood flow, NO may be formed by a NOS independent mechanism. It was found that both purified and tissue containing XO catalyze the reduction of nitrite to NO. This redox reaction requires NADH as an electron donor, and is oxygen independent. The inhibitory profiles suggest that reduction of nitrite takes place at the molybdenum center of XO. These findings suggest a role for XOR as a source of NO derived from endogenous nitrate and nitrite under ischaemic conditions ranging from sub-normoxia to anoxia when NO-synthase does not function [11, 12].

Physiological Importance of NO

Nitric oxide (NO) has, in only the past 20 years, become recognized as a very, very important compound in human physiology. This period of time turned out to be very important for two reasons: (a) the extensive research and accompanying publicity on the relationship between nitrite and cancer resulted in firmly entrenched perceptions of cured meat as a contributor to human cancer that continue to this day, and (b) the discovery of endogenous nitrite in the body was the forerunner to a subsequent major breakthrough in biology [2]. The breakthrough came in 1986 when it was shown that nitric oxide was a major biological messenger molecule responsible for regulation of blood pressure and blood flow, neurotransmission and brain function, immune system function, wound healing, vasodilation, inhibition of platelet aggregation, neurotransmission and cytotoxic host defense mechanisms. NO itself is antimicrobial and cytotoxic, and it is further involved in the generation of an array of reactive molecules and even more potent antimicrobial substances (including, potentially destructive RNS, such as peroxynitrite), which makes NO a defensive molecule against various pathogens, tumor cells and alloantigens. This turned out to be such a momentous discovery that the 1998 Nobel Prize for Physiology/Medicine was awarded to the three researchers who identified the critical biological role of nitric oxide [13].

Consequently, the current hypothesis is that tissue and blood nitrite provides a low-oxygen source of NO because it is easily formed from nitrite. To test this hypothesis, researchers have been studying the effects of dietary nitrite on tissue concentrations of nitrite and on induced heart attacks in mice. They have found that dietary nitrite significantly reduced injury and increased survival from heart attacks. They further suggested that dietary nitrite may be a critical component for cardiovascular health. This is a complete, 180-degree change in thinking about nitrite and human health. Thus, nitrite has an important role in physiology and dietary nitrite appears to be protective against cardiovascular disease and injury [1]. However, in contrast to the large body of knowledge regarding NO in animal cells, the physiology and biochemistry of NO in milk has been obscure. Therefore, knowledge of the *in vivo* concentration of NO is very much needed in order to explore physiological roles of NO in mammalian milk.

XANTHINE OXIDOREDUCTASE (XOR)

(XOR) is a complex molybdoflavoprotein. The fully constituted enzyme is a dimer, each subunit of which contains one molybdenum atom, one FAD and two non-identical iron-sulfur redox centers. Although XOR interacts with a wide range of reducing and oxidizing substrates, its conventionally accepted role is in purine catabolism, catalyzing the oxidation of hypoxanthine and xanthine to uric acid [14, 15].

Mammalian XOR exists in two interconvertible forms, xanthine dehydrogenase (XDH, EC 1.1.1.204), which predominates *in vivo*, and xanthine oxidase (XO, EC 1.1.3.22). While both forms of the enzyme reduce molecular oxygen, only XDH can reduce NAD^+, which is its preferred electron acceptor. Reduction of oxygen leads to superoxide anion and hydrogen peroxide and it is the potential to generate these reactive oxygen species (ROS). There is increasing evidence that XOR has additional physiological functions associated with its synthesis of ROS and reactive nitrogen species (RNS), which have important roles in inflammation and host defense. Although XDH is the predominant form found in normal cells and tissues, XO appears to have an important role in cell and tissue injuries. Various forms of stimuli induce the conversion of the XDH to the XO form, presumably resulting in intensive synthesis of ROS and RNS [14].

On the basis of above properties, a role for XOR has been proposed in innate immunity. Innate immunity is composed of: (1) surface epithelia that provide local physical and molecular barriers, (2) inflammatory reactions and the activation of conserved cell-signaling pathways, (3) numerous systemic protective molecules and (4) various phagocytotic cells. All of these components work together to resist and prevent the action of toxic molecules and the rapid spread of potentially fatal pathogens. The protective functions of XOR in innate immunity are, as at the cellular level, linked to its detoxification reactions, its synthesis of uric acid and, particularly, its synthesis of numerous ROS and RNS. XOR activity and uric acid are generally found in the blood plasma of many mammalian species and levels are particularly high during numerous disease states. ROS and RNS perform, at low levels, numerous cellular and physiological functions as second messengers but, at high levels, can act as microbicidals. XO has also been implicated in protective antiviral responses by catalyzing the conversion of retinaldehyde to retinoic acid. Derivatives of retinoic acid can inhibit viral replication, thus potentially preventing the spread of viral diseases [15].

Proposed mechanisms of pathophysiological involvement of XOR are largely based on the well-known properties of the bovine milk and rat liver enzymes, and although results obtained in experimental animal systems are commonly extrapolated, at least implicitly, to humans, remarkably little is known about the human enzyme.

More recently, purification of XOR from human milk has been described. Human milk XOR exhibits NADH-oxidase activity that is fully equivalent to that of the bovine milk enzyme, demonstrating the integrity of the FAD center of the human enzyme as compared with bovine counterpart. Human milk XOR, while showing physicochemical properties very similar to those of the bovine milk enzyme, has only approximately 5% of the activity of the latter towards xanthine and related substrates suggesting dissimilarities between the bovine and human enzymes at the molybdenum and iron-sulfur centers. Comparison of the Mo contents and XO activities of human and bovine XOR allowed estimation of activities corresponding to 100% Mo content. This gave estimates of 59% and 55% content of inactive Mo-containing enzyme for human and bovine XOR respectively. XOR purified from human milk was shown to contain 0.04 atoms Mo per subunit. The human enzyme was approximately 30% deficient in iron-sulfur centers. Clearly, there are also significant differences between the enzymatic activities of human and bovine milk XORs, particularly in their potential for generating ROS and RNS, which are of major clinical interest. Resulfuration experiments, together with calculations based on enzymatic activity and Mo-content, led to an estimate of 50-60% desulfo-form [15]. Thus, human milk XOR is not only grossly deficient in Mo but is also substantially lacking in iron-sulfur centres. Thus, it seems clear that bovine and human XORs contain similar demolybdo-forms of the enzyme that are deficient in Fe/S. The relative molecular weights of the two enzymes are experimentally indistinguishable from each other and correspond to the values (including all cofactors) derived from the deduced amino acid sequences. Similarly purified XOR from human milk was shown to contain approximately 15 fold lower molybdenum content and enzymic activity [15]. The essential difference is that the content of this demolybdo-from is much higher in the human case and an important question is why should this be so? It is likely that the human milk samples used are usual in that the donors came from Mo-deficient area. Thus, in human and bovine milk XOR also exists in enzymatic inactive demolybdo form.

With regard to mammalian milk XOR generally, unoccupied Mo sites are not confined to the human enzyme. Preparation of XOR from goat and sheep milk contain only 0.09 and 0.18 atoms Mo per subunit respectively and, although purified bovine milk XOR is clearly much richer in Mo, it is still 40% deficient. It is far from what advantage might derive from this. It is of interest that, while XOR plays a key role in the process of milk secretion, this does not require active enzyme, depending rather on XOR protein [14, 15]. Moreover, the specific activity of human XOR has been shown to peak dramatically in the first few weeks postpartum, possibly to coincide with an antimicrobial role in the neonatal gut. Thereafter, specific activity rapidly falls to consistently low levels, probably, when an antimicrobial function of milk is less critical.

Thus, XOR is best known as an evolutionary conserved housekeeping enzyme, as mentioned above, with a principal role in purine catabolism. By generating mice with a targeted disruption of XOR, it was discovered the additional role of XOR as an essential protein for milk fat droplet secretion from the lactating mammary gland, i.e., these findings add further support to the idea that XOR protein plays a physiological role in milk equal in importance to its catalytic function as an enzyme [15].

Uric Acid As a Potential Antioxidant

Uric acid, a product of XOR reactions in animals has been recognized as a potential ROS scavenger. Uric acid may be oxidized nonenzymatically by ROS to form allantoin. Uric acid is an effective inhibitor of ROS at levels found in human plasma and also significantly protects against another strong oxidant, peroxynitrite, whereas other antioxidants are protective at concentrations exceeding those usually found in blood plasma. Thus, XO is peculiar enzyme ironically producing both toxic superoxide and the potential antioxidant uric acid [10].

MOLYBDENUM DEFICIENCY

It has been claimed that molybdenum status influences susceptibility to certain forms of cancer and that the high incidence of esophageal cancer among the Bantu in Transkei (South Africa) is associated with a deficiency of this element in locally available food. Studies in Henan province, China, suggest that a high incidence of esophageal cancer is associated with lower than normal contents of molybdenum in drinking water and food as well as in serum, hair and urine. Esophageal cancer tissue also had lower molybdenum content than normal. It may well be relevant that inclusion of 2 or 20 μg of molybdenum/g in the diet of rats has been found to inhibit esophageal and stomach cancer following the administration of N-nitrososarcosine ethyl ester. Molybdenum in the drinking water of rats at a concentration of 10 mg/l inhibited mammary carcinogenesis induced by N-nitroso-N-methylurea [6, 7, 8].

Molybdenum deficiency has not been identified in free-living animal species. It has, however, been identified in a single subject receiving total parenteral nutrition and can be achieved in animal studies. Animals can be made molybdenum deficient by feeding them diets containing high amounts of tungsten or copper. Both tungsten and copper are molybdenum antagonists. Molybdenum deficiency has also been produced experimentally in goats by feeding them purified diets, very low in molybdenum. In goats, a molybdenum deficient diet was associated with reduced fertility and increased mortality in both the mothers and the offspring. Molybdenum deficiency in animals results in retarded weight gain, decreased food consumption, impaired reproduction and a shortened life expectancy [16].

The high dietary Mo contents did not reduce the growth of animals and after Mo-administration the highest Mo levels were found in liver and kidney. However, molybdenum levels in milk of Mo-administrated animals was not studied. No recommended dietary allowance (RDA) has been established for molybdenum. The estimated range recommended by the Food and Nutrition Board as safe and adequate is 75-250 micrograms per day for adults [16]. The results indicate that supplemental Mo in the amount of 10 mg/L of drinking water inhibited mammary carcinogenesis [17].

EXOGENOUS PHOSPHOLIPIDS INCREASE THE ACTIVITY OF MILK XOR

Milk is essential for mammal newborns, providing nourishment and protection. Milk is the only diets whose sole function in nature is food. Milk is a white or yellowish natural emulsion in which lipids are present as droplets called Milk Fat Globules. Phospholipids and sphingolipids of milk form an integral part of Milk Fat Globule Membrane (MFGM). It is a protein-lipid biopolymer and surrounds fat globules in milk [16]. The MFGM is expected to be inhomogeneous with significant amount of proteins in the membrane (Figure 1). One of the main MFGM phospholipids attributing to the biological role of Milk Fat Globules is Sphingomyelin, along with Phosphatidyl Choline and Phosphatidyl Ethanolamine. It acts as a natural emulsifying agent thereby preventing flocculation and coalescence of fat globules in milk and protecting the fat against enzyme activity. Milk Fat is a combination of both saturated and unsaturated fatty acids. The phospholipids and sphingolipids in milk are gaining interest due to their nutritional and technological qualities. Sphingolipids and their derivatives are highly bioactive compounds with anti-cancer, bacteriostatic and cholesterol-lowering properties. In phospholipids, the head group consists of a phosphate residue that esterified with a second alcoholic compound such as ethanolamine, choline, serine and inositol. Phospholipids form a bilayer in which the nonpolar regions of the lipid molecules in each layer face the core of the bilayer and their polar head groups face outward, interacting with aqueous phase on either side [18].

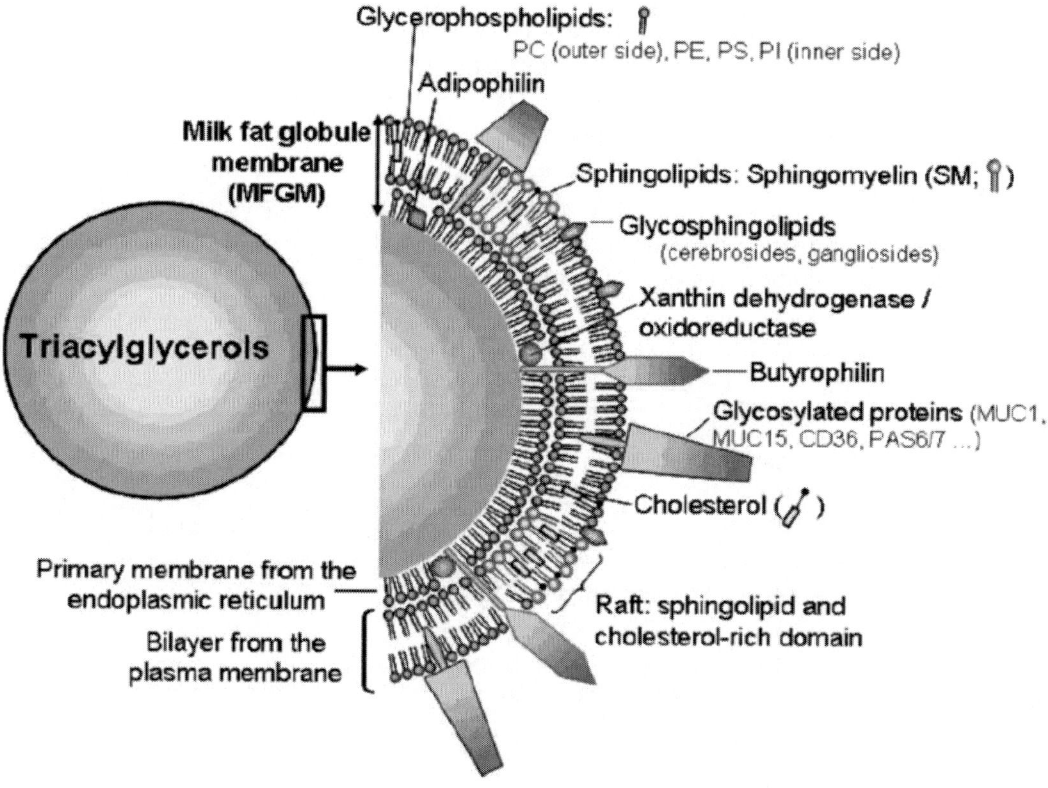

Figure 1. Structure of Milk Fat Globule Membrane (MFGM).

The major protein components of the MFGM layer are butyrophilin and xanthine oxidase (XO) along with at least 30 identified proteins (Figure 1). The enzyme was found to represent more than 8% of the intrinsic protein of the bovine MFGM. XO is present between the monolayer and bilayer and inactive. The enzyme can be released into the plasma by various treatments. Phospholipids were found to release the free XO from the fat-globule membrane [18]. The process of emulsification of hydrophobic fat globules by the detergent action of phospholipids in the gut breaks the globules down to mixed micelles. The hydrophobic moieties (fatty acid chain) of phospholipids are inserted into the hydrophobic fat globules and the hydrophilic polar head groups interact with and face the water. The formation of small micelles from large fat globules greatly increases the surface area available for the action of XO (Figure 1), in essence forming a monomolecular layer around the fat.

CONCLUSION

In contrast to the large body of knowledge regarding NO in animal cells, the physiology and biochemistry of NO-production in milk has been obscure. Therefore, knowledge of the in vivo concentration of NO is very much needed in order to explore physiological roles of NO-production in mammalian milk by XOR. The evidence for above hypotheses must come from further studies aimed at understanding the precise roles of molybdenum administration in the reduction of nitrate and nitrite by milk XOR and formation of physiological important NO. The conditions suitable for initiating the incorporation of dietary molybdenum in milk XOR remains elusive, requiring further research. Until now, there was no conclusive data available to prove whether exogenous phospholipids increase the activity milk XO and NO-production.

REFERENCES

[1] M.T. Gladwin, A.N. Schechter, D.B. Kim-Shapiro, R.P. Patel, N. Hogg, S. Shiva, R.O. Cannon III, M. Kelm, D.A. Wink, M.G. Espey, E.H. Oldfield, R.M. Pluta, B.A. Freeman, J.R. Lancaster, M. Feelish, J.O. Lundberg. *Nature Chemical Biology,* 1(6), 308-314 (2005).

[2] N.S. Bryan, *Free Radical Biology & Medicine* 41, 691-701 (2006).

[3] S. Correia, M.Barroso, Â. Barroso, M. F. Soares, D. Oliveira, C. Delerue-Matos. *Food Chemistry.* 120 (4), 960 (2010).

[4] A. Milkowski, H. Garg, J. Couglin, N. Bryan. *Nitric Oxide.* 22, 110 (2010).

[5] N.G. Hord, J.S. Ghannam, H.K Garg, P.D Berens, N.S. Bryan. *Breastfeed Med.* 6(6), 393 (2011).

[6] W.J. Blot. J.Y. Li, P.R.Taylor. *J. Natl. Cancer Inst.* 85, 1483 (1993).

[7] G.J. Brewer, R.D. Dick, D.K. Grover. *Clin. Cancer Res.* 6, 1 (2000).

[8] C.D. Seaborn, S.P. Yang. *Biol. Trace Elem. Res.* 39(2-3), 245 (1993).

[9] J.D. Reynolds, G.S. Ahearn, M. Angelo, J. Zhang, F. Cobb, J.S. Tamler. *Proc. National Acad. Sciences* 104(43), 17058 (2007).

[10] Z. Alikulov, N.P. L'vov, V.L. Kretovich. *Biokhimia.* 45(9), 1714 (1980) (Russian).

[11] T.M. Millar, C.R. Stevens, N. Benjamin, R Eisenthal, R. Harrison, D.R. Blake. *FEBS Letters.* 427, 225 (1998).

[12] Z. Zhang, D. Nauthon, P.G. Winyard, N. Benjamin. *Bioch. Biophys. Res. Comm.* 249, 767 (1998).

[13] N.S. Bryan, K. Bian, F. Murad. *Frontiers in Bioscience.* 1 (2009).

[14] Rr. Mendel. *Biofactors.* 11, 147 (2000).

[15] B.L. Godberg, G. Schwarz, R. Mendel, D. Lowe, R. Bray, R. Eisenthal, R. Harrison. *Biochemical Journal Immediate Publication.* 1 (2005).

[16] J.R. Turumland, W.R. Keyes, G.L. Peiffer. *Am. J. Clin.Nutr.* 62(4), 790 (1995).
[17] A. Berssnyi, E. Berta, I.Kadar, R. Glavits, M. Szilagyi, S. G. Fekete. *Acta Veterinaria Hungarica,* 56, 55 (2008).
[18] R. Mesilati-Stahy, K. Mida, N.J. Argov-Argaman. *Agric Food Chem.* 59(13), 7427 (2011).

In: News in Chemistry, Biochemistry and Biotechnology ISBN: 978-1-63117-273-1
Editors: G. E. Zaikov, G. Nyszko, L. P. Krylova et al.

Chapter 20

THE PROSTOR AND FERM KM-1 COMPLEX PROBIOTIC ADDITIVES: INNOVATION BIOTECHNOLOGICAL PREPARATIONS FOR ENHANCING THE QUALITY OF DOMESTIC FISH MIXED FEED

D. S. Pavlov*[*,1], *N. A. Ushakova*[1], *V. G. Pravdin*[2], *L. Z. Kravtsova*[2], *S. A. Liman*[3] and *S. V. Ponomarev*[4]

[1]A.N. Severtsov Institute of Ecology and Evolution, Russian Academy of Sciences, Russia, Moscow,

[2]The "NTC BIO," LLC, Russia, Belgorod Region, Shebekino Town

[3]The "Agroakademia," LLC, Russia, Belgorod Region, Shebekino Town

[4]The "Bioaquapark" Innovation Centre – The Scientific Centre of the Aqua-Culture at the ASTU, Astrakhan, Russia

ABSTRACT

The ProStor and Ferm-KM-1 complex probiotic preparations based on solid-state fermentation of beet pulp with a probiotic association (three strains of *Bacillus subtillis, Bacillus licheniformis*, a complex of lactic acid bacteria) have been developed. A *Cellulomonas* microorganism has been additionally introduced to the Ferm KM-1 probiotics composition.

Some fish mixed feed formulations with use of the preparations have been developed. In experiments, the efficiency of new mixed fodders for the young of carp and sturgeon has been demonstrated.

Keywords: Probiotics, biofilm, phytobiotics, feed, fish farming

[*] E-mail naushakova@gmail.com.

INTRODUCTION

The prerequisite of effective economic management in modern industrial fisheries is increasing their productivity that is directly linked to the use of complete and cost- effective feed. The most important task is to create and use in fish farming the feed, which:

- is maximally needed for the body to ensure its vital functions
- has growth/development stimulating properties as well as prevention and anti-stress characteristics
- ensures the environmental safety of feed produced.

In order to improve the quality of fish mixed feed, some enzyme, probiotic, prebiotic and probiotic/enzyme combination feed additives are used, as well as complex probiotic preparations enriched with phytocomponents.

Probiotic fodder preparations are regarded as a potential alternative to feed antibiotics, so the use of probiotics is considered an essential point of obtaining ecologically clean feed [1-5]. Probiotic preparations balanced with phytochemicals, show an enhanced biological activity due to the combination of the actual probiotic effect and the action of a phytobiotic.

Probiotics are live microbial supplements that have a beneficial effect on the body by improving the intestinal microbial balance, and stimulate the metabolism and immune processes. Probiotics are widely used in mixed feed for fish [6-9]. In themselves, probiotics do not provide a significant amount of nutrients for producing more products. But their biological potential improves fish health, enhances productivity levels, and better use of feed.

The determining factor of the probiotics efficiency is, in many ways, the technology of formulating these preparations. Modern biotechnology approaches to the development of probiotic preparations imply, firstly, the use of different types of microorganisms in certain combinations, and, secondly, their production in a form allowing their long-term storage at normal temperatures, and granulation.

The technology for production of the biologically active complex probiotic preparations ProStor and Ferm KM-1 is based on a partial solid state fermentation of beet pulp with a probiotic association. The final product includes biomass of probiotic bacteria forming a biofilm on the surface of a phyto-carrier, products of their metabolism, phytosubstrate biotransformation products, prebiotics - pectins of beet, and phytocomponents. The bacterial composition of the preparations contains vegetative cells of three strains - *Bacillus subtillis, Bacillus licheniformis*, and a lactic acid bacteria complex. The ProStor preparation contains in the probiotic association a unique strain *Bacillus subtillis* – a producer of hydrolase class enzyme, which has anti-inflammatory and antiviral effects, stimulates the immune reactions of the body. A cellulolytic *Cellulomonas* microorganism is additionally introduced to the Ferm KM-1 probiotics composition and capable of both synthesizing enzymes that break down cellulose, and producing lysine, the essential amino acid. Depending on the type of fish and their food, the effect of biological action of bacteria varies. Therefore the preparation Ferm KM-1 increases the digestibility of all feed components, and, to the upmost degree, of fiber in case when the feed mix contains a lot of fiber, which is important, for example, for the carp. For the sturgeon on the protein diet, the preparation increases the digestibility and protein digestibility of feed.

Probiotic bacteria have an enhanced viability and resistance to adverse environmental conditions for they are in the form of a biofilm on a phyto-carrier (Figure 1). The preparations contain enzymes: cellulase, amylase, complex of proteases, lipase, as well as organic acids, biologically active substances, vitamins, amino acids, immunoactive peptides – products of probiotics metabolism. The preparations comprise phyto-particles that are a cellulose microsorbent.

The preparations are featured with combining probiotics and prebiotics (mannans and glucans on cell walls of yeast *Saccharomyces cerevisiae*), and phytobiotics of the medicinal plants - e*chinacea purpurea* and *holy thistle*. Echinacea has immunomodulatory properties. Echinacea preparations exhibit antibacterial, antiviral and antifungal properties. When intaking the echinacea preparations at metabolic disorders, at the impact of different chemical compounds of toxic nature, contained in the feed (heavy metals, pesticides, insecticides, fungicides), a stimulation of the immune system has been observed.

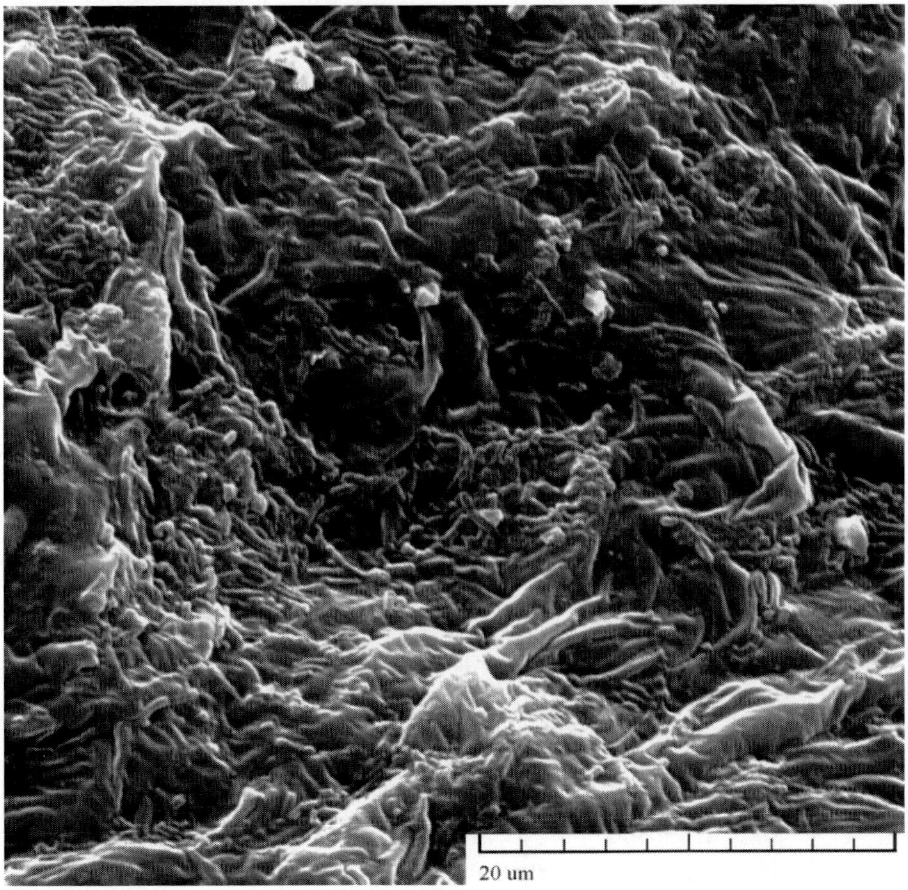

Figure 1. Microphotograph of the fermented sugar beet pulp with biofilm of probiotics.

Holy thistle is used for prevention of various liver affections. Preparations of holy thistle increase protective properties of liver to infection and poisoning, stimulate the formation and excretion of bile. The positive effect of the plant also affects the liver, and the whole digestive tract.

The special feature of the ProStor and Ferm-KM-1 products is the presence of yeast cell walls. They contain mannanoligosaccharides and beta-glucans, which effectively bond and absorb in the gastro-intestinal tract different pathogens. Beta-glucans have a stimulating effect and optimize the immune system. The preparations increase the digestion and feed efficiency, growth rate, optimize the productive indices of fish, effective in the treatment and prevention of parasitic diseases. The preparations that are hi-tech, bulk products of brown colour, with slightly specific odor, which makes it easy to mix them with compound feed components. They tolerate forage production processes without loss of biological activity.

Warranty storage period of preparations - six months from the production date, subject to +30°C temperature and relative humidity up to 75%.

Table 1. The efficacy of the ProStor preparation for carp and sturgeon juveniles

Index	Carp		Bester (starlet + beluga cross)	
	Experiment, 1.5 kg ProStor/t mixed feed KM-2M	Control, mixed feed KM-2M	Experiment, 2 kg ProStor/t mixed feed OT-7	Control, mixed feed OT-7
Absolute weight gain, g % of control	8.9 142.4	6.25 100.0	14.2 215.1	6.6 100.0
Average 24-h weight grow rate, %	6.78	5.65	11.2	7.84
Food expenses, units % of control	1.8 81.8	2.2 100.0	1.2 63.1	1.9 100.0
Survival rate, %	100	100	100	87

The ProStor and Ferm KM-1 preparations are used in the feed for the young and adult fish (the carp and the sturgeon). The preparation is administered in the feed in the feed mills or farms, by mixing. They are applied daily to feed on recommended zootechnical dosage rates (for the carp 1.0-1,5 kg per ton of feed, for the sturgeon 1.5-2,0 kg per ton of feed). Side effects and complications at the use of preparations at the recommended doses have not been observed. There are no contraindications. Fish products after the use of preparations can be used without restrictions. The efficacy of the ProStor preparation for fish is demonstrated in an experience with carp and sturgeon juveniles (Table 1).

The preparation in an amount of 1.5 kg per ton of feed was introduced to the feed KM-2M for the carp, and in the amount 2.0 kg per ton of feed OT-7 for the sturgeon. The underyearlings were kept in aquaria in groups of 15 animals. Fish breeding and biological indices of young carps and sturgeons as for absolute weight gain and average daily growth rate were higher than the ones of the control carp group, respectively, by 45% and 25%, and for control sturgeons - respectively, by 120% and 45%. The experimental sturgeon fry survival rate demonstrated was by 13% higher than the index of the control fish. The cost of 1 kg of growth gain of the experiment carp was 23.4 rubles, which is 13% lower than in the controls (26.95 rubles). The cost of 1 kg of growth gain in the experiment bester equaled 21.2 rubles, which is 35% lower than in controls (33.0 rubles).

Table 2. Fish breeding and biological indices of growing for two-year old sturgeon hybrids

Indices	Experiment versions	
	Control	Experiment with *Cellulomonas*
Weight, g: initial final	250.6±19.17 292.5±22.4	243.8±20.86 3042±31.2
Fullton's condition factor, %	0.35(100%)	0.39 (111%)
Absolute weight grow rate, g	41.9 (100%)	60.4 (144%)
24 h grow rate, g	1.32 (100%)	1.95 (148%)
24 average 24 h grow rate, %	0.50(100%)	0.72 (144%)
Weight gain coefficient, unit	0.031(100%)	0.045 (145%)
Food coefficient	1.2 (100%)	1.0 (83%)
Survival rate, %	100	100

Feed cost indices (1.2 units) of pilot feed line with similar values of better feed foreign companies. In experiments on the cultivation in a closed water supply for young sturgeons on the Ferm-KM-1 diet at 0.1% in the production OT 7 feed of for young sturgeons, the condition factor, as well the absolute and average daily weight gain coefficient significantly increased (Table 2).

The results of checking the efficiency of the incorporation of the ProStor and the Ferm KM probiotic preparation to the mixed feed for the sturgeon demonstrate higher industrial productivity rates for Russian-Siberian sturgeon hybrid. The obtained data as fishery/biological indices allow to recommend the use of the ProStor and Ferm KM-1at the large-scale mixed fodder production for they provide higher productivity figures, lowering the costs for feed and stable health conditions for the fish cultivated.

REFERENCES

[1] L. I. Bychkova, L. N. Yukhiменко, A. G. Khodaк et al. *Fish farming*, № 2, 48 (2008) (in Russian).

[2] V. D. Pokhilenко, V. V. Perelygin: *News of medicine and pharmacy*, 18(259), 56, (2008) (in Russian).

[3] S. Harbarth, M. H. Samore: *Emerg. Infect. Dis.*, 11, 794 (2005).

[4] A. D. Pickering: Stress and Fish. A. D. Pickering (ed.). London-N.Y.: *Acad. Press*, 1, (1993).

[5] T. Matsuzaki: *Immunol Cell Biol.,* 78 (1), 67 (2000).

[6] Yu. N. Grozesku, A. A. Bakhareva, E. A. Shulga: *News Bulletin of Samara Scientific Center, RAS*, 11, 1(2), 42 (2009) (in Russian).

[7] B. T. Sariev, A. N. Tumenov, Yu. M. Bakaneva, N. V. Bolonina: ASTU News Bulletin. *Ser. Fish Farming.* 2, 118 (2011) (in Russian).

[8] V. V. Panaseнко: *Fish farming*, 1, 74, (2008) (in Russian).

[9] S. V. Ponoмaryov, Yu. N. Grozesku, A. A. Bakhareva: Industrial fish farming. Moscow: *Kolos*, 2006. 320 p. (in Russian).

In: News in Chemistry, Biochemistry and Biotechnology ISBN: 978-1-63117-273-1
Editors: G. E. Zaikov, G. Nyszko, L. P. Krylova et al.

Chapter 21

COMMON LICORICE *GLYCYRRHIZA GLABRA* AS AN EXAMPLE OF THE USE OF PLANT EXTRACTS AND BIOLOGICAL COMPONENTS OBTAINED FROM THE PLANTS OF AN ARID ZONE

***O. V. Astafyeva*[1], *M. A. Egorov**[*,2] *and L. T. Sukhenko*[3]**
Federal State Budgetary Institution of Higher Professional Education
Astrakhan State University, Astrakhan, Russia

ABSTRACT

Use of phytogenic components or preparations instead of their chemical counterparts is a vital direction. As a rule, natural preparations obtained from plants have a slower and milder effect and do not accumulate in organism, do not cause side effects, i.e., they are free from the drawbacks which are often observed when purely chemical substances are implemented.

The unique and outstanding character of the obtaining and production of biologically active extracts with antibacterial and phytoncide properties from the Astrakhan Region plants are determined by the local environment conditions: high insolation, high temperature and low humidity, that contribute to accumulating increased concentrations of biologically active substances.

These conditions promote antimicrobial, bactericidal, immune defensive and antioxidant activity of the obtained plant extracts.

Keywords: Biologically active substances, antibacterial activity, common licorice *Glycyrrhiza glabra*

[*] Phone: (8512) 25-17-54; Fax: (8512) 25-17-18; E-mail egorovs.mail@gmail.com.

INTRODUCTION

The substances extracted from plants possess certain advantage over their chemical counterparts. As a rule, natural preparations obtained from plants have a slower and milder effect and do not accumulate in organism, do not cause side effects, i.e., they are free from the drawbacks which are often observed when purely chemical substances are implemented. The plants, as a food item for humans and animals and an inherent part of the living world, contain complexes of biologically active substances, which have been adapting to impact the living organisms for millions of years, and it is a very important advantage of the natural preparations with favourable therapeutic action.

Hence, use of phytogenic antibacterial preparations instead of their chemical counterparts is a vital direction in medicine, pharmacology and cosmetology [1]. Plant extracts and biologically active substances obtained from them are showing promise in this field. Activity of the extracts is mainly predetermined by the presence of chemical substances in them [3,5]. These primary biologically active substances have different composition and are related to different classes of chemical compounds: vitamins, flavonoids, terpens, hormones, alkaloids etc. [5]. A special place among biologically active substances is taken by terpens and their derivatives (terpenoids, saponins, glycosides etc.). Substances, classified as plant hormones, e.g., brassinosteroids, which perform the important regulatory functions in plants [4] are also related to terpens. It is the presence of these chemical substance groups that is responsible for various properties of the natural specimens: antibacterial [7], antioxidant [6], antifungal [2] etc.

This research is vital as it is aimed at extracting biologically active components from plants of arid zone and their use as biological preparations for the purposes of cosmetology and pharmacology as well as other industries. The unique and outstanding character of the obtaining and production of biologically active extracts with antibacterial and phytoncide properties from the Astrakhan Region plants are determined by the local environment conditions: high insolation, high temperature and low humidity, that contribute to accumulating increased concentrations of biologically active substances. These conditions promote antimicrobial, bactericidal, immune defensive and antioxidant activity of the obtained plant extracts [9].

EXPERIMENTAL PART

The paper is aimed at proving the possibility of using plants from arid zone and biologically active components obtained from them (with common licorice as an example) as biological preparations and their components with antibacterial properties. Antibacterial activity of the common licorice root extracts and fraction obtained from them, containing 1-2 chemical biologically active components, has been researched.

50% ethanol extracts of *Glycyrrhiza glabra* root and fractions obtained from them served as objects for the research.

Antibacterial activity was studied on nonpathogenic test microorganisms *Staphylococcus aureus* RCIO (Russian Collection of Industrial Organisms) B-1899, *Escherichia coli* CC (circadian culture) RCIO B-1911 and *Bacillus subtilis* RCIO B-1919.

Complex of biologically active substances from *Glycyrrhiza glabra* root was extracted with 50% ethanol in 1:5 proportions.

Biologically active substances of the obtained extracts were separated by fractionation as monocomponents. Fractions were obtained as a result of column liquid chromatography. Elution of biologically active substances of *Glycyrrhiza glabra* was carried out with 1% alcoholic solution (90% ethanol) of ammonia in the column (h=25 sm, d=2.5 sm).

Antibacterial activity of extracts and fractions was defined employing the method of direct diffusion into agar media, seeded with microorganisms, with measuring of the diameter of zone of inhibition (DZoI) and method of serial dilution of the preparations under consideration in growth media aimed at measuring the minimum inhibitory concentration (MIC) [8].

These studies have been carried out jointly with the colleagues from Ca'Foscari University of Venice (Venice, Italy) for several years.

RESULTS AND DISCUSSION

As a result of fractionation of the extracts – 50% ethanol extracts of biologically active substances of *Gl. glabra* root, fractions were obtained and the influence of these fractions on the nonpathogenic test microorganisms were studied employing the method of direct diffusion into agar media with the use of lunulae and method of serial dilution (MIC). Method of direct diffusion into agar media allowed obtaining the qualitative characteristics (DZoI - diameter of zone of inhibition) of the antibacterial activity of the fractions under consideration. Method of serial dilution permitted to get the most consistent results of test microorganism activity inhibition with the minimum concentrations of active substances of different obtained fractions of common licorice root extracts, as well as quantitative characteristics of antibacterial activity, expressed in MIC – minimum inhibitory concentration.

Antibacterial activity of the common licorice root extracts and the fractions of biologically active components separated from them has been studied. Table 1 shows the results of the study.

Table 1. Antibacterial activity of the common licorice root extracts and the fractions separated from them (DZoI method)

Fractions (1mg/ml)	*E.coli*	*St. aureus*	*B. subtilis*
	DZoI, mm		
E1	0	0	0
E2	16.0±3.4*	0	6.5±3.1*
E3	14.5±2.3*	0	6.6±2.8*
E4	15.8±0.7*	6.7±1.2*	12.1±1.3*
E5	0	8.3±0.9*	0
Extract	14.1±3.6*	11.5±2.3*	10.5±6.8*
50% Ethanol	0	0	0

Note: DZoI is the zone of inhibition diameter; * - differences with control sample are true at $p \leq 0.05$; 0 is the absence if antibacterial activity.

As Table 1 proves, fractions E4 and E5 exercised antibacterial (bacteriostatic) activity on *St. aureus* culture. Moreover, fraction E4 exercised antibacterial activity on the three test cultures (*St.aureus*, *E. coli*, *B. subtilis*). Fractions E3 and E2 showed the same level of antibacterial activity on strains of *E. coli* and *B. subtilis*.

Table 2. Antibacterial activity of the common licorice root extracts and the fractions separated from them (MIC method)

Fractions	*E.coli*	*St. aureus*	*B. subtilis*
	MIC, mcg/ml		
E1	1000.0±0.0	1000.0±0.0	1000.0±0.0
E2	62.5±9.5*	1000.0±0.0	1000.0±0.0
E3	125±19.4*	1000.0±0.0	62.5±13.9*
E4	62.5±11.3*	62,5±8,9*	1000.0±0.0
E5	1000.0±0.0	250.0±21.3*	1000.0±0.0
Extract	333.3±24.3*	500±14.9*	500±17.5*

Note: MIC is minimum inhibitory concentration, * - differences with control sample are true at $p \leq 0.05$.

But minimum inhibitory concentration (MIC) of E2 fraction is higher applied to *E. coli* (250 mcg/ml), and E3 fraction – applied to *B. subtilis* (500 mcg/ml) (Table 2).

CONCLUSION

Among the fractions of the ethanol extracts of common licorice root only one revealed stronger antibacterial activity than 50% extract when applied to *B. Subtilis.* Fractions of common licorice extract possess higher inhibitory activity compared to the common licorice extract itself when applied to *E.coli*. When applied to *St. aureus,* the extract showed the highest antibacterial activity compared to monocomponents in the form of fractions.

The obtained results prove the possibility of use of the common licorice root extracts and biologically active substances separated from them as fractions under consideration as biological preparations and their components with antibacterial properties.

At present ASU-based Laboratory of Biotechnologies has developed common licorice extract with anti-tuberculosis properties and phytobalms “INSOFIT”, which contain extracts of plants of arid zone possessing antibacterial properties. The formulation of these phytobalms includes common licorice root extract.

REFERENCES

[1] Astafieva, O. V., Novichenko, O. V., Egorov, M. A. The Possibility of Use of Extracts from Higher Hydrophytes and Geophytes of the Astrakhan Region for the Needs of Cosmetology // *Biochemistry and Biotechnology: Research and Development Binding*: *Hardcover,* USA, 2012, pp. 147-152 (in English).

[2] Bataeva, Y., Dzerzhinskaya, I., Egorov, M., Magzanova, D., Astafyeva, O. Growth-Promoting and Fungicide Characteristics of Cyanobacterial Communities from Ecosystems of the Astrakhan Region // *Journal of Agriculture and Food Technology*, 2 (12), 2012, pp. 184-187 (in English)

[3] Burlando, B., Verotta, L., Cornara, E., *Bottini-Massa, Herbal principles in cosmetics. Properties and mechanisms of action* / Boca Raton*: CRC Press,* Taylor and Francis group, 2010. – pp .226-231 (in English)

[4] Egorov, M.A. *Brassinosteroids as Possible Nanoregulators of Biological Systems. Biochemistry and Biotechnology: Research and Development.* USA, New York State, Nova science publishers, Inc., 2012. pp. 143-146_(in English)

[5] Muravieva, D. A., Samilina, I. A., Yakovlev, G. P. *Pharmacognosy* / Moscow: *«Medicine»,* 2002. - 656 c. (in Russian).

[6] Sabrina Fabris, Mariangela Bertelle, Oxana Astafyeva, Elena Gregoris, Roberta Zangrando, Andrea Gambaro, Giuseppina Pace Pereira Lima, Roberto Stevanato Antioxidant properties and chemical composition relationship of europeans and brazilians propolis // *Pharmacology & Pharmacy*, 4, 2013, pp. 46-51 (in English)

[7] Sukhenko L. T., The Biotechnology of Phased Drinking Water Purification in the Conditions of Astrakhan Region // *Biotechnology, Biodegradation, Water and Foodstuffs,* USA, New York State, Nova science publishers, Inc., 2009, pp. 143-145 (in English)

[8] Sukhenko, L.T. *Laboratory and practical training on microbiology with elements of inframicrobiology: recommended practice*. Part 1. Astrakhan: ASPU publishing house, 1999, 17 p. (in Russian).

[9] Sukhenko, L.T. Prospects of Extraction of Anti-Microbial Biologically Active Substances from some Wild Plants of the Astrakhan Region //*Vestnik Orenburgskogo Universiteta,* No 4, 2011, P.56-62.

In: News in Chemistry, Biochemistry and Biotechnology ISBN: 978-1-63117-273-1
Editors: G. E. Zaikov, G. Nyszko, L. P. Krylova et al.

Chapter 22

THE STUDY OF MORPHOGENETIC PECULIARITIES OF WINTER RAPE (*BRASSICA NAPUS* L.) PRIMARY EXPLANTS *IN VITRO* CULTURE

O. L. Klyachenko and N. V. Nikiforova*
National University of Life and Environmental Sciences of Ukraine, Kyiv, Ukraine

ABSTRACT

The paper presents the results of experiments performed for obtaining of virus-free plant regenerants of four rape (*Brassica napus* L.) cultivars through the callusogenesis and direct organogenesis. Morphogenetic peculiarities in primary explants cultured on optimally selected medium for callusogenesis and regeneration was studied. Comparative analysis of the direct and indirect plant regenerants morphogenesis was made. The primary effectiveness of the winter rape studied varieties' indirect morphogenesis was proved.

Keywords: Rape, morphogenesis, explant, callus, rhizogenesis, plant regenerants

Rape (*Brassica napus* L.) is the major oilseed crop and one of the most important sources of vegetable oil in a food and industrial usage, as well as high-protein feed. In terms of produced oil volume, rape is on the third place in the world after soybean and cotton [1,2]. At the present stage of scientific development for the selective process intensification the usage of biotechnological methods, direct tissue culture, cell selection and genetic engineering techniques are effective and enable the rapid multiplication of high-producing material. The method of tissue and organ culture is important for the processes of cells and tissues regeneration empowering, especially during the rational system of seed production organization. In this way we can multiply genetically valuable plants, heterotic hybrids, as well as sterile and parthenocarpic genotypes.

* National University of Life and Environmental Sciences of Ukraine, 03041, Kyiv, Heroiv Oboronu str., 15. Email: nikiforova.n.v@mail.ru.

Genetic variability that is common for callus and cell cultures allows obtaining regenerants that have new hereditary traits and is future-oriented on a new parent material obtaining. Regeneration of many Brassicaceae genus representatives is well studied, but specific genotype is not always characterized by regeneration according to the protocol described for the culture [3].

The effect of ploidy level, peculiarities of cultivars and explants on the multiplication factor of rape culture in vitro is studied incompletely, causes of poor bud formation and root development in particular genotypes are not completely determined.

The aim of our study was to investigate morphogenetic peculiarities of different winter rape cultivars in vitro culture.

MATERIAL AND METHODS

The study involved the kind of winter rape Ukrainian selection *Aliot*, *Nelson*, *Syntetic*, *Antariya*. Investigations were conducted in several ways. The intensity of callus formation and explants regeneration rape and the ability of root development of winter were studied.

The experiment began with the seeds sterilization according to our scheme, based on existing techniques with the following planting on the non-hormonal culture medium according to Murashige and Skoog (MS) [4]. Cotyledonary leaves were used as explants. For the callusogenesis induction was used MS medium laced with adenine 10 mg/l, gibberelic acid (GA) 0.05 mg/l, 6-benzylaminopurine (6-BAP) 0.5 - 1.5 mg/l naphthaleneacetic acid (NAA) 0.5 mg/l, kinetin 2.5 mg/l, 2,4-dichlorophenoxyacetic acid (2,4 D) 2.5 mg/l and sucrose 20 g. Obtained initial callus was transferred on a fresh medium for keeping growth. For the investigation of callus morphogenesis it was transferred to MS medium laced with growth regulators - kinetin 0.25 mg/l, 6-BAP 1.5 - 3 mg/l, NAA 0.5 mg/l and cultured under illumination (3000 - 4000 lux) and in absolute darkness (in thermostat), at a temperature of + 24-26°C and relative humidity 70-80%. Periodically calluses were examined and defined according to color, texture and growth rate.

Sprouts of plant regenerants were grown on the light in the growth chamber at a temperature + 24-26°C with 16 hours photoperiod. Sprouts rooting on a rhizogenous medium, which was laced with the half concentration of macro- and micronutrients MS, MS vitamins, 0.5 mg/l adenine and 20 g of sucrose, was examined on a weekly basis.

RESULTS AND DISCUSSION

Morphogenetic potential of cultured plant cells is determined by their genotype and culture conditions. One of regeneration frequency rising method is based on the artificial selection of culture media and conditions for in vitro cells growing for each particular genome [5].

Callus induction was observed on MS medium laced with adenine at a concentration 10 mg/l, concentration of GA 0.05 mg/l, 6-BAP 0.5 - 1.5 mg/l, NAA 0.5 mg/l, kinetin 2.5 mg/l, 2.4-D 2.5 mg/l and sucrose 20 g. Cotyledonous and true rape leaves were planted on the medium and cultured under the illumination and in absolute darkness at a temperature + 24-

26°C and relative humidity 70-80%. During cultivation of explants on the surface of a culture medium during the first three days significant swelling of the winter rape initial explant of all studied cultivars was observed. On the seventh day of cultivation the beginning of callusogenesis was marked, and up to 16 day callus formed in almost all samples, both on the coup injury (Figure 2c, d), and on the side surfaces of explants (Figure 2a, b).

Obtained initial calluses differed in morphology and weight gain. On the medium MS1 (MS + 10 mg/l adenine, 0.05 mg/l GA, 0.5 mg/l 6-BAP, 0.5 mg/l NAA) was obtained the biggest number of winter rape calluses of *Aliot* and *Synteti* cultivars, varieties *Nelson* and *Antariya* showed better results on the medium MS2 (MS + 0.05 mg/l GA, 0.5 mg/l BAP, 0.5 mg/l NAA). The overall trend of morphologically correct callus formation and biomass growth had been observed on the MS1culture medium (Table 1).

For the somatic embryogenesis induction in winter rape callus tissue obtained calluses were transferred to a modified MS medium laced with growth regulators - kinetin 0.25 mg/l, 6-BAP 1.5 - 3 mg/l, NAA 0.5 mg/l. Tubes were transferred to a growth chamber with 16-h photoperiod (3000 - 4000 lux), at a temperature + 24-26°C, humidity 70-80% and morphological changes were observed.

Figure 1. Peculiarities of a rape (*Brassica napus* L.) direct morphogenesis in vitro cultue: *a* – *Aliot* variety, *b* – *Nelson* variety, *c* – *Syntetic* variety, *d* – *Antariya* variety.

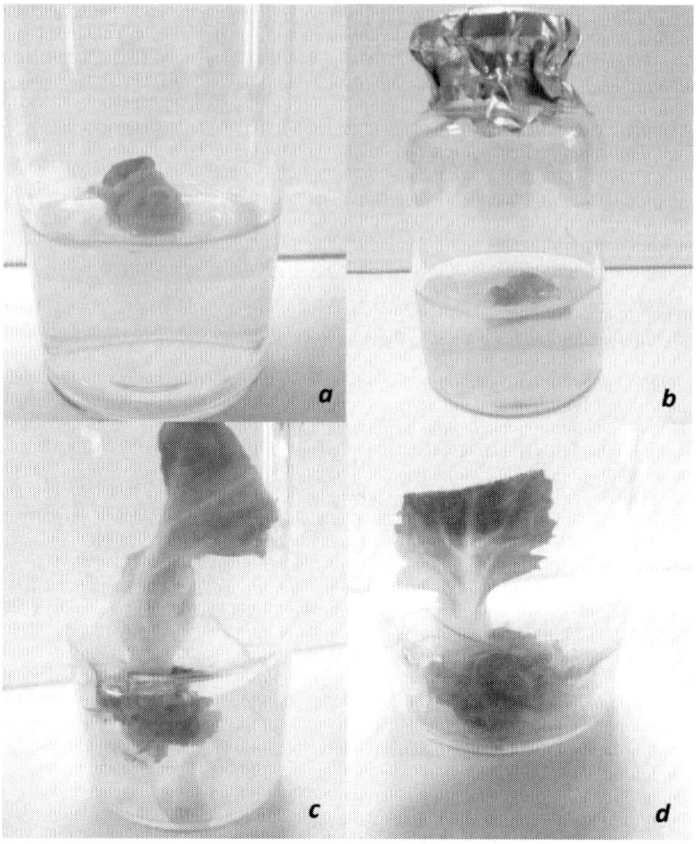

Figure 2. Peculiarities of rape callusogenesis (*Brassica napus* L.) in vitro cultue: *a* – *Aliot* variety, *b* – *Nelson* variety, *c* – *Syntetic* variety, *d* – *Antariya* variety.

After 14 days of cultivation on the surface of callus formation of green indurations was observed, which in a few weeks evolved into morphologically correct microrosettes and sprouts (Figure 2a). High results in morphogenic callus formation of *Aliot* and *Syntetic* winter rape varieties observed on medium MS2.1, *Nelson* variety showed idetical results on two medium - MS2.2 and MS2.3, and *Antariya* variety distinguished by the best callusogenesis on MS2.2 medium.

A high regenerative ability of the winter rapeseed plants of all varieties on MS2.3 nutrient medium was marked (Table 2).

Parallel experiments on the cultivation of four winter rape varieties for direct morphogenesis induction on MS medium laced with 0.25 mg/l kinetin were conducted. All material was grown in a growth chamber at a temperature + 24-26°C, relative humidity 70-80%, duration of the photoperiod was 16 hours. After 3 weeks of cultivation cutting grafting of obtained plant regenerants was made and further usage of microcuttings (1 - 2 cm) with a shortcut leaves that had axillary meristems. On the second week the rapid regeneration of sprouts and axillary buds was marked. But the number of regenerated plants through the direct morphogenesis significantly infrared to the number of regenerated plants obtained by the indirect morphogenesis.

Also during the cultivation process of four rape varieties' regenerants were characterized by differences in the rate of development, color, and plant formation (Figure 1).

Table 1. Peculiarities of winter rape (*Brassica napus* L.) callusogenesis

No	Medium composition	Variety	Quntity of cotyledonary leaves tranfered on the medium for callusogenesis , pcs.	Quantity of obtained calluses	Callus description
1	MS 1.1 MS + adenine 10 mg/l, GA 0.05 mg/l, 6-BAP 0.5 mg/l, NAA 0.5 mg/l	*Aliot*	30	26	Light green, compact with a slight villosity
		Nelson	30	18	
		Syntetic	30	24	
		Antariya	30	21	
2	MS 1.2 MS + GA 0.05 mg/l, 6-BAP 0.5 mg/l, NAA 0.5 mg/l	*Aliot*	30	28	Light brown, friable, agranular
		Nelson	30	20	
		Syntetic	30	23	
		Antariya	30	22	
3	MS 1.3 MS + 6-BAP 1.5 mg/l, NAA 0.5 mg/l, kinetin 2.5 mg/l, 2,4-D 2.5 mg/l	*Aliot*	30	18	Light brown, friable, hard-coarse granular
		Nelson	30	18	
		Syntetic	30	20	
		Antariya	30	19	

Table 2. Peculiarities of sprouts morphogenesis and regeneration in winter rape (*Brassica napus* L.) callus culture

No	Medium composition	Variety	Calluses quantity, pcs.	Morphogenic calluses, %	Regeneration ability, %
1	MS 2.1 MS + kinetin 0.25 mg/l	*Aliot*	30	60	22
		Nelson	30	52	24
		Syntetic	30	58	22
		Antariya	30	54	24
2	MS 2.2 MS + 6-BAP 3 mg/l, NAA 0.5 mg/l	*Aliot*	30	62	36
		Nelson	30	54	30
		Syntetic	30	60	24
		Antariya	30	57	32
3	MS 2.3 MS + 6-BAP 1.5 mg/l, NAA 0.5 mg/l	*Aliot*	30	50	36
		Nelson	30	54	38
		Syntetic	30	58	38
		Antariya	30	48	42

The essential stage in the process of meristematic plant obtaining, which are ready for planting in the ground, is the stage of sprouts rooting obtained from the isolate. With the purpose of rooting microrosettes and rape sprouts were transferred on the rhizogenous medium laced with the half concentration of macroelements MS, microelements MS, vitamins MS, 0.5 mg/l adenine and 20 g of sucrose. Well-rooted plants with developed leaf laminas and dark green petioles were taken out from the tubes for the adaptation. The root system was washed from agar traces with distilled water and rinsed with a 1% solution of potassium permanganate. Regenerated plants were planted in sterile soil, pre-fried in a dry-air sterilizer, and covered with a glass cylinder. Periodically, plants watered with a solution of macro- and microsalts according to Murashige and Skoog laced with sucrose 30 g/l. Three weeks later they were transferred on the field.

Conclusion

Therefore, morphogenetic peculiarities of initial explants during the cultivation on optimally selected medium for callusogenesis and regeneration were studied. The primary effectiveness of the winter rape studied varieties' indirect morphogenesis was proved. Developed complex of biotechnological techniques can be successfully used in traditional breeding and provide a basis for development of new varieties and hybrids, breeding material multiplication.

References

[1] O. L. Klychenko, I. D. Sytnik, O. K. Galchinska: *Winter and spring rape seed. Biology. Selection. Biotechnology*. Kyiv, 2012. 245 p. (in Ukrainian).

[2] O. L. Klychenko, I. D. Sytnik: *Agrarian education and science,* 2, 9 (2002). (in Ukrainian).

[3] G. P. Kushnir, V. V. Sarnackaya: *Microclonal propagation of plants: theory and practice*. Kyiv, 2005. 270 p. (in Russian).

[4] T. Murashige, F. Scoog: *Physiol. plant.*, 15, 473, (1962).

[5] S. G. Kolumbaeva, S. A. Jokobaeva, K. K. Boguspaev: *Biotechnology. Theory and practice*, 1, 56, (1996). (in Russian).

In: News in Chemistry, Biochemistry and Biotechnology ISBN: 978-1-63117-273-1
Editors: G. E. Zaikov, G. Nyszko, L. P. Krylova et al.

Chapter 23

CYTOLOGICAL CHANGES IN SPERMIA OF THE RUSSIAN STURGEON (*ACIPENSER GUELDENSTAEDTII* B.) AFTER CRYOPRESERVATION BASED ON THE COMPOSITION OF CRYOPROTECTIVE MEDIUM

G. V. Zemkov and T. I. Pochevalova*
Astrakhan State University, The Laboratory of Biotechnologies, Astrakhan, Russia

ABSTRACT

The article presents some experimental results concerning the problem of genetic conservation of valuable and endangered animal species. In our own investigations we studied the cryoprotective properties of glycerin, dimethyl sulfoxide and heparin of different proportions on the example of the spermatic fluid of sturgeon Acipenser guldenshtadti (Brandt). These results were compared with the cryoprotectors of well-known composition. In addition, freezing and storage of the spermatic fluid were executed in ultralow temperature. Cold tolerance of the sperm cells has been linked to the composition of cryoprotectors. It was estimated by the sperm mobility and in accordance with the results of the morphological analysis of cells by the light microscopy method.

Keywords: Cryopreservation, cryoresistance, defrostation, sperm cells, cryoprotectors, Russian sturgeon, cytomorphological analysis

The gene pool preservation of valuable species of animals and plants has become an urgent problem due to the growing man-made environment in the last 30-40 years. Many species of the ecosystem have already disappeared and others are on the brink of extinction. Reserve management and conservation of valuable commercial species, as traditional ways of protecting them from extinction, can not fully solve the problem of the gene pool preservation

* E-mail: akimochkinatanja@mail.ru.

of wild fauna and flora now. Experience in cryobiology helped to develop conservation biotechnology of long-term storage of reproductive cells of animals in extremely low temperatures for later use, aimed at restoring the population level of animals (Veprintsev, Rott, 1984).

The issues of cells protection from being destructed by water crystals which are formed during freezing of biological objects have been and remain very important for the research in this area. These ice crystals have a destructive effect when biological material is frozen (Lozina-Lozinskiy, 1972). There are a lot of researches in this area; scientists are trying to modify the existing cryoprotective mixtures and develop new ones.

The objective of the experimental study we carried out was to find all the specific features of morphological distortions in spermia of the Russian sturgeon after storage of spermatic fluid at extremely low temperatures and with different composition of cryoprotective medium at the same time.

Materials and Methods

Spermatic fluid of reproductive male specimens of the Russian sturgeon of later run was taken at fish-breeding plants of the Astrakhan region, employing the method of decantation after pituitary injection. 15 samples were taken from fifteen male specimens. 5 variants of different cryoprotective media were tested (Table 1).

Table 1. Qualitative and quantitative composition of cryoprotective mixtures

Cryoprotective medium	Composition of cryoprotective medium	Authors
Composition № 1	80% of stock solution, 10 mM sucrose,(1.71 g/l) 10 mM marmitol (0.98 g/l); 10% of egg yolk, 10% of dimethylsulfoxide	Ananiev, Andreev and others, 1998
Composition № 2	80% of stock solution; 20 mM sucrose (6.84 g/l); 10% of egg yolk; 10% of dimethylsulfoxide	Our modification
Composition № 3	80% of stock solution; 20 mM marmitol (3.94 g/l); 10% of egg yolk; 10% of dimethylsulfoxide	– " –
Composition № 4	80% of stock solution; 10% of egg yolk; 10% of dimethylsulfoxide	– " –
Composition № 5	0.05 ml of glycerin; 0.05 ml of DMSO; 0.04 ml of heparin	New composition that we developed

Before freezing samples of spermatic fluid were placed in a refrigerator for 30 minutes at a temperature of 8 degrees and then they were frozen with the help of step-by-step method in liquid nitrogen. Spermatic fluid was in the fridge at -196°C in liquid nitrogen. Quality of native and defrosted spermatic fluid was determined by morphological integrity of spermium.

RESULTS OF THE RESEARCH

During the study we found out that all samples of native spermatic fluid contained spermium with well-marked morphological features. The head part had the normal elongated form with extension at the base that is typical for sturgeon spermium (Ginsburg, Detlaf, 1975), tail was extended and could be easily seen (Figure 1).

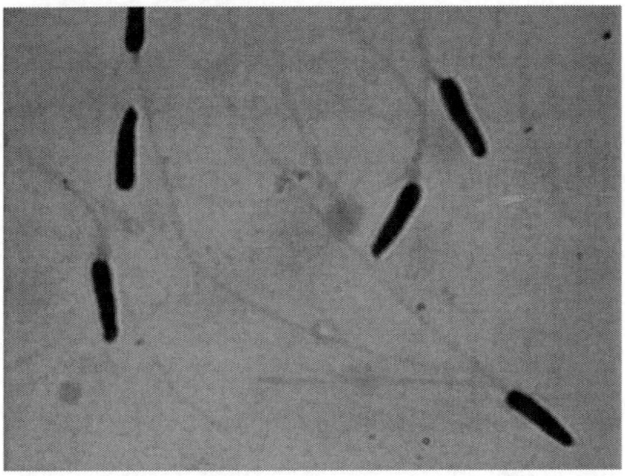

Figure 1. Control sample. Smear of native spermatic fluid before freezing. Sperm cell head of a regular form, the length of the tail is proportional to its head, no signs of fragmentation or curvature. Azure-eosin Romanovsky's stain. Magnification x 100.

Wide variation in the degree of distortion of the structural organization of the Russian sturgeon spermia depending on the composition of the cryoprotective medium was found during the experiments. After freezing the spermatic fluid in liquid nitrogen with cryoprotective medium №1 added, in some cases, spermia acquired a rounded shape and they were randomly distributed in the mass, it was also difficult to see the tail (Figure 2A). In other cases, we could observe the swelling of the spermatic fluid smears, hypopigmentation and a significant increase in the size of the round-shaped head that had obvious signs of destruction. We could hardly see the tail (Figure 2B).

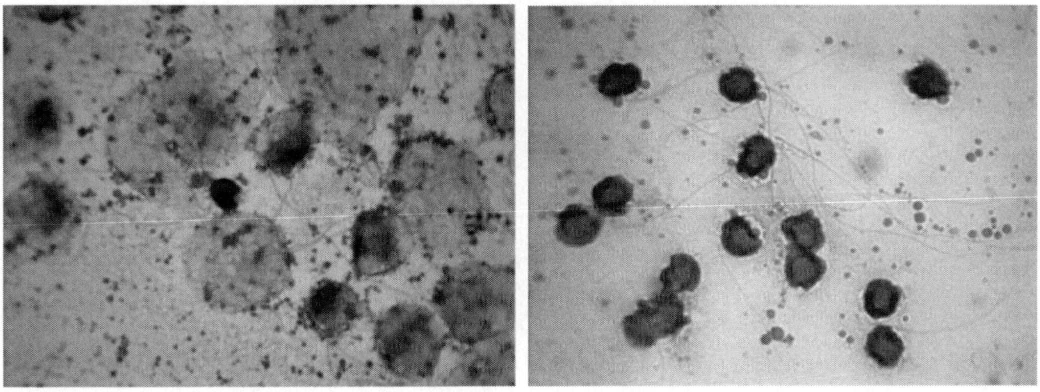

(A) (B)

Figure 2. A smear of defrosted spermatic fluid of Russian sturgeon with the addition of cryoprotector № 1. A and B – different cases. Azure-eosin Romanovsky's stain. Magnification x 100.

When we used a cryoprotective medium №3 with mannitol without sucrose, we saw the spermia contraction and reduction of their sizes (Figure 3A); the introduction of mannitol instead of sucrose (medium №2) caused schistocytosis (Figure 3B). We could see the tail clearly after introduction of sucrose instead of mannitol.

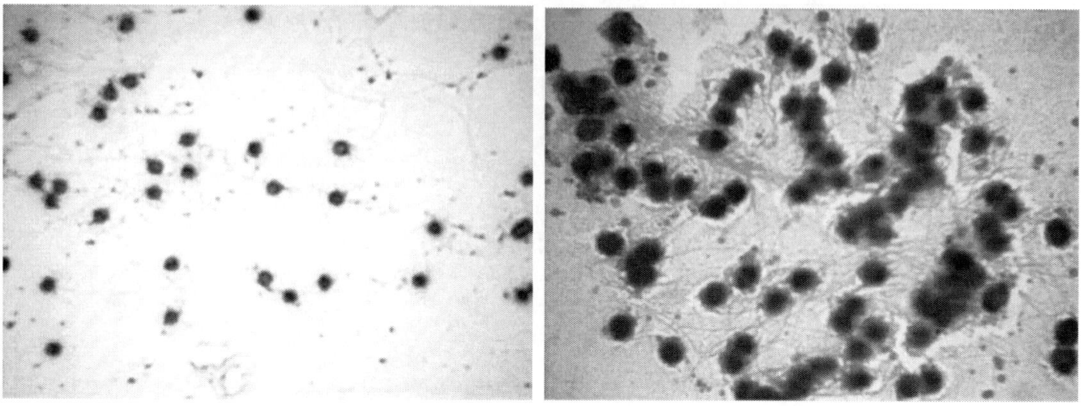

(A) **(B)**

Figure 3. Defrosted spermatic fluid of Russian sturgeon sugared before freezing: manna sugar d (A) and sucrose (B). Azure-eosin Romanovsky's stain. Magnification x 100.

When we used a cryoprotective medium №4 without introduction of sucrose and mannitol, in some cases we could see aggregation in certain places of smears but the head part had a normal shape. We could clearly see the tail (Figure 4). Consequently, we should use this medium without introduction of mannitol and sucrose.

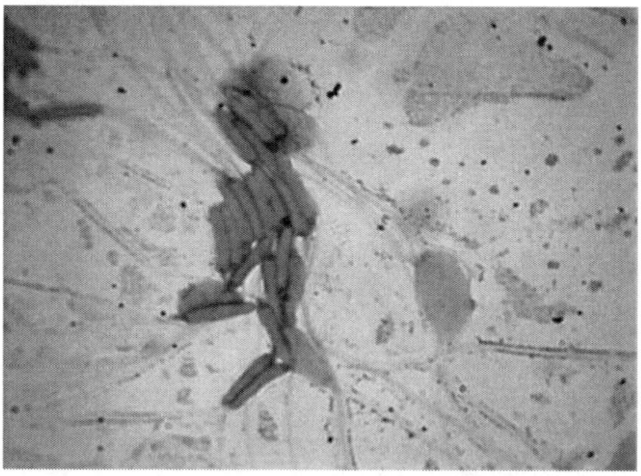

Figure 4. A smear of defrosted spermatic fluid of Russian sturgeon frozen with cryoprotector № 4, no added sugar. Azure-eosin Romanovsky's stain. Magnification x 100.

When we added our modified cryoprotective medium №5 to the spermatic fluid, almost in all cases, we could hardly see the tail of the spermia after defrosting, but we saw the head part clearly, it remained unscathed and had the normal shape (Figure 5A), which is very close to the morphology of the control samples of spermia. In rare cases, when we used this medium, we observed swelling of the spermium heads, some cells with clear sings of degradation (Figure 5B), which is also an indicator of individual fish cells cryoresistance.

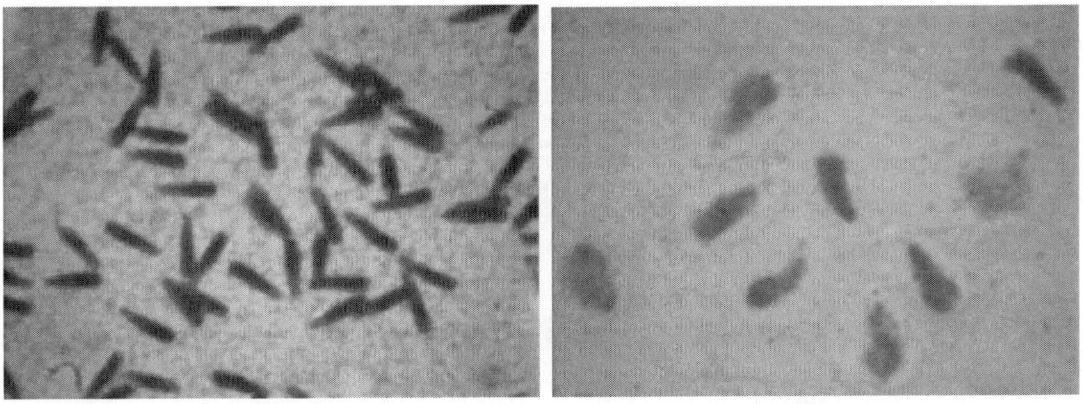

(A) (B)

Figure 5. Defrosted spermatic fluid of Russian sturgeon with the addition of cryoprotector № 5. A and B – different cases.Azure-eosin Romanovsky's stain. Magnification x 100.

The above data first of all show dependence of morphological and functional organization of spermia on the composition of cryoprotective media that should protect cells from destruction when we freeze spermatic fluid. These data served as the basis for the development of an electronic database "Morphological Distortions of Fish Spermia at Different Conditions of Freezing" (Certificate of state registration of the database № 2008620286) and the search for effective composition of cryoprotective media providing a high survival rate of cell material from the processes of cryopreservation - defrosting.

REFERENCES

[1] V.I. Ananiev, A.A. Andreev, T.S. Golovanov,N.N. Petropavlov, L.I. Tsvetkova Experience in cryopreservation of inconnu and beluga sperm. - *Fisheries. Ser. Aquaculture*. - M., 1998 - Vol. 1. - P. 25 – 32

[2] B.N. Veprintsev, N.N. Rott. Conservation of genetic resources. The problem of gene pool preservation. Pushchino, 1984. – p. 48

[3] L.K. Lozina-Lozinskiy – Essays on Cryobiology. Leningrad, *Science*, 1972. P. 288.

In: News in Chemistry, Biochemistry and Biotechnology ISBN: 978-1-63117-273-1
Editors: G. E. Zaikov, G. Nyszko, L. P. Krylova et al.

Chapter 24

DEVELOPMENT OF NONTOXIC METHODS OF RODENT POPULATION CONTROL AS AN ALTERNATIVE APPROACH FOR BIG CITIES

V. V. Voznessenskaya* and T. V. Malanina
A. N. Severtzov Institute of Ecology & Evolution, Moscow, Russia

ABSTRACT

Rodents cause considerable economic damage to agriculture and industry production. In the urban area in addition to economic losses, human health and safety from rodent transmissible zoonoses are of concern. Highly toxic methods are applied currently in Russia to manage rodent populations, which are not safe for humans and other mammalian species. Major pitfalls: high toxicity to humans and other non-target species; environmental pollution; development of avoidance behavior and rodenticide resistance in rodents. Biological activity of Felidae family pheromone L-Felinine has been described in the house mouse and Norway rats.

Keywords: Rodents, reproduction, population control, nontoxic repellency, reproductive inhibitors, steroid hormones

INTRODUCTION

Rodents cause considerable economic damage to field and fruit crops on annual basis.

In the urban area in addition to economic losses, human health and safety from rodent transmissible zoonoses are of concern. Highly toxic methods are applied currently in Russia to manage rodent populations, which are not safe for humans and other mammalian species (Voznessenskaya et al., 2004). Major pitfalls of current approaches: high toxicity to humans and other non-target species; environmental pollution; development of avoidance behavior

* A. N. Severtzov Institute of Ecology & Evolution, 33 Leninski prospect, Moscow, 119071, Russia, fax: (495)9545534, email: veravoznessenskaya@gmail.com.

and rodenticide resistance in rodents. Moreover, individuals, survived after rodenticide treatment exhibit reproductive outbreaks (Rylnikov et al., 1992). Zoonoses attributable to rodents are exacerbated during periods where their population erupts. Methods that can dampen these irruptive population cycles would prove highly desirable. It is our goal to develop a product that will dampen the amplitude of these rodent population cycles. Rodent population size is regulated by external as well as internal (zoosocial) factors. Predator population density is the most influential external factor (Hentonnen et al., 1987). Major internal regulating factor is the rodent population density itself. Our investigation is aimed to develop a product based on a number of substances involved in the regulation of rodent population density under natural conditions. Reproductive control in wildlife species who comes into conflict with humans has received increasing attention as a humane method for managing wildlife populations. Moreover, modeling studies show that contraception as a tool to management populations is best suited for species with high population turnover, i.e., short generation time and high reproductive output. Thus, rodents are ideal targets for this management tactic. Predator urine is used as a wildlife management tool to repel herbivorous animals from areas. Fundamentally, avoidance of predator urine by potential prey, and by implication the areas where predators frequent, is presumably evolutionary advantageous because it lowers the risk of predation. Potential prey can discriminate predator urines as opposed to that of other herbivores on the basis of the urine's odor. One consequence of a high meat diet is the presence of sulfurous compounds in the urine. These compounds result from protein digestion and metabolism. When sulfurous compounds are removed from predator urines by mercury treatment herbivorous animals are no longer repelled by the urine's odor (Nolte et al., 1994). Our previous research showed the effects of predator odors on various aspects of rodent reproductive behavior and reproductive output (Voznessenskaya et al., 1992; 2004; 2006; Sokolov et al, 1992; Kassesinova, Voznessenskaya, 2009). Felinine is a unique sulfur-containing amino acid found in the urine of domestic cats (Rutherfurd et al., 2002). Felinine is unstable in water solution and exists in the form of mixture of amino acid and sulfur-containing volatile compounds. One of four of them: 3-mercapto-3-methyl-1-butanol has a characteristic cat odor and believed to be a pheromone. Miyazaki et al., 2006). We now present evidence to support bioactivity of L-felinine with rodents.

MATERIALS AND METHODS

Test Subjects

Test subjects were 3-4 month old Norway rats (Rattus norvegicus) and 4-6 month old mice (Mus musculus); both from an outbred laboratory population. Before the start of the experiments, females were housed in groups of 3-4, and males were housed singly. Experimental rooms were illuminated on 14:10 hours light:dark schedule, and maintained at 20-22°C. Food and tap water were provided *at libitum*. Virgin females in proestrus/estrus, as determined by vaginal cytology, were chosen for the mating experiments. Sexually experienced males that were not mated in the 14 days before the test were used as sires. The morning after pairing, the females were checked for successful mating, as indicated by the presence of a vaginal plug. Successfully mated females were then housed singly.

Reproductive Output

For each experimental group, the total number of offspring was counted as well as number of pups per female; sex ratio was determined. In addition, we also weighed pups at the time of weaning (21-st day after birth).

Collection of Urine

Urine from domestic cats (*Felis catus*) was used as a source of predator chemical cues. These cats normally hunt for mice and have mice as part of their diet. If needed, additional meat was added to their diet. Freshly voided urine was frozen (-22°C). Once defrosted, urine was used only once. Non-predator urine was obtained from guinea pigs. Individuals of these species were placed into metabolic stainless-steel cages overnight, and urine was collected and stored using the method described above. Urine was collected and stored at -22°C.

"Open Field" Test with Added Stress (Rough Handling Conditions)

An "open arena" (D=0.7 m) with bright lights was used. Pregnant females were placed for 15 minutes in the centre of the arena on 1-st, 3-d , 5-th and 7-th day of gestation. During the test, we also used a buzzer, which made a loud noise, every 5 minutes. In addition, mice were handled roughly to physically induce stress. Blood samples from sublingual vein were drawn after each test for progesterone and corticosterone assay.

Assay for Progesterone and Corticosterone

Animals within each treatment were randomly assigned to one of four cohorts. Blood samples (100 μl) were obtained from sublingual vein every second day for each cohort for each of the treatment for the first seven days of gestation. This minimizes the handling and sampling of individual mice, while allowing a detailed study of changes in hormonal pattern as a function of time and treatment. Our experience shows that this method of repeated blood sampling has no long-term effect on visible scarring associated with traditional tail sampling technologies (Miller et al., 1997). Samples were centrifuged and the plasma frozen at -20°C until subsequent analysis. Plasma progesterone and corticosterone were assayed (in duplicate) by enzyme immunoassay (EIA) method (DRG, USA).

Assay for Fecal Corticosterone Metabolites

In small animals like mice, the monitoring of endocrine functions over time is constrained seriously by the adverse effects of blood sampling. Therefore, we used noninvasive technique to monitor glucocorticoids with recently established 5a-pregnane-3ß,11-ethol,21-triol-20-one enzyme immunoassay (Touma et al., 2004) to assess adrenal activity in mice under conditions of long lasting exposures to predator odors. Mice were exposed to L-Feinine (0.05%) on everyday basis for period of two weeks. On completion of exposures fecal material was collected from each animal over 24 hours. Extraction procedure was performed with 80% methanol. Concentration of corticosterone metabolites was measured with spectrophotometer (Spectramax340, Molecular Devices, USA) at 450 & 670 nm. Specific antibodies were received from prof. E. Möstl laboratory (University of Veterinary Medicine, Vienna).

Immunohistochemistry Assay

To visualize activated neurons on olfactory bulbs sections in response to stimulation, Fos protein immunohistochemistry was used (Flavell and Greenberg, 2008). Fos protein is a product of c-fos known as immediate early gene which is induced quickly by different stimuli including cell depolarization (Sheng and Greenberg, 1990). Labeling Fos provides a physiological marker of neurons activated in response to specific stimuli. Half life span of protein Fos is two hours: depending on specific characteristics and neural cell localization optimal exposure time for maximal Fos detection may range from 45 to 90 min (Voznessenskaya et al., 2010). To stimulate main and accessory olfactory system mice were exposed L-Felinine (0.05% in water) for 40 min using half duty cycle (one minute—specific odor, one minute—clean air). Immediately after exposure mice were perfused with 3% paraformaldehyde in phosphate buffer. Olfactory bulbs were removed and postfixed in paraformaldehyde for 16hours. We used standard procedure for fixation of olfactory bulbs, cryoprotection and immunohistochemical staining of olfactory bulbs sections (DellaCorte, 1995). We used indirect avidin/biotin method; horseradish peroxidase was used as enzymatic label, diaminobenzidin (DAB) was used as chromogen. Sections were made at 20 μm using cryostat Triangle Biomedical. Immunostaining was made according to standard three day protocol using primary antibodies Santa Cruz Biotechnology (USA):c-fos (4) sc-52, dilution 1: 500. For visualization and counting of Fos positive cells we used Nikon©Eclipse E400 microscope with camera Nikon©Coolpix 990. For picture analyses we used ImageJ (NIH).

Experimental Design

The experimental method consisted of applying 0.2 ml of a test solution (urine of 0.05% L-felinine) to the bedding of pregnant rats or mice every other day for different time durations. This application maximised the likelihood of physical and odour exposure of the test stimulus to the female. In experiments, three treatment levels were used:

(1) tap water (WAT), as a negative control;
(2) urine from guinea pigs maintained on a vegetarian diet (vegetables, grains and water *ad libitum*), as a urine control (GPU);
(3) urine from domestic cats maintained on a feral mouse diet (CU), as a model stimulus representing unadulterated predator urine. Cats were maintained on the feral mouse diet for 14 days before urine collection;

After mating, females were randomly assigned to treatment groups: WAT, GPU, CU. Mean differences among treatment groups were determined in separate analyses for the number of pups and sex ratios using software STATISTICA8..

RESULTS

Exposure of mated female mice to intact cat urine provoked block of pregnancy in 30-70 % of cases depending on the season. At the same time average percent of pregnancy block in control animals did not exceed 15% (n=16, p<0.001, Fisher test). In autumn-winter season exposure to L-felinine (0.05%) provoked pregnancy block in 70% of mated female mice while in control group we observed only 20% of females with block of pregnancy (n=20, p=0.043, Fisher test). In spring-summer analogous exposures to L-felinine provoked pregnancy block in 62.5% of mated females compared with 12.5% in control (n=9, p=0.046, Fisher test). In felinine treatment group (figure 1) number of pups per fertile female was 2.5± 1 while in control – 5.70±1.00 (n=28, p=0.046 Mann-Whitney U test). Sex ratios in mice also were affected in favor of males by both treatments: cat urine (p< 0.001) and L-felinine (p<0.01). Data presented in figure 2, 3. Exposure of pregnant rats to L-felinine did not affect significanty litter size though we observed significant reduction in cat urine treatment group. On the contrary sex ratio in rats was affected in both treatment groups in favor of males: urine (p=0.0007), L-felinine (p=0.0007).

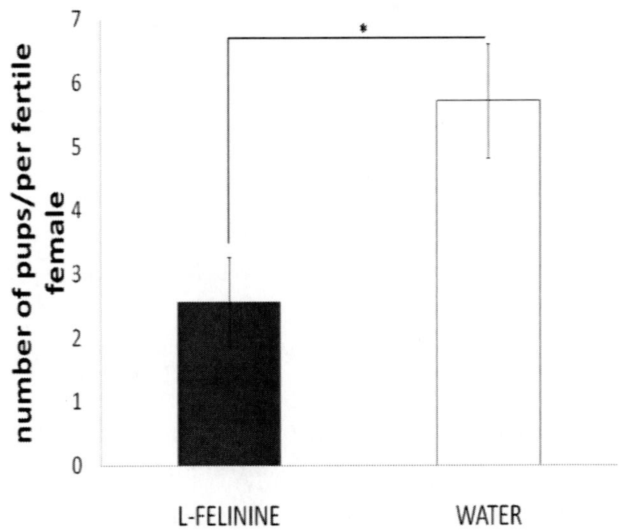

Figure 1. The influence of exposures to L-felinine (0.05%) during gestation on reproductive output in house mouse *Mus musculus* (Mann-Whitney U Test *p≤0, 05, n=28, ┬ - SEM).

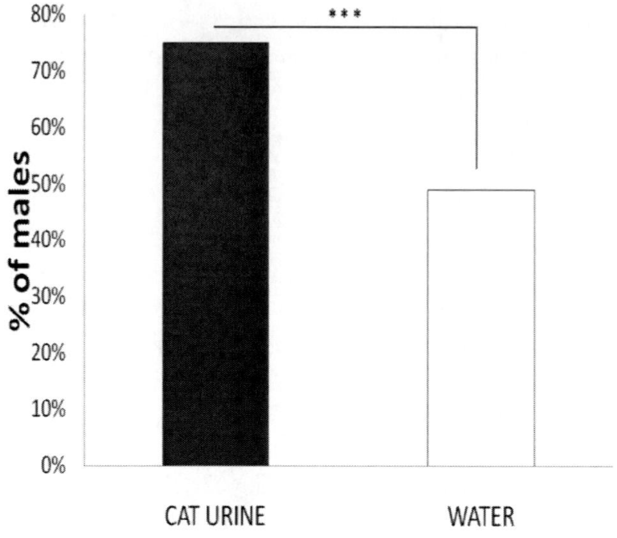

Figure 2. The influence of the cat urine *Felis catus* exposures during gestation on sex ratio in house mouse *Mus musculus* (***$p \leq 0{,}001$, n (cat urine)=52, n(water)=118, Fisher test).

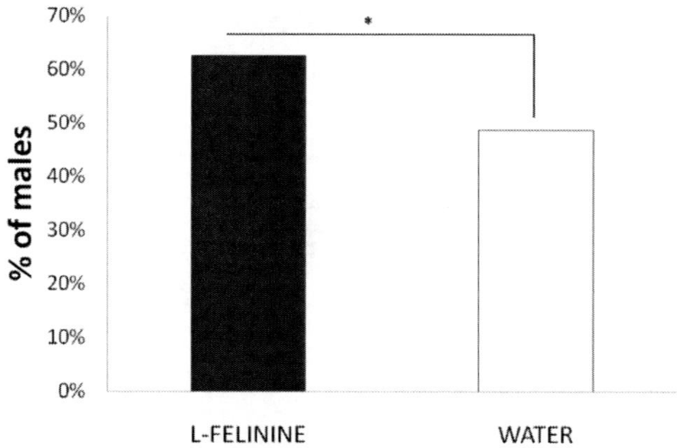

Figure 3. The influence of the L-felinine (0.05%) exposures during gestation on sex ratio in house mouse *Mus musculus* (*$p \leq 0{,}01$, n(L-felinine)=72; n(water)=160, Fisher test).

Table 1. The influence of exposures to cat urine *Felis catus* on plasma corticosterone in house mouse *Mus musculus*

	Plasma corticosterone (ng/ml)		
	1 -st day	3-d day	5-th day
Cat urine	681,25±135,16	706,25±123,63	716,25±105,55
Open field with "added stress"	371,25±175,05	183,75±86,34	96,25±34,61
Guinea pig urine	278,87±96,91	204±26,98	168,75±25,87
Water	92,75±43,51	77,88±22,8	84,25±17,7

(M±SD; n=8, each group).

We observed clear elevation of plasma corticosterone ($p<0.001$, $n=8$, Tukey test) in response to felinine in mice (table 1). As positive control we used “open arena” test with “added stress”. Mice responded to this kind of treatment with elevated corticosterone but we observed habituation during the course of consecutive placements (days 1-5). At the same time mice did not habituate to consecutive exposures to felinine. We also observed such a habituation in mice introduced to other novel stimulus – guinea pig urine. To explore for how long predator chemical cues may provoke elevated corticosterone we exposed mice to L-felinine for two weeks. On completion of exposures fecal glucocorticoid metabolites were measured for each animal. In control group concentration of corticosterone metabolites was 203, 85 ± 47, 74 ng/ml, in felinine treatment group – 702, 15 ± 122, 24 ng/ml ($n= 13$, $p<0.001$, t-test). The response of laboratory naive animals to predator scents and failure to habituate to the stimulus indicate the innate nature of the response. Chronically elevated cortocosterone may be responsible for the induction of pregnancy block.

We did not observe any differences in plasma progesterone for cat urine/felinine treatment groups and control animals. Immunohistochemical studies revealed neural activation in response to stimulation with L-felinine at the level of main olfactory bulb as well as at the level of accessory olfactory bulb indicating the involvement of both systems (main olfactory and vomeronasal) in detection of L-felinine which is important if practical applications are considered. In solution L-felinine is unstable; exists in form of mixture of amino acid and sulfur-containing volatile compounds. Most likely that 3-mercapto-3-methyl-1-butanol (felinine derivate) binds to receptors in main olfactory epithelium.

Discussion and Conclusion

Reproductive traits in rodents are affected by a number of environmental, social and chemosensory factors, e.g., the nutritional status of females will influence ovulation rate and litter size (Hamilton and Bronson 1985), as will exposure of females to other rodents of various social status (Steiner et al. 1983; Huck et al. 1988). Other well-described influences include synchronization of ovulation amongst female cohorts (Whitten 1956), acceleration or delay of puberty (Vandenbergh 1969; Lombardi and Vandenberg 1977), pregnancy block owing to stress, and failure to implant blastocysts when female rodents are exposed to the odor or urine of strange males (Bruce 1959).

The majority of these studies on reproductive inhibition have focused on intraspecific influences of semiochemicals and how they influence reproductive output and behavior in females. A few studies have focused on between-strain influences or interspecific influence, although the source odor generally is still confined to rodents.

During our investigation on the effects of predator odor on rodent reproduction and repellency, we found that female rats exposed to cat urine during pregnancy had reduced litter sizes at parturition (Voznessenskaya et al, 2004). Exposure to predator odor also caused disruptions of the oestrous cycle (Voznessenskaya et al. 1992). These effects bear striking similarities to the studies of the effects of rodent urine odor on intraspecific rodent reproduction. If such similarities are broadly based, then similarities in mechanisms of perception, reproductive physiology, and chemical nature of stimulus might be anticipated.

We do not believe that reduction in litter size is attributable to an adaptive response by rodents to predator odors. Rather, we propose the following interpretation. Urine contains information about the identity of individuals, reproductive status, and dominance status. We postulate that urine also contains information about environmental quality as reflected by nutritional status. Investigation of urine from a variety of sources would serve as an efficient way to integrate environmental information. During times of food depletion, an individual could assess the nutritional status of the population. If food becomes limiting, rodents will begin to catabolise their own muscle protein and the urine will contain larger amounts of protein degradation products. These signals could serve to trigger mechanisms that would affect reproduction. Given that the generation time of rodents is short, complete reproductive inhibition may not be adaptive. However, reduced reproduction may be beneficial. Reduced reproduction would relieve energetic constraints on lactating females that might otherwise jeopardize survival if a full litter size were attempted.

Litters are biased toward producing males when predator or rat catabolic urine is used as a stimulus. This is consistent with theory on reproductive value. Even with reduced litter size, females may still experience lower survival probabilities during reproduction and lactation in food limiting environments because of energetic constraints. However, males would be less constrained by such energetic considerations. Thus, their survivorship probabilities may be higher than females, and by implication their value in contributing to fitness would also be higher. So then, why should rodents reduce reproduction when presented with predator urine? Predators on rodent diets would produce urine with many of the same rodent-derived metabolic products. It is only coincident that the two urines produce the same effect.

The proposed method utilizes naturally derived compounds that pose no environmental hazard. In nature, predators are one of the most powerful extrinsic factors affecting prey population cycles (Hentonnen et al. 1987; Klemola et al. 1997). At the same time, high population density in rodents is the most powerful intrinsic factor for regulation of population density. Our method utilizes combination of intrinsic and extrinsic factors regulating population density under natural conditions. One of the most serious advantages of this method is lack of habituation to repeated exposures to such types of compounds. At the time we keep a sixteenth generation of rats in our laboratory under persistent exposures to predator odors (Voznessenskaya et al., 2006). These animals still responding to predator urine exposures with reduced litter size. The proposed method should prove useful in reducing our reliance on pesticides with less favorable environmental properties while achieving the goal of reducing rodent populations.

ACKNOWLEDGMENTS

This research was supported by grants from Russian Foundation for Basic Research #07-04-01538a, 10-04-01599a and 14-04-01150a, Russian Academy of Sciences, Program "Zhivaya Priroda" and MK-709.2012.4.

REFERENCES

Dellacorte C.: Experimental Cell Biology of Taste and Olfaction: Current Techniques and Protocols, Boca Raton: CRC, 1995, p.145.

Touma C., Palme R., Sachser N.: *Hormones and Behav.*, 45, 10, (2004).

Nolte D. L., Mason J. R., Epple G., Aronov E. V., Campbell D. L.: *J. Chem. Ecol.*, 20, 1505, (1994).

Kassesinova E., Voznessenskaya V.: *Chem. Senses*, 34 (3),E35, (2009)

Hamilton G. D., Bronson F. H.: *Amer. J. Physiol.*, 250, 370, (1985).

Hentonnen H., Oksanen T., Jortikka A., Haukisalmi V.: *Oikos*, 50, 353, (1987).

Bruce H. M.: *Nature* (London), 61, 157, (1959).

J. G. Lombardi, J. G. Vanderbergh: *Science*, 196, 545, (1977).

Vanderbergh J. G.: *J. Endocrinol.*, 84, 658, (1969).

Rutherfurd K. J., Rutherfurd S. M., Moughan P. J., Hendriks W. H.: *J. Chem. Ecol.*, 19, 1405, (2002).

Miyazaki M., Yamashita T., Suzuki Y., Soeta S., Taira H., Suzuki A.: *Comp. Biochem. Physiol. B. Biochem. Mol. Biol.*,145, 451, (2006).

Sheng M., Greenber M. E: *Neuron*, 4, 477, (1990).

Bacon S. J., McClintock M. K.: *Physiol. Behav.* , 56, 359 (1994).

Flavelll S. W., Greenberg M. E.: *Annu. Rev. Neurosci.*, 31, 563, (2008).

Klemola T., Koivula M., Korpimaki E., Norrdahl K.: *J. of Animal Ecology*, 66, 607, (1997).

Huck U. W., Pratt N. C., Labov J. B., Lisk R. D.: *J. Reprod. Physiol.*, 83, 209, (1988).

Rylnikov V. A., Savinetskaya L. E., Voznesenskaya V. V.: *Soviet Journal of Ecology.*, 23 (1), 46, (1992).

Sokolov V. E., Voznessenskaya V. V., Zinkevich E. P.: In: R. L. Doty, D. Muller-Schwarze (Eds): *Chemical Signals in Vertebrates 6*. Plenum Press, New York, 267, (1992).

Voznessenskaya V. V., Wysocki C. J., Zinkevich E. P.: In: Doty, R. L., Muller-Schwarze, D. (Eds): *Chemical Signals in Vertebrates 6*. Plenum Press, New York, 281, (1992).

Voznessenskaya V. V., Krivomazov G., Voznesenskaia A. E., Klyuchnikova M. A.: *Chem. Senses*, 31 (5), A84, (2006).

Voznessenskaya V. V., Klyuchnikova M. A., Wysocki C. J.: *Current Zool.*, 56(6), 813, (2010).

Voznessenskaya V. V., Naidenko S. V., Clark L., Pavlov D. S.: In: Zaikov G. E. (Ed): *Biotechnoogy and the Environment Including Biogeotechnology*, Nova Science Publishers, NY, 59, (2004).

Voznessenskaya V. V., Voznesenskaia A. E., Klyuchnikova M. A.: *Chem. Senses*, 31 (8), E43, (2006).

Whitten W. K.: *J. Endocrinol.*, 13, 399, (1956).

In: News in Chemistry, Biochemistry and Biotechnology ISBN: 978-1-63117-273-1
Editors: G. E. Zaikov, G. Nyszko, L. P. Krylova et al.

Chapter 25

ANTIOXIDANTIVE ACTIVITY OF FOREST AND MEADOW MEDICINAL HERBS

Z. G. Kozlova[*]
Emanuel Institute of Biochemical Physics of the Russian Academy of Sciences (RAS), Moscow, Russia

ABSTRACT

Quantitative data for fat- and water-soluble antioxidants in medicinal herbs were obtained. These data vary for fat-soluble antioxidants in the range from 1.0×10^{-3} to 9.9×10^{-3} M/kg and from 1.7×10^{-3} to 1.8×10^{-2} M/kg for the total content of fat- and water-soluble AO.

Keywords: Antioxidants (AO), oxidation, medicinal herbs, antioxidant activity (AOA)

INTRODUCTION

The study of health-giving herbs attracts interest because most of these plants are used in medicines. Our ancestors have used them for centuries for treating various ailments and interest in them continues to this day. Their broad health-giving properties make it possible to successfully use them in practically all therapeutic spheres. The vegetable kingdom is rich in phenol compounds which possess antioxidant properties and biological activity allowing them to participate in regulating the oxidizing processes in the human organism. The aim of the present work was to study and evaluate the antioxidant activity (AOA) of a number of medicinal plants. Eight forest and meadow herbs (Coltsfoot leaves, St. John's wort herb, Burdock leaves, Clover inflorescences, Dandelion leaves, Plantain leaves, Tansy inflorescences and Nettle leaves) were investigated. It should be noted that each of these plants has a specific health-giving action.

[*] Corresponding author: Z. G. Kozlova. Emanuel Institute of Biochemical Physics of the Russian Academy of Sciences (RAS), 119334, 4 Kosygin St., Moscow. E-mail: yevgeniya-s@inbox.ru, Fax: (495) 137-41-01.

Characteristics of Forest and Meadow Medicinal Herbs

NETTLE (leaf) (URTICA DIOICA L.) contains up to 269 mg% of vitamin C, carotene and other carotenoids (up to 50 mg%), vitamins of B and K groups, formic, pantothenic and other organic acids. In the leaves there has been found up to 5% chlorophyll, more than 2% tanning agent, gum resin, protoporphyrin, sitosterol, iron, quercetin, coffee, p- coumare, feryl acids, acetylcholine. Nettle preparations are taken internally to stop bleeding, intensify contraction activity of the uterus and improve coagulation of blood. In folk medicine nettle is used mainly to stop bleeding, as a diuretic and fever- reducing preparation, in treating rheumatic fever, liver and gall-bladder illnesses.

PLANTAIN (leaf) (PLANTAGO L.) contains glycoside, aucubin, ferments – invertin and emulsion, tanning agents, mucilage, carotene, ascorbic acid, little vitamin K and alkaloids.

A tincture of PLANTAGO L. leaves has an expectorant action and is used as an auxiliary means in treating bronchitis, whooping-cough, bronchial asthma and tuberculosis. The juice of fresh PLANTAGO L. leaves is effective in treating chronic gastritis, stomach and duodenal ulcers when the gastric-juice acidity is normal or below normal.

CLOVER inflorescences (TRIFOLIUM PRETENS L.) contain glycosides – trifoline, izotrifline, alkaloids, essential oil, vitamins C, B, carotene, resinous substances, fatty oil, etc.

Expectorant diuretic and antiseptic means are helpful in treating bronchial asthma, against anemia, collapse breakdown, malignant tumour.

BURDOCK (leaf) (ARCTIUM LAPPA L.). In leaves there is tanning agent, mucilage, essential oil, vitamins (particularly ascorbic acid).

Burdock used to stimulate metabolism and as a diuretic, sudorific for gastritis stomach ulcer.

COLTSFOOT (leaf) (TUSSILAGO FARFARA L.) contains bitter glycoside tussillagin, sitosterol, gallic, apple, tartaric acids, saponins, carotenoids, ascorbic acid, inulin, and dextrin. Coltsfoot used for treating dropsy, scrofula, pulmonary tuberculosis, hypertonic.

ST. JOHN'S WORT (herbs) (HYPERICUM PERFORATUM L.) has flavanoids – hyperozid, rytin, quercetin, essential oil, tanning agents (to 10%), carotene, ceryl alcohol, vitamin C, and an insignificant amount of choline and traces of alkaloid.

Herbs are used for morbidity indigestion, inflammation of the liver, nephritis.

TANSY – floscule - (TANACETUM VULGARE L.) contains ketone tuion, flavanoids, tanning (4.5%) agent, alkaloid (to 0.5%), quercetin, bitter substance, gum, resinous substance, of vitamins A and C. Tansy is used for fever and aching bones, shattered nerves, joint articulation, rheumatic fever, headache. DANDELION (leaf) (TARAXACUM OFFICINALE WEB) consists of vitamins A, B_1, B_2, C, carotene, phosphorus, and iron. It is recommended for atherosclerosis, neurosis, sleeplessness and normalizes digestion [4].

Method of Experiment

The method of investigation is to determine AOA directly and is based on the study of the kinetics of chain oxidation of model hydrocarbon isopropyl benzene (cumene) initiated by azo-bis-isobutyronitrile.

The most universal indicator is the rate of absorption of oxygen by oxidizing matter (oxidation rate). It reflects the total result of the main reaction occurring in the system.

The AOA of a preparation is characterized by a drop in oxidation rate and is determined by the induction period. The method is very sensitive, exact and informative. The analysis permits to investigate fat- as well as water-soluble AO. To investigate water-soluble AO, the analysis was performed in a mixture of polarized and non-polarized hydrocarbon solvents: to cumene there was added a mixture of hexane, dimethylsulfoxyde and water [1-3].

RESULTS AND DISCUSSION

The object of investigation was the 8 aforementioned dried wild plants. The material in cumene was extracted during 24 hours. The antioxidant content of these products was determined.

As an example, the Figure 1 shows the kinetic dependences of oxygen absorption in a model reaction of initiated cumene oxidation in the absence of antioxidant (straight line 1) and in the presence of Burdock (curve 2), Plantain (curve 3), Clover (curve 4), Nettle (curve 5).

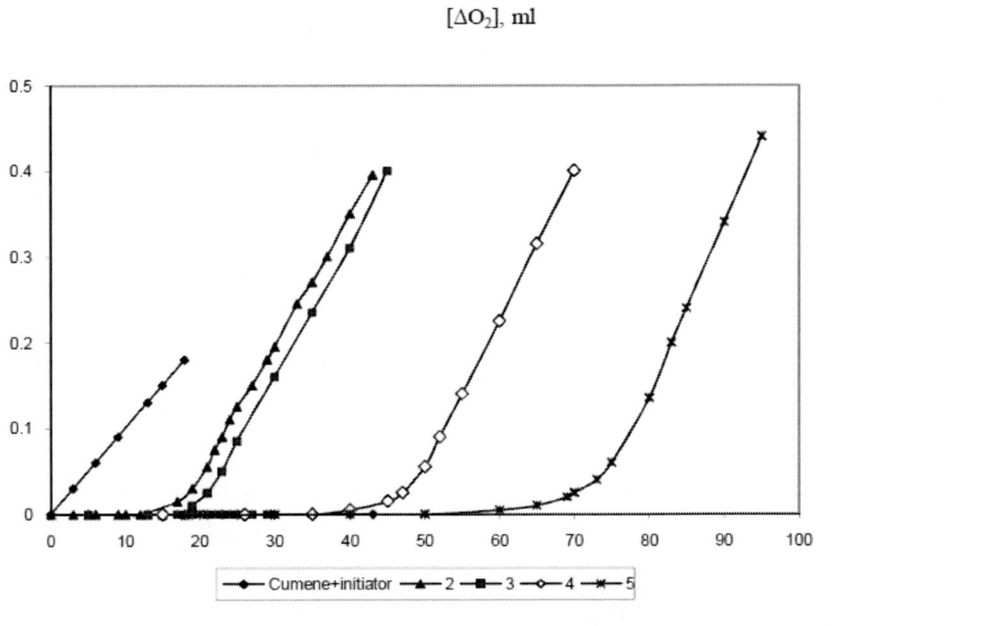

1 – hydrocarbon (cumene) + initiator AZO-bis-IZOBUTYRONITRILE, 1 mg),
2 – with Burdock added (17.5 mg), τ = 18 min,
3 – with Plantain added (6.3 mg), τ = 20 min,
4 – with Clover added (30.1 mg), τ = 47 min,
5 – with Nettle added (16 mg), τ = 74 min.

Figure 1. Kinetic Dependences of Oxygen Absorption (1 ml of hydrocarbon, 1 mg of initiator, t = 60^0 C).

It can be seen from the figure that in the absence of the additive hydrocarbon oxidation proceeds at constant rate (straight line 1). When the preparation is added, the oxidation rate at the beginning is strongly retarded but begins to increase after a certain period of time. This is indicative of the presence of AO in the additive. The rise in reaction rate is due to the expenditure of AO. When it is used up, the reaction proceeds at the constant rate of an uninhibited reaction.

Table 1. Bioantioxidant Activity of Investigated Matter

№	Matter	Antioxidant Content in Dry Matter (M/kg)		
		Fat-soluble part	Water-soluble part	Sum of fat- and water-soluble part
1	Nettle (leaf)	9.9×10^{-3}	8.5×10^{-3}	1.8×10^{-2}
2	St. John's wort	5.7×10^{-3}	3.1×10^{-3}	8.8×10^{-3}
3	Plantain (leaf)	6.5×10^{-3}	1.9×10^{-3}	8.4×10^{-3}
4	Clover	4.1×10^{-3}	3.0×10^{-3}	7.1×10^{-3}
5	Burdock (leaf)	2.1×10^{-3}	1.0×10^{-3}	3.1×10^{-3}
6	Coltsfoot (leaf)	1.6×10^{-3}	1.3×10^{-3}	2.9×10^{-3}
7	Dandelion (leaf)	1.0×10^{-3}	1.5×10^{-3}	2.5×10^{-3}
8	Tansy (floscule)	1.0×10^{-3}	7.0×10^{-4}	1.7×10^{-3}

Diagram Illustrating Results in Table

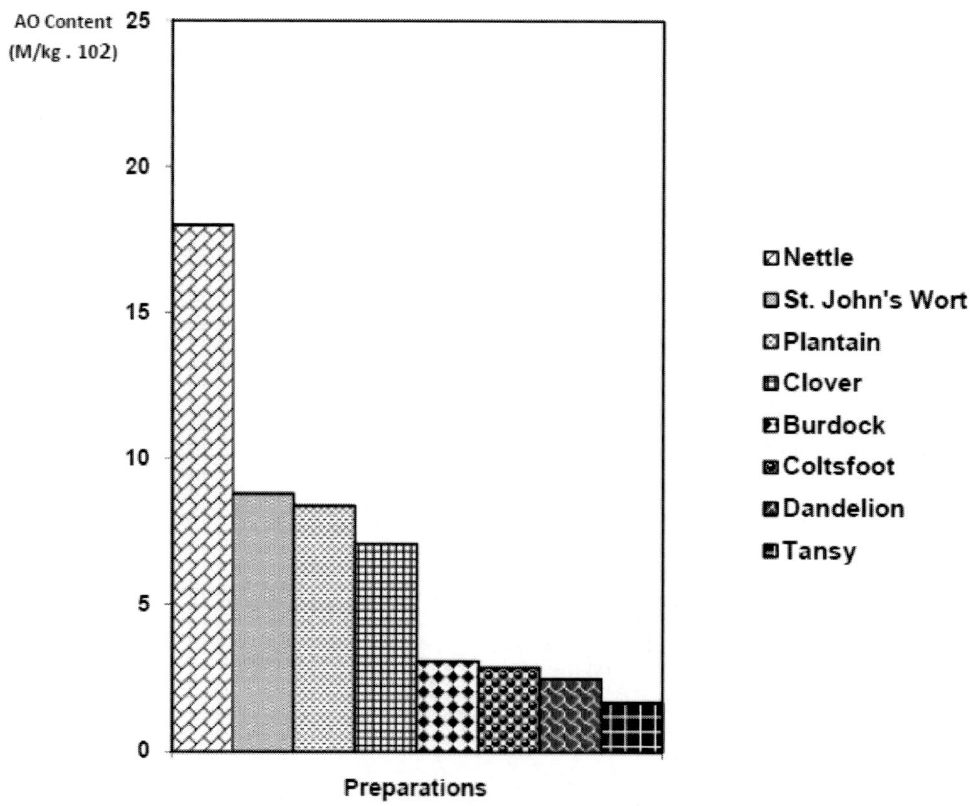

Data on the antioxidant content of investigated matter are presented in the Table 1. These data are illustrated by the Diagram.

Quantitative data for fat- and water-soluble AO in medicinal herbs were obtained. These data vary for fat-soluble AO in the range from 1.0×10^{-3} to 9.9×10^{-3} M/kg and from 1.7×10^{-3} to 1.8×10^{-2} M/kg for the total content of fat- and water-soluble AO. In the main, all samples containing more than 1.0×10^{-3} M/kg are potential sources of AO for the organism. The relatively high content of AO in medicinal herbs was found in the following gradation: Nettle > St. John's wort > Plantain > Clover > Burdock > Coltsfoot > Dandelion > Tansy.

Obtained AO values for wild herbs correlate well with our earlier obtained values for medicinal herbs. This permits to raise the status of wild herbs to that of Medicinal herbs long known [5].

REFERENCES

[1] Tsepalov, V. F., et al.: *A Method of Quantitative Determination of Inhibitors*, Certificate № 714273 dated 15.10.1979. (in Russian).

[2] Kharitonova, A. A., Kozlova, Z. G., Tsepalov, V. F., et al.: Kinetic Analysis of Antioxidant Properties in Complex Compositions by means of a Model Chain Reaction, *Kinetika i Kataliz. J.*, 1979, Vol. 20, № 3, pp. 593-599. (in Russian).

[3] Tsepalov, V. F.: A Method of Quantitative Analysis of Antioxidants by means of a Model Reaction of Initiated Oxidation in the book *"Investigation of Synthetic and Natural Antioxidants in vitro and in vivo"*, Moscow, 1992. (in Russian).

[4] Solovyova, V. A.: *Russia's Medicinal Herbs*, S.-Petersburg, 2006. (in Russian).

[5] Kozlova, Z. G., Kharitonova, A. A., Tsepalov, V. F., and Nevolina, O. A.: Quantitative Evaluation of Antioxidants in Spice-Aromatic and Medicinal Herbs, *Republican scientific conference "Spice-Aromatic and Medicinal Herbs: Outlook for their Use"*, Minsk, 1999. (in Russian).

INDEX

B

C

D

E

F

G

H

I

N

Q

R

S

T

U

V

W